The Institutional Position of Seaports

The GeoJournal Library

Volume 51

The titles published in this series are listed at the end of this volume.

The Institutional Position of Seaports

An International Comparison

by

HENRIK STEVENS
*Zeeland Seaports Authority,
Vlissingen, The Netherlands*

KLUWER ACADEMIC PUBLISHERS
DORDRECHT / BOSTON / LONDON

A C.I.P. Catalogue record for this book is available from the Library of Congress

ISBN 0-7923-5979-8

Published by Kluwer Academic Publishers,
P.O. Box 17, 3300 AA Dordrecht, The Netherlands.

Sold and distributed in North, Central and South America
by Kluwer Academic Publishers,
101 Philip Drive, Norwell, MA 02061, U.S.A.

In all other countries, sold and distributed
by Kluwer Academic Publishers,
P.O. Box 322, 3300 AH Dordrecht, The Netherlands.

Printed on acid-free paper

This is a translation of the original work in Dutch
"De Institutionele Positie van Zeehavens: Een internationale vergelijking"
published by Uitgeverij Eburon
Translated by Kathy Owen

Printed in the Netherlands.

To Corianne and Gemma

Contents

Illustrations

Figures

Maps

List of Graphs

List of Tables

Preface

The phenomenon of international seaport administration is the subject of this book. As a Ph.D.-student at the Delft University of Technology (period 1993 - 1997) I had the opportunity to develop and exercise my hobby on a full time base. The result was a Ph.D.-dissertation which was defended in December 1997. Unfortunately, these research results were published in Dutch while the majority of the interviewees and employees in the world of international seaport administration are English speaking people. Both for the reason of high international relevance of the results I felt the necessity to get this Ph.D.-research translated and published in English.

With the excellent help of my promotor Prof.dr. W.G.M. Salet I found Prof.dr. H. van der Wusten prepared to cover this study on international seaport administration in the Kluwer GeoJournal Library series. I thank Mr. Van der Wusten for giving me this opportunity. But also due to the outstanding help of my current employer 'Zeeland Seaports Authority' the funding for the translation became very quickly possible. I thank the Managing Director of Zeeland Seaports Authority, Mr. J.M.H.G. Philippen, and the Commercial Director, Capt. J. Verkiel, for their interest and wonderful help in getting this study translated. And of course my sincere thanks go to Katy Owen who actually made this dream come true.

Finally I would like to remind myself that this book has become what it is due to the love and patience of my wife Corianne and our little girl Gemma.

CHAPTER 1

Introduction to the Study

'The fundamental issues of the world and of people are not bound to time' (Eucken, 1950:4)

1.1.   Impetus for the Study

The *reconsideration of tasks and responsibilities* by central governments is an international phenomenon. A trend towards increased market mechanism (privatisation and deregulation) and decentralisation (territorial and functional) has become widely accepted since the early eighties under the leadership of the former political leaders of the United States (Reagan) and Great Britain (Thatcher). Although the political doctrines were harmonised, the methods of implementation were highly diverse. In the United States initially the decentralisation opportunities presented by 'New Federalism' were emphasised where local government had to rely once more on its own vitality (Agranoff, 1982). At the same time the opportunities offered by privatisation were primarily geared to contracting out (Kettl, 1993). In Great Britain from the outset government discontinued its interest in many sectors which were then left to the mercy of market forces (Swann, 1981). In the Netherlands where there were fewer state-owned industries to sell, options for reducing the scope of the public sector were sought through deregulation and privatisation.

Over the years it became clear that the greatest obstacle to this 'back to core business' policy was not so much bureaucratic resistance but the enormity of the task for government of taking up a new position vis-à-vis other social actors and applying this consistently (Henig, 1990). The administrative reorganisations proved in practice to lead to a *fixation on resources* (budgets, personnel, regulation) and they were seldom *integrated into a coherent framework* (Salet, 1994). This sometimes resulted in a public monopoly being transformed into a private monopoly or a former public service being swallowed up by the market. It also sometimes occurred that regulation was used by public authorities even more than in the former situation in order to guarantee quality (Swann, 1981) or that when contracting out the public authorities got insufficient value for money (Kettl, 1993). It must be concluded that due to the multitude of interactions between various public authorities and social actors it is difficult to determine how the two actors will behave towards

1

each other in these changed circumstances. This makes it difficult to predict what impact new divisions of authority will have in the long run.

In this book the term 'reorganisation' will be used in its broader sense. Although the forms may vary from territorial decentralisation to the sale of public services, the motivation remains the same, i.e. easing the operational burden of (central) public authorities. In its broader sense, the term 'reorganisation' refers to the redefining and transferring of government tasks and powers to organisations situated between the state and the market. This term is less politically loaded than such terms as 'liberalisation' and 'privatisation' which moreover tend to mean different things in different countries. It also includes the forms which are involved in functional decentralisation.

It appears that the common characteristic of reorganisation proposals is that they take shape from a strategic perspective with the intention of reducing the overall size of the public sector (Salet, 1994). Under the slogan 'less government, more market', policy makers were charged with achieving efficiency improvements. What were termed the 'Major Operations', however, did not always have the desired effect. In the early nineties, a more critical political wind began to blow in the Netherlands which relativised the expectations based on political ideology. This led to the disappearance of deregulation and privatisation from the political agenda and less drastic forms of privatisation (agencies and functional decentralisation) being given a more prominent role. This change in political attitude towards reorganisation proposals seems to have been motivated by the failure of deregulation (due to the requirements of policy neutrality) and of a number of privatisation operations.

Given the political nature of the reorganisations, it is logical that the desired outcomes cannot always be realised. After all, making cuts in one's own department is always a painful business and will be given shape as cautiously as possible. The political nature of the reorganisations also means, however, that in a political climate like that of the Netherlands which is based on consensus, reorganisation proposals get stuck at the level of specific political targets and opportunities (Salet, 1994). This can result in a stack of proposals but no account being taken of the effects a proposal will have on the general position of the civil service and the how the various proposals will affect each other. In some sectors it remains extraordinarily difficult to change or even question the responsibility of the public authorities. The fact is that in practice these are long-standing relationships in which every participant has played a part. The question remains of whether the established system still leads to the desired outcomes and whether it would not be more sensible to question the division of responsibilities once more and subsequently design a new system in accordance with agreed principles (cf. Chubb and Moe, 1988).

The analytical framework of this book will be built on the idea that without a normative basis there will, in the long term, be a lack of clear definition and safeguarding of responsibilities between the provincial and local authorities and between those authorities and social actors. A lack of clarity in the delineation of responsibilities can mean that bearing responsibility becomes a risky business and for this reason is consequently avoided.

Based on the theory that in reorganisation proposals a lack of political or market controls is created, this study focuses on how the reorganisation itself might be rationalised. The position the state occupies in a society or in a specific sector forms the subject of our research. This is operationalised through the specific division of responsibility between public authorities and social actors. The study assumes that life in modern western states requires high quality standards but that the state, due to its limited scope for action, cannot implement all the requirements itself and thus appears compelled to hand over matters to (quasi-)markets. Furthermore, the state production of goods and services entails undesired non-market failures (Wolf, 1979 and 1993) which might be curbed by certain forms of privatisation. In this study, the possibility of governments setting limiting conditions on social actors is a relevant issue.

Particularly in modern economic systems, due to the intertwinement of state and market and the increasing quality standards, it remains interesting in an academic sense to look at the various relationships which can exist between public authorities and social actors and the conditions needed to enable forms of governance to come into force. Although many theories have already been published about governance and regulation, this study wants to demand continuing political consideration for standardising the position of various public authorities within society before embarking on concrete action. This might appear to be a very general question which has no relevance today. Far from it. As we have said, the pliability of society is being called into question more than ever, yet public authorities are expected to solve social problems. Proposals for the reorganisation of public authorities are directly linked to the capacity of other social actors to solve problems and the quality standards which they set for those solutions. For a long time this dimension of reorganisation proposals seemed to have been forgotten and it looked as if insufficient account had been taken of the specific position of the authorities or the new position of the reorganised body. The theoretical objective of this study is to provide an insight into the division of control in a number of state-market relationships. By focusing on the question of control, we can examine whether the reorganised governmental task or agency is able to meet the required expectations.

1.2.   International Ports as Empirical Field of Study

For various reasons the technical and empirical side of international ports was chosen for this study. International seaports demonstrate a wide variety of administrative forms as a result of their own historical and cultural development and their position within their own political structure (Goss, 1979). They are managed by administrative bodies (port authorities) which generally occupy a relatively independent position between state and market and whose administrative structures can vary greatly. Furthermore, they maintain relations with central and local government and with private enterprise. In the second place, due to the direct involvement of public and private actors, issues pertaining to the division of responsibilities and regulation are an ongoing feature of seaports. Third, little if anything is known about the academic rationale underlying the administrative structure of international seaports. Decisions regarding the establishment of new activities or the influencing of existing activities by the authorities are often dependent on the incidental opinions of local political leaders or the director of the port authority. The general frame of reference for such assessments is unclear. What little academic literature exists is often dated or does not deal directly with the issues relevant to this study. Finally, the available literature on the administrative structure of seaports was written by economists and is highly descriptive in nature (Goss, 1979). Recommendations are based on the existing and desired division of ownership and scale size (Goss, 1990 a-d). Such administrative criteria may be important at a simplistic level but they provide no insight into the contextual relationships within which port activities take place. These criteria deliver no insights on the relative responsibilities of public and private parties. Despite these shortcomings this line of reasoning still prevails in the international seaport world.

The international privatisation of ports has been under discussion for a number of years. The European Commission, for example, has had various studies conducted into the administrative and financial situation in the European ports. The results of these studies, however, are not fully available to the (academic) public. The privatisation discussion is also explored at the level of international consultation between port authorities (International Association of Ports and Harbors) but this discussion has become bogged down at the very broad stage of case comparison and has been unable to rise above the level of individual rationale. For these reasons it seems logical to reconsider the administrative structure of important activities in international seaports but from an administrative perspective. The way in which responsibilities with regard to port activities are normalised is deferred to a subsequent study. The empirical objective of the present study is to map the administrative structure of a number of major international seaports and to assess the format of port activities in the light of normative principles. In contrast to the usual *success indicators* used for ports, e.g. the number of people employed there and the tonnage handled, this study uses a completely different *assessment framework*. Reasoning from the normative principle that underlies the structure of a particular port activity implies that concrete actions will be successful if they are generated from this normative basis. In other words, it is not only what is expected of a port activity, but also the achievement of that activity which constitutes success.

1.3. Problem Definition and Normative Framework

The social problem addressed in this study is the difficulty we have described in predicting the consequences of the new division of responsibilities caused by governmental reorganisations. This complicates any attempt to gain an overall view of the success of a reorganisation. Seaports, due to their administrative variation and the current international discussion on privatising them, form an ideal location for identifying and studying various types of responsibility divisions or control mechanisms. As a normative point of reference we assume the need for giving clear responsibilities an administrative structure. Unclear relationships between subjects muddy the interpretation of positive and negative signals, thus creating a reduction in *learning capacity* which adversely affects the success of an organisation or activity in the long term.

Public administration has a need for knowledge regarding the positioning of organisations situated between state and market while at the same time there is a dearth of literature on the administrative structure of international seaports. The objective of this book is therefore twofold: first, to develop a theoretical analytical framework to analyse the diversity of port activities and to provide an insight into the conditions needed for activating the learning capacity of port activities; and second, to generate empirical knowledge about seaports.

The premise underlying this study is that the chance of creating a sustainable and successful learning capacity in public and private seaport organisations is only present if their internal and external functioning is governed by a consistent system of checks and balances. Working on this assumption, we examine which expectations can be formulated about the necessary conditions for the development of learning capacity in the diverse institutional contexts of international seaports. For finding an answer to this normative question empirical research is required. In order to avoid any misconceptions about the scope of this study, it should be made clear at once that empirical research into the learning of a large number of international seaports cannot be addressed in this book. As with the above-mentioned normative issue, the empirical research pertains to the institutional conditions and examines which conditions actually underlie the various control relationships of international seaports.

This general empirical hypothesis first of all requires the formulation of both a regulatory typology of control relationships and also of the highly diverse port activities. In developing the first typology the emphasis should lie on the specific institutional conditions whereas the typology of port activities must be internationally manageable. Furthermore, this hypothesis requires that the main differences in managing seaports should be addressed and that attention be devoted to their similarities. Finally, this hypothesis ensures that lessons are drawn from the empirical cases using the normative framework.

1.4.   Delineating the Subject of Study

The theoretical study of the administrative structure of a seaport has to contend with the 'problem' that ports form part of the international economic system and thus encounter numerous outside influences. A clear-cut delineation of the physical feature which constitutes a 'port' is easily made in an administrative sense because a port authority has the formal control over a legally delineated port area. However, such a description would not do justice to the current port authority practice of undertaking activities outside the formal *port area,* the influence of competition between ports (cf. Goss, 1990a) and the involvement of private enterprise in a port, even though this is not directly found in the port area itself (cf. Fleming 1991; Van Klink, 1995). The opportunities for and threats to a port are not only found within the port itself and should thus be expressed in some way in the institutional model of the port.

If international economic regulation were taken as sole point of departure for assessment, however, the regional and national (socio-)economic functions would be accorded too little weight. The criterion of scale which is thus addressed is a highly fluid concept as far as administrative structure is concerned. The objective of the study is to identify various forms and mechanisms of control. The normative basis and the administrative structure of the relevant market segments of the most important port activities have been chosen as the empirical subject of this study. A number of port activities are then examined to discover the specific role of the port authority within the formal area. This does not alter the fact that relevant relationships between the port authority and public and private actors, whether or not these occur in other ports, may also form part of the analysis. Such relationships are considered relevant in so far as they affect the administrative structure of activities in the port, e.g. the influence of central government, or provincial and local government, on port planning or on nautical management regulations. At the same time, attention is paid to the influence of competition from other ports and the possible cooperation with areas in the port's hinterland or even with other ports.

1.5.   Methodology Used

The approach chosen in this book is that of an *international comparative study*. There were a number of reasons for this. It is interesting from an academic viewpoint to examine the extent to which various political structures and cultures generate different solutions to the control problem. Furthermore, ports form a part of the international economy, they have frequent contact with each other and endeavour to advise each other on various areas. For equally legitimate reasons it is interesting to look at how port activities are given different shape internationally. The dearth of international literature on the institutional position of seaports provides an extra incentive for conducting international research. And, as mentioned above, the available literature takes an institutional-economic viewpoint (division of ownership and scale size) and does not include the individual control mechanisms and principles which are central to this study.

Case Selection
The ports studied were chosen on the basis of their differences in administrative structure. International comparative study involves 'replacing the names of countries by variables'. An international comparative study must be interdisciplinary. The most important dependent variable is the way in which a control relationship and/or the relation between public and private actors is operationalised. Such a question has relevance for every structure and culture, as attested to by the enthusiasm and approval for the subject shown by many of those interviewed. The institutional typology which is developed in chapter two, based on theoretical principles from law, economics and sociology, is not bound to specific structural or cultural factors and thereby offers an instrument for an international comparison. Furthermore, use is made of a typology of port activities which can be found in every port in the world.

This study meets the general success standards of international comparative research. The theoretical framework used is of an *interdisciplinary* nature and the number of units to be studied is not overlarge. This applies both to the number of ports and the number of variables to be studied in the ports. *Several levels of analysis* are used (ranging from the market planning level to that of specific port activities and plans). Differences between the ports are not only *qualitative* (the nature of the port as regards economic significance and type of goods transhipment) but also *quantitative* (different sizes, annual tonnage handled, etc.). It is the intention to very tentatively lay some *causal links*. International ports are directly influenced by international economic activity and are dealt with on the basis of cost comparisons by internationally operating companies. In other words, international economic regulation exerts a strong influence on the institutional structure of seaports. By addressing the consistency of the role division between public and private actors in a port and comparing it with international economic regulation, the study can formulate expectations about the learning capacity of that port. Finally, international comparative research is aimed at *the development of types* and here a twofold approach will be used. First, a typology will be developed to indicate public roles and second, port activities and the institutional structure of international ports will be categorised.

Although, theoretically speaking, the study contains all the elements necessary to stand the test of criticism, it must be admitted in all fairness that the empirical depth and detail vary from case to case. In this respect it is more difficult to talk of a comparative study. Furthermore, institutional structures formed the central focus so that little empirical material about the equally important informal side of port organisations was produced.

Methods

In all the ports, interviews were held with political representatives, chief executive officers and managers of the port authority's financial, marketing and legal/administrative departments. They were well informed as to what was being organised in the port and how. Since the management of ports, in the sense of public control regarding port affairs, is central to the study, the port authorities formed the most important source of information. Nevertheless, the study also looked at the input from the private side. Sometimes there was an opportunity to talk to representatives from the local Chamber of Commerce, a specific lobby from the business community or specific companies in the port. At the same time, important reports and brochures were obtained from the port authority.

Another part of the material was compiled via contacts with the International Association of Ports and Harbors. An important aspect of this was attendance at the biennial conference of the International Association of Ports and Harbors in Seattle in 1995. In addition to the interviews with the local port authority, the conference provided the opportunity to arrange contacts with other ports still to be visited. Travelling to the various ports meant it was also possible to visit the local libraries (sometimes specialised port libraries), large bookshops and the maritime and transport faculties of universities.

1.6.   Guideline

The structure of the book is as follows. Chapter two deals with the theoretical analytical framework, followed in chapter three by a more general impression of relevant port issues and literature. Chapters four to eleven are mainly of interest to people working in port management and each can be read independent of the others. Chapter twelve presents a direct comparison between the various ports and the general findings are addressed. The diagram below shows the structure of the book.

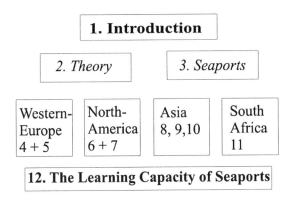

Figure 1.1. Structure of the Book

CHAPTER 2

The Regulation of Responsibilities

'The objects of our desires and aversions are not objects but relations. (...) *The goals we seek are changes in our relations or in our opportunities for relating; but the bulk of our activity consists in the "relating" itself'* (Vickers, 1965:33).

2.1.   Introduction

The transfer of tasks and powers from central government bodies to decentral bodies and (quasi-)market organisations implies the creation of different *control relations*. Based on the observation in the previous chapter that it is difficult to predict the consequences of institutional changes, this study seeks various *control mechanisms* which show the conditions under which the state and social actors will interact with each other after reorganisation[1]. Control mechanisms denote the principles which determine how the control of the decision making is organised. At the same time, these principles determine how the coordination between supply and demand will occur. This chapter will explain why an instrumental approach, which could result in the detailed analysis of various control methods, was not chosen. Based on the difference between *organising* and *steering* control relations, theoretically different control dimensions are revealed. By means of this distinction, an individual definition of the *institution concept* and its position in relation to other disciplines and the individual field of study is determined (section 2.2).

After first focusing on the concept of 'economic regulation' in a more general sense and the variety of positions and roles in which public authorities may manifest themselves in an oriented market economy (section 2.3), section four addresses the normative starting point and touchstone for this study, i.e. the argument in favour of clear control relations. The theoretical operationalising of the various control dimensions subsequently occurs by focusing the analysis on three different but interrelated levels of abstraction. Section five addresses the *constitutional level* which will be dealt with by means of insights from *regulation theory*. At the level of various organisations, ideas from *control theory* are reviewed (section six). By means of a typology for the individual activities level, section seven

---

[1] The concept of 'control' is understood to mean: 'the way in which accountability is regulated'.

examines how the control is divided. These three sections provide the building blocks which jointly lead, in section 2.7, to a *typology of the various control relations between state and market*. Each type is characterised by its own control mechanism. The final paragraph contains a summary.

2.2.   Regulation versus Management of Relationships

The point of departure for the theoretical analytical framework is the distinction between regulating relations between state and market as against managing them. The *regulation of relations* (through legislation) involves seeking and laying down *sustainable principles or norms* on which long-term expectations with regard to each other's behaviour can be based. Regulating a relationship between subjects is linked with seeking *the nature and existential conditions* of the relation concerned. When a public authority pursues *regulatory policy* it formulates the principles on the basis of which the division of responsibilities between state and society will occur. Regulating is the standardising of relations between subjects. The position of a public authority can be determined based on these norms.

The *steering of relations* is aimed at the goal- and results-driven influencing of relations between subjects, determined by the demands of the moment. It comprises intervening in existing relations in order to gain particular results. Steering is aimed at solving problems. Subsidising companies to use clean technology or the temporary aid given to Philips in the form of a *technolease* are examples of government steering behaviour. Political aspirations are the guiding principle underlying steering behaviour: there is a policy to be pursued.

Regulation and steering are directly interlinked and frequently flow into each other, making analysis very awkward at times. As normative point of departure, we follow the current theories on legislation and regulation from economic public law based on regulation as fixed point of reference. In accordance with the *coordination requirement* of economic public law, steering should be referred to the chosen regulation. In this line of reasoning, steering has no justification purely on the grounds of current objectives because it will only be effective if it correlates with existing standardised expectations with regard to government behaviour. Of course, the nature of law cannot be wholly laid down beforehand and regulatory legislation needs to be constantly reinterpreted and updated as new circumstances occur (see Nonet and Selznick, 1978). But one can never let the steering issues stand alone. The interaction between the regulatory and steering aspects must constantly be sought. Nonet and Selznick select a methodology which reasons from the situational rationality towards common principles (i.e. from the bottom up). This working method is certainly quite feasible in a *common law system*, but it is important to ensure that the law is not ultimately seen as an *instrument* while overlooking the other function of the law, i.e. as a safeguard which is based on normative foundations. In fact, the interaction between regulation and steering is always central.

From the political regulation angle a government should always first regulate economic and social activities by creating sustainable responsibility principles and thus link its own (steering) position to *identifiable* margins. The distinction between the regulating of relations and the pursuit of goal-oriented policy (the influencing of those relations) is essential because totally different considerations underlie it and the effects are correspondingly varied. Regulation is not usually concerned with the direct achieving of particular policy aims but with conditioning the different positions of the parties involved in such a way that a relative equilibrium is maintained in which they can pursue their own aims (Salet, 1987:8). This distinction makes it clear that opportunities for effective intervention by public authorities are linked to *institutional boundaries*. Particularly at a time when the steering capacity of public authorities is being reconsidered, it appears more desirable than ever to consider this distinction. Nevertheless, regulatory politics is not about an organisation principle of a higher order, as Salet (1994) explains. 'The fundamental legal code is not based on a principle of prudent organisation but on a tacit "moral bond of reciprocity". It does not seek marginal benefit but binds the legal persons and the government which pursues policy to a moral code of what is and is not appropriate. In this sense "regulatory politics" is not an organisational instrument of a higher order. (...) The role of legal rules based on values should therefore not be organised from goal-rational motives but be founded in the normalisation of the intercourse between legal persons' (Salet, 1994:118).

The concept of '*institution*' is reserved in this study for the domain of the sustainable normalisation of relations between subjects. An institution is characterised by the fact that it either *sustainably* limits certain opportunities for action or facilitates them. These are patterns of rules and/or norms and principles on which long-term expectations with regard to each other's behaviour may be based. Compare in this context the difference between the concepts of institution and organisation. An institution lays down or comprises the limiting conditions for interaction between subjects whereas an organisation is set up in order to achieve results with a view to certain objectives. An organisation is expected to conform to the institutional norms laid down by the general rule.

A PublicAdministration Framework for Analysis and a Sociological.Paradigm
The analytical method employed is that of a public administration framework, i.e. it incorporates legislative, sociological and economic insights on the control issue (the regulation of state-market relations). The general paradigm used for the analytical framework is taken from sociology, the *theory of social relations* (Salet, 1994). This paradigm makes it clear that social behaviour involves giving shape to useful relations between a number of subjects. Social behaviour, in whatever form and under whatever conditions (e.g. power relations), including the norms and rights which are the result of this, is understood as inter-subjective behaviour. Subjects are involved with each other at *varying intervals* and, as a consequence of the specific purpose and reason for the relationship, know what to expect of each other. The expectations can be more and less (formally) explicit. In modern societies, public authorities have numerous relations with other social actors. They are guardian of the public interest, contracting party, arbiter, producer, consumer, etc.

As a result of this diversity of tasks, responsibilities and public roles, it is not always clear which relationship is the main one. The disadvantage of this is that the expectations of social actors (citizens and private organisations) with regard to the behaviour of public authorities towards specific aspects of policy are very muddled. The public authorities concerned get the impression that the right hand no longer knows what the left hand is doing and this can lead to unreliable behaviour. When relations are clear, and thus transparent, subjects know where they stand with each other and good relations can be safeguarded. It will be argued that clear relations benefit the control and consequently the viability of the organisation or the relations. Section 2.4 deals in more detail with the normative viewpoint and the operationalisation of hybridity.

Relationship with Existing Public Administration and Other Disciplinary Insights
A public administration framework for analysing the control issue lends itself to the use of insights from other disciplines. Two criteria were discussed above which will be used to pinpoint other insights into regulation. The first reasons from the regulation of relationships. The concept of "institution" is reserved in this study for sustainable patterns of values and norms. The second criterion is the relations between subjects, which are central because social behaviour is understood as inter-subjective behaviour. '*Institutional analysis*' is used here to mean the search for the (normative) preconditions for sustainable relations between subjects.

Within public administration, current thinking about regulation usually takes place at organisation development level. When cooperation between public and private actors is potentially more efficient for achieving aims and solving problems, then cooperation becomes the norm. Contemporary public administration appears to recognise the formation and implementation of objectives as the main target of research. From this point of view, public authorities increase their legitimacy when they solve problems with those involved. By commencing the reasoning from interdependent positions, relations between public authorities and other social actors are horizontalised. This concerns governance within what are termed 'policy networks'.

Regulating relations on the basis of goal rationalisation is often short-lived because the rationality of the moment does not transcend the situation. In time (when policy is piled on policy and the call for clarity is heard once more) it becomes necessary to devise a new scheme of responsibilities. The question is thus whether the legitimacy of government policy allows itself to be organised through such a streamlining of interests or whether it finally boils down to the more fundamental choices which underlie the policy format and which reflect the normative relations between the state and society.

Sociology, economics, law and political science each have their own individual interpretations of the concept of institution. A more detailed and topical treatment of the various interpretations of this concept in economics, sociology and political science and their subdisciplines can be found in e.g. Scott (1995). The differences are linked on the one hand with the present *subject of research* (the nature of the regulatory relations) and on the other hand with the *scope of the theory* (at the regulatory or governance level). Sociology may claim to be the mother of the institution concept. In this branch of science, the concept is used as 'the pattern of *sustainable* values and norms'. In sociology, institutions are seen as the result of human behaviour, but not as consciously designed by it. For the design of human behaviour the concept of 'organisation' is used. An organisation is a functional interplay for achieving aims and results. If the subject of research concerns the normative relationship between one organisation and another, however, then the term institution is again used (Vickers, 1965).

In modern states the law functions as a source and means for establishing a sustainable value system in the intercourse between individual citizens and between the state and society. This study concerns the latter relationship. The position of the state (boundaries and capacity for action) is regulated through legislation (constitutional law and administrative law). Legislation comprises the formal reflection of the solidified normative patterns in inter-subjective intercourse. The explicit reflection of specific normative patterns often has its roots in the past and carries through (whether consciously or not) to the present. The choice of a democratic constitutional state and its specific use in the past is not fundamentally called into question in the Netherlands, for example. Since the state has the exclusive power to draw up and maintain the normative, legal framework for relations with society, its position is essentially different from that of other social actors. The design and concrete implementation of the democratic system (powers of state bodies, decision-making procedures, etc.) ensure that this exclusive position is not abused. The law derived from the state is not detached but is reciprocated in existing and changing social relationships. Within the political framework a hierarchical relationship thus exists between the state and other social actors (citizens, companies, interest groups, etc.) in which the hierarchy is the expression of legitimacy (Salet, 1994). The current emphasis in public administration on the reciprocal dependence between state and society in solving social problems, increases the chance that too little value will be attached to the clarifying effect of hierarchical elements.

More specifically as regards economic relationships, such as those which are to be found in seaports, the relations and reciprocity between law, state and society are expressed in the insights on economic law. This subdiscipline is concerned with the tension inherent in government regulation. On the one hand this refers to the sustainable harmonisation between the *law as economic instrument* (for influencing economic relationships) and the *law as safeguard* (the aggregate of formal and material principles to which the public sector and other economic subjects are bound by the choice of ends and means). On the other hand it involves the shaping of a mixture of economic and legislative relations in various sectors and market segments in as consistent a way as possible. In particular it is the emphasis which this discipline lays on the normative foundations as point of departure for governance which makes its insights so interesting. The following sections explore this in more detail.

Although interesting methodical insights on regulation and control are supplied from other (sub)disciplines, no use is made of these since a conscious choice has been made to base the analysis on the *regulating of relations*. The general characteristic of an economics approach is namely that it assumes *scarcity* and the desire for *optimising* towards solutions. In institutional economics the nature of the regulatory relationship is expressed in the *division of rights of ownership* (Berle and Means, 1932) in the *size of transaction costs* (Coase, 1937; Williamson, 1975) or in the *degree of information asymmetry* between principal and agent. Due to the current trend towards optimisation (to reduce the information asymmetry) by e.g. the 'staging' of a particular division of ownership, the proposals for making savings on transaction costs (search and information costs, negotiation costs etc.), or by a combination of the two, these analyses take place at the goal-rational level.

In political science, institutions have traditionally been defined as the principle and formal expression of constitutional law. In the sixties, the emphasis shifted to systems theory and attention was focused on the effects of political behaviours. These studies are located at goal-rational level because they are aimed at clarifying the behaviour of the electorate or of politicians. Nowadays, the focus has shifted to normative political theory and the design of various rule systems.

Goal-rational considerations also come into play in the subdisciplines formed in the interface between economics and law (economic law) and between economics and political science (public choice). Economic law provides economic explanations for the existence of the law and analyses the efficiency of legal rules. The public choice approach comprises the economic analysis of public decision making and the behaviour of public administration. Its various research subjects range from the decision-making rules at constitutional level to the design of public administration (e.g. Downs, 1967).

Theoretical Analysis at Different Empirical Abstraction Levels
The economic and political game rules which are formulated at society level have consequences for the way in which concrete organisation structures, conventions and actions can be used. New practices also affect the existing administrative structures and become as it were *embedded* in a greater institutional whole which is dynamic in the long term. Although reciprocity occurs over time, for the sake of analytical clarity it is important to distinguish a number of abstraction levels.

The institutional-theoretical search for control dimensions for various relations between state and market will be elaborated by means of an analysis at three different levels of abstraction. In the first instance, insights from *regulation theory* are used and concepts are generated to describe economic legal regulation (the constitutional level). Assuming an *oriented market economy*, we examine how accountability can be institutionalised for the various sorts of organisations within it. To this end insights from control theory (in particular Vickers, 1965) are reviewed. The third level of abstraction deals with the regulation of the consumer's control (as individual or organisation) over various specific tasks and objectives.

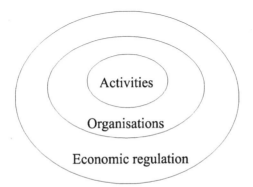

Figure 2.1 Theoretical Analysis at Three Levels of Abstraction.

2.3.   Economic Regulation and Various Government Roles

The theoretical analytical framework starts from regulation theory which addresses the general division of participation between state and market. It concerns the institutional structure of economic regulation. How centralised is the decision making on the deployment of the means of production and how are supply and demand coordinated? Central to this is the general economic role division between state and private sector which is taken for granted to such an extent it is no longer disputed, but functions as an important background frame. In international seaports, too, where the public authorities and the business community have a high level of direct involvement, the general role division is often no longer disputed. It only becomes important to those involved when one of the parties announces a change of course or makes this clear through concrete action.

Economic regulation concerns the way in which the economic process is normalised; this concerns the norms for relations within which the process of production and the division of goods for satifsying demand take place, as well as the governance and coordination mechanisms pertaining to these. The concept of economic regulation reflects the regulatory framework for economic behaviour and is distinguished from the behaviour itself, such as: producing, consuming, investing, saving and so forth. Or as Hellingman and Mortelmans so cogently express it: it concerns 'the pattern of rules that determines who takes the decisions regarding production and the destination of goods and services and how these are coordinated' (1989:25). The fact that economic regulation is no longer disputed can imply, however, that it is taken so much for granted that policy makers think that they need no further ratification. This line of argument carries the potential for the deterioration of the joint frame of reference because it does not lead to integration and conscious change.

To what extent the state lays down or tries to influence the limiting conditions for economic activity depends on the current political ideology (Lindblom, 1977). Eucken (1950) has made it clear that the economic process can be coordinated in principle by three different actors: the state, groups and the market.

- *Leadership by the state* in a centrally-controlled economy was considered by Eucken to be undesirable for a number of reasons. Disadvantages are the concentration of economic power, the amalgamation of political and economic power, the uncertainty and social dependence, and the threat to the constitutional state and to freedom from the potential abuse of power. Furthermore, a centrally-controlled economy in an increasingly open international economy can only be maintained at enormous cost. Finally, the coordination of supply and demand by public authorities leads to arbitrariness and uncertainty because public bodies cannot be held accountable for their output (cf. Wolf, 1979).

- Another form of social regulation of the economic process distinguished by Eucken is *leadership by groups*. This form is also not preferred since it results in an unstable equilibrium and tends towards imbalance due to individual groups pursuing their own interests and excluding others by entering into agreements. 'Serving the public interest is the last thing these groups have in mind' (Eucken, 1950:6).

- Eucken thus concluded that the group anarchy solution and the corporation can only exist temporarily and that it finally comes down to a choice between central government management of important sections of economic activities and *management by the market* and where the latter, according to Eucken, was the most favourable form which remained and could lead to respectable outcomes. He based his normative argument on the liberal-economic value of the guarantee of maximum freedom. Empiricism reveals hybrids everywhere, however: compare, e.g., the various gradations in self-regulation (Gupta and Lad, 1983).

The control mechanism differs fundamentally in these three basal regulatory models. In management by the state, control over economic activities is in the hands of the politicians. Decision making about the allocation of the means of production and the coordinating of supply and demand takes place through the *budget mechanism* and *political and administrative procedures*. In the management by the market model, the coordinating principles are the *price mechanism* and the *competition between suppliers*. These two controversial issues and their various aspects are usually contrasted with each other as shown below.

Table 2.1. Eucken's Models of Economic Regulation
(borrowed and abridged from Hellingman and Mortelmans, 1989:27)

Regulatory models	Barter relations	Centrally controlled economy
Typical regulatory basis	■ market ■ exchange	■ planning ■ estimation and assignment
Coordinating principle	■ price mechanism ■ competition	■ budget mechanism ■ political control
Determination of objectives	■ individual preferences ■ own interest	■ communal choices ■ collective interest
Decision making	■ individual consensus ad idem	■ political and administrative procedures
Power over economic subjects	■ autonomy ■ private ownership	■ prescription ■ collective ownership
Legal form	■ contractual	■ statutory regulations, consensual

Hellingman and Mortelmans (1989) do not distinguish management by groups as a separate form of regulation because Eucken designated the two other models as *internally consistent*. They interpret management by groups as '*consultation economics*' but cannot equate consultation as a decision-making and coordination mechanism with the price mechanism and the budget mechanism in the two other models (ibid. p. 33). Geelhoed (1985), however, does distinguish between the three forms of regulation although he uses the term '*économie concertée*' rather than *consultation economy*. Some authors distinguish more than three basic types of economic regulation but it is doubtful whether this contributes anything to the analysis. Geelhoed attempts to clarify the distinction between Eucken's three forms of management by distinguishing between different social spheres: *imperium*, for which the exercise of state authority is assumed and *dominium* whereby the state can function as one of many market parties. In free trade relations, both spheres of action are legally separated and public authorities function with clearly separated roles. In the économie concertée the lines between the exercise of state power and the functioning as a market party become blurred, whereas in the centrally controlled economy the state monopolises both spheres of action.

Presenting various models of economic regulation is not done with the intention of providing a minutely detailed account of the ins and outs of an economy. The dichotomy as deliberately presented by Hellingman and Mortelmans (1989) does not do justice to numerous intermediary organisations which, in modern economies, may be located between state and market. Furthermore, the attention devoted to economic regulation often fails to take account of characteristics such as the degree of development and rate of innovation, the relationship between various economic sectors and the degree of openness and dependence on foreign markets, whereas these are often just as important to a description of an economy as the degree of centralisation. And of course, these two models ignore the countless tools which a government has at its disposal to influence the market without obstructing spontaneous market relations. The purpose in analysing economic regulation is that it forms the *point of departure* for economic policy and gives an insight into general control mechanisms such as the price/competition mechanism for the market and the budget mechanism for the public sector. The disadvantage of an emphasis on elements of the economic structure, such as the degree of development, is that it undermines the economic viability of a system. The indicators of the economic structure tend to become values in themselves (aims are elevated to values) without their primary mechanisms being challenged. By first emphasising economic regulation in a general sense, the deployment of more intricate control mechanisms can then be distilled from it.

2.4.   Intermezzo: Clear Relations as Normative Point of Departure

The normative point of departure for this study was to promote clear relations between state and market. The underlying reasoning is that the *effectiveness of control and/or participation* and the *capacity to learn* are linked to the clear-cut division of responsibility and the clear-cut delineation of checks and balances. In short, when relations have a clear ordering principle, then subjects know where they stand with each other and what they can expect of each other over time. It will then be possible to interpret positive and negative information from a fixed point of reference and to convert it into effective action.

Unclear relations, in terms of the unclear delineation of responsibilities, are defined by the term *hybrid relations*. Hybrid implies the existence of mixed or bastardised forms. There are a number of types of hybrid relations. Well-known indicators are e.g. *recurring problems with divisions of authority problems*, the creation of a *power vacuum, excess profits* made by (quasi-)monopolies, *corruption, unclear judicial process, unfair competition*, etc. In the longer term, hybrid relations can have (serious) adverse effects on the functioning of the actors involved. The following section first describes some views on the assessment of hybridity and subsequently examines how it is operationalised within the framework of this study.

2.4.1. INTERPRETATIONS OF HYBRIDITY

Hybrid relations are a real headache for many administrators because they are a combination of two different coordination mechanisms of human behaviour which need each other at market regulation level but which are not compatible with each other in direct reciprocity. 'It's bootless to try to harmonize commerce and guardianship into one joint system of morality. Trying to do it can't produce harmony - quite the opposite. The contradictions are innate. We have no way to escape them' (Jacobs, 1994:106).

Jacobs (1994) terms the two different coordination mechanisms '*trading*' and '*taking*' and these refer to exchange and authority/power relations respectively[2]. The concept of 'taking' stands for the behaviour of those actors in the system who have the legal and/or legitimate authority and position to enable them to 'enforce' willy-nilly the affairs of other actors (Jacobs calls those positions the 'guardians' of the system). This is why the concept of 'taking' is associated with authority and power relations.

---

[2] Jacobs (1994) makes the perceptive remark that human beings are the only social animals to have two coordination mechanisms thus effectively parrying comments from critics who felt two was far too few.

The two systems of norms are not equal. With regard to productive activities, the system of norms for the market (exchange relations) is more effective in the long term than that for collective action. Van Dinten (1991) makes it clear that 'helping' each other (one interacts because of the joint motivation around a central idea) is no match for 'using' each other (one interacts primarily to achieve one's own interests). 'What happens when the "We help each other" system is in operation and the "We use each other" system appears on the scene. Behaviour will then be introduced whereby accounts are settled as quickly as possible. It creates a 'what's in it for me' attitude. Those who continue to work in groups in accordance with the "We help each other" model will lose because in the first place they do not think of settling accounts until later, and in the second place they become the ones who have to ensure the results of the group and so have to work harder. In consequence they will also decide to switch to "We use each other" and thus this system becomes dominant and will supplant "We help each other"' (Van Dinten, 1991). A plea for clear relations between state and market would appear to be called for but it is quickly forgotten when concrete aims or problems arise.

In the short term, a state-market relation is not restricted to its specific institutional character and the path to specific goal-driven solutions appears clear. A greater threat is posed to the original position if these goal-oriented solutions take structural root in some way and impinge on the original character and position of the relationship. As long as (financial) risks can be shifted to others, no-one will have any problem with these changes. As soon as responsibilities are redressed, however, a whole range of supervisory conditions immediately appear in order to be able to maintain the former position as far as possible. What should be done, for example, with the contractual part of a professor's salary if he no longer generates any private funding? And what about a faculty which no longer attracts students but which can support itself with private funding? The examples are simply endless but relate in every case to considerations concerning the original regulatory principle of the relationship.

Although internationally there is little empirical literature available on hybrid relations, academic views regarding the existence and handling of hybridity are divided. Furthermore, it should be noted that the *subject of analysis* of these studies can vary. In the current public administration literature, hybridity is applied both to *relations* between state and market as well as to hybrid *organisations*. In this study, the assumption is that the hybrid features of an organisation will affect the way in which relations with public and private organisations are established and therefore insights into both forms of hybridity will be addressed. A second reason for distinguishing 'hybrid relations' lies in the definition of the institution concept.

Academic assessment of the 'hybridity' phenomenon varies considerably. Some scholars take the view that hybrid forms are necessary in a dynamic and complex society and they advocate seeking answers to the question of how public authorities should deal with hybridity. They seek flexible solutions and goal-oriented aid for the hybridity issue and they come up, for example, with *norms for government behaviour*, which is an attempt to restore the original power relation, or they pursue *function transformation strategies* because hybrid organisations always tend to go from one hybrid to another. The group of public administration scholars who accept hybridity hold the view that solutions can be devised for the problems relating to hybridity (unreliability, opportunism). Another group of scholars are of the opinion that hybrid relations, due to the disparities in the systems of norms, will never ever be controllable. The starting point of this second group of scholars is that relations between subjects must be pure and clear and that specific goals and problems must be tested against the nature of the sustainable relations and expectations.

The standpoint of this second group appeals to me. The central theoretical assumption of this study is that organisations will only respond to changes in their environment in an innovative and alert way if they can be effectively called to account. Effective accountability or being answerable for actions can only occur, in my view, if relations are clear (second theoretical assumption). In the following subsection I want to clarify why the learning capacity of hybrid organisations or hybrid relations is of a different order to the learning capacity of clear relations. The final subsection will then address the operationalisation of this concept.

2.4.2. HYBRIDITY AND LEARNING CAPACITY: DOES THE END JUSTIFY THE MEANS?

In their seminal work '*Organizational learning: a theory of action perspective*', Argyris and Schön (1978) distinguish three forms of organisational learning: single loop, double loop and deutero learning. '*Single loop learning*' concerns the generating of creative solutions within existing organisational frameworks and strategic objectives. '*Double loop learning*' goes one step further and refers to the generating of solutions by modifying the strategic frameworks of the organisation which then form the point of departure for policy and administration. And finally they distinguish '*deutero learning*' which means the organisation of the individual learning process. A collective memory is assembled by keeping a record of specific incidents and other forms of historiography (keeping a logbook of the organisation) so that past experiences can be referred to in new situations.

The interpretation which Argyris and Schön give to their *'learning concepts'* can be situated at goal-rational level. This has consequences for the meaning and import of double loop and deutero learning in particular. On page fourteen of their book they define the term 'norms' which according to them are turned entirely upside down in double loop learning (obvious from conflicts within the organisation). To illustrate their point they take the example of an organisation in the cane sugar processing industry.

'The norms, strategies and assumptions embedded in the company's cane-growing practices constitute its *theory of action* for cane growing. There are comparable theories of action implicit in the company's ways of distributing and marketing its products. Taken together, these component theories of action represent a theory of action for achieving corporate objectives. This global theory of action we call "instrumental". It includes *norms* (italics HS) for corporate performance (for example, norms for margin profit and for return on investment), strategies for achieving norms (for example, strategies for plant location and for process technology), and assumptions which bind strategies and norms together (for example, the assumption that maintenance of a high rate of return on investment depends on the continual introduction of new technologies)' (Argyris and Schön, 1978:14-15).

In Argyris and Schön's description, double loop learning appears to occur at the instrumental, goal-rational level and does not involve interpreting the concept of norms in terms of altering sustainable relations with the institutional environment. I would like to use the concept of double loop learning, however, for an existential reconsideration at regulatory level. This includes introducing changes in the mix of success standards. This takes place, for instance, when market conditions change (new markets become available through new technologies, e.g.). Using this new definition I want to discuss the learning capacity of hybrid organisations.

According to the advocates of the flexible solution to hybridity problems, hybrid organisations rely heavily on deutero learning. Because they tend towards one hybrid form after another, they need to have a 'supply' of explanations in stock for why they are now using (yet) another new approach. Reasoning from the axiom that relations should be clear in order to link expectations to recognisable margins, I want to defend the standpoint that hybrid organisations can learn in a single loop way but that the reference points for actualising double loop learning are lacking.

Hybrid organisations are exceptionally able to solve goal-oriented problems within existing frameworks. Initially, they are labelled as extremely 'clever'. This is not surprising because they are able, when it suits them, to change hats and allow their own goal-oriented objectives to prevail in interactions. In the long term, however, they are accused of opportunism and unreliability, and for outsiders the superficial attraction of their cleverness soon palls. Due to their goal-oriented modifications, the phenomenon of double loop learning is not really applicable to hybrid organisations: existing strategic organisation frameworks are not being essentially changed but swing back and forth between forms of 'trading' and 'taking'. Compare, for example, the opportunistic behaviour of public authorities in the Netherlands in their approach to soil contamination and noise pollution. The public authorities swing from 'wise father' type behaviour, when they appear superior to everyone else, to 'after all we're all equal' behaviour when it suits their purposes. There are no sustainable solutions for the changes in the organisational framework because the position of the organisation is constantly in dispute.

Furthermore, the organisation gleans little from its own learning process (deutero learning) because opportunities and threats are seen as isolated incidents, '*opportunity windows*' (Kingdon, 1984). It is forced to act on the basis of situational rationality because otherwise the tension between taking and trading would become too great and its position would be called into question by the outside world. It comes down each time to trying to find a satisfactory '*trade-off*' between taking and trading for which the unique situation sets the limiting conditions. It never reaches deutero learning because each trade-off is unique. As soon as a trade-off is no longer unique, one of the two coordination mechanisms gets the chance to become institutionalised at the cost of the other (this is usually trading) in which case there is no longer hybridity.

The inability to achieve double loop learning due to the lack of fixed points of reference, and the reduced necessity to organise deutero learning because of the uniqueness of the trade-off situations, makes it possible for hybrid organisations to keep putting their own situationally-determined interests first in their interaction with others. Initially, the hybrid organisation is accused of being unreliable but its hybrid behaviour can also lead to risks being passed on to society. It is anticipated that the generally expensive ad hoc steering will achieve very little in the long term. Finally, it is expected that hybrid organisations, due to their intangible character, will become too self-absorbed and no longer react adequately to stimuli from the environment. They take their own goal-rational explanation too far, so that innovations and quality improvements fail to materialise.

2.4.3. OPERATIONALISING HYBRIDITY: TESTING INSTITUTIONAL CONSISTENCY

Operationalisations take place within the chosen general research framework. Savas (1982 and 1987) defines hybridity as the *piling up of various institutional arrangements* in order to produce goods and services. A voucher in combination with a production subsidy and a unique franchise is an example of hybridity.

In this study, hybridity is operationalised as follows. Taking the requirement of *cohesion in regulatory politics* as a starting concept, the économie concertée as such is not a hybrid form but it becomes one if it is not consistently disaggregated into organisational initiatives with the adoption of their accompanying tasks and goals. There is generally a market-oriented system, however, in which the state chooses to separate the dominium and the imperium. When governments in such regimes do not adhere to this separation in implementing concrete initiatives, then hybridity is found. A recent example of this is the operation of a private property by a public organisation where competition is closed to certain market segments as was the case for a long time in the telecommunications sector.

The discussion has recently got underway on the privatised and self-regulating government departments which, in addition to performing their public task, also enter the market. Now that the game rules for companies are being tightened via the prohibition system under the new Competition Act, there is increasing social demand for tight statutory regulations for these 'moonlighting' public departments which operate at arm's length. Thus a whole range of limiting conditions have been devised, aimed at preventing competition disturbance by agencies and independent administrative bodies. The provision of a limitative enumeration of activities in which the public service is allowed to compete with private providers and/or the introduction of a ban on transferring cross-subsidies to subsidiary companies, are examples of such limiting conditions. However logical such limiting conditions appear at first sight, they inspire little confidence in the long term since no institutional safeguard is created between the imperium and the dominium. After all, how is it possible to guarantee effective supervision of something as intangible as paying cross-subsidies (accountancy tricks)?

The existence of hybrid relations and hybrid organisations cannot be denied: ports in particular demonstrate a variety of forms. This is why in the final part of this section we will turn to the question: in what institutional way can hybrid organisations be activated to furnish accountability.

Clear (historical) reasons, arguing in favour of maintaining hybrid positions, can usually be pinpointed. For example, public authorities want to develop initiatives in the market which would not otherwise get off the ground. The value which is safeguarded in such cases is the *guarantee of long term investments* and the development of large-scale projects. Another possibility is that the combination of public and private sphere can have a synergetic effect. This might be the case, for example, in the cooperation between nautical management and port planning. The value which is guaranteed is *expertise*.

In principle, there are an abundance of motives which argue in favour of directly linking both spheres of one organisation. The only problem is, however, whether in the long term they can preserve the value underlying the motive. Devising institutional solutions to the problem of hybrid organisations aims at guaranteeing the specific configuration of public and private values for which the organisation was created. Those guarantees must be implemented in such a way that they constantly call for the combining of the two spheres.

Although I agree with Jacobs (1994) that a combination cannot easily tolerate such directness in the long term, I want to outline how I understand the institutional interaction with this type of organisation. The assumption is that it is not appreciated (politically or administratively) if the two spheres are separated because a certain *surplus value* is expected. One option for activating the learning capacity of hybrid organisations for large-scale projects, for example, is the *political appointment of a council of independent and unpaid experts* for a certain period which would be invested with far-reaching sanctioning powers with regard to strategic policy and finance. Appointments should not be for an extended period since otherwise there is a chance that the council might lose its objectivity towards the organisation. Another option is the creation of strong political reserve powers such as the approval of annual budgets and reports, the option to discharge a director, to set up an independent committee of inquiry in the event of serious conflicts, etc.

An important general element for activating the learning capacity is that there should be heavy investment in the financial controls of the hybrid entity because a great deal of autonomy can be derived from acquiring various resources. The automatic acquiring of monies must always be able to be questioned. As regards the guarantee of expertise, in some cases a form of *benchmarking* can be developed. This concerns the competition for delimited components between different public services or with other private providers. The fact that there is another, more or less artificially built-in provider, ensures that there is more of a handle on a comparable performance (the benchmark). This construction assumes the comparability of services, uniform performance criteria and the same competition conditions.

2.5.   Control Relations at Organisation Level

In modern societies, numerous relations between state and market can be found. Port authorities are mainly active in hybrid positions because they are responsible both for the public management of the port (safety and access) and the private operation (site leasing and superstructure). The 'many shades of grey' which are also found in other sectors of society make it difficult to discover how the accountability in concrete organisation formats is actually regulated. This section will deal with the institutional conditions which can help in getting to grips with the tangled system of control over organisations. The analysis concerns both the role of the port authority as well as the role of organisations involved in other port activities.

The Control Relation Between Supraregulator and Regulator: the Content
The institutional analysis of accountability by organisations concerns the formal and informal control relations with the relevant representatives of the 'outside world'. A control relation is the relationship between the controller and the controlled. A more positive description is that it concerns the relationship between the organisation and those to whom the organisation feels itself obliged to account for its actions. Vickers (1965) refers in this context to a 'supraregulator' and a 'regulator' respectively in order to express this theme of accountability in a broad sense, whether it pertains to an internal (hierarchical) relationship or to an interorganisational relationship. The term 'regulator' denotes the *policy freedom and relative independence* which an organisation (also called *agent* in economic theory) possesses. It 'regulates' and/or controls its own situation up to a point. This can be a consequence of the transfer of resources to others (Berle and Means, 1932) or of making general demands on policy makers (Lipsky, 1980). Finally, thinking in terms of supraregulator and regulator makes the relationship between internal and external organisation more overt. The formal and informal way in which an organisation is held accountable by the outside world has consequences for the internal division of tasks.

The core of the control issue can be broken down into three separate aspects. For the supraregulator, it is first of all important that he has an understanding of the way in which the regulator selects, processes and presents information about the state of the system. He then needs to discover the regulator's own standards against which he is testing the situation. The final question is how the supraregulator selects and formulates an answer to the regulator's actions. Stated in this way it is not assumed that the supraregulator has total control over the regulator and consideration is given to the regulator's individual desires and problems.

In order to steer effectively, the supraregulator needs to have an insight into the *success standards* which (should) apply to the regulator. The supraregulator is always faced with the problem of *information asymmetry* which carries over in two crucial ways[3]. First, it is not possible to get all the information about the preferences and capacities of the regulator on the table, so the supraregulator cannot precisely estimate the value of the information he has obtained. Second, the regulator can conceal certain actions and transfer the costs to others. These two factors deprive the supraregulator of a full insight into the regulator's conduct which means that the answer to the first two conditions mentioned above is always incomplete. '*Fine tuning*' is absolutely out of the question. The more the regulator is able to take up a position independent of the supraregulator, the more likely it becomes that the success standards which the organisation itself considers important will diverge from those which are given a positive assessment by the supraregulator. This is particularly the case if it is difficult to measure output directly or to assess the outcome of an action. A regulator's ability to inspire confidence and live up to expectations plays the most important role in such cases.

Vickers (1965) makes it clear that it is possible to view *conditions for success* (external expectations) and *success standards* (internal motivation criteria) as relations. This provides a concrete interpretation of the valid external expectations of the supraregulator with regard to the performance and conduct of the regulator. This method gets round the information asymmetry problem by emphasising the more generic expectations with regard to the outcomes of that performance. The consequences also become clear when external and internal expectations regarding the position and performance of the regulator do not tally. This means not only that the supraregulator is not getting *value for money* but also that an urgent look needs to be taken at institutional guarantees which will promote compliance with the success standards. The degree to which the internal success standards correspond to the external conditions for success determines the need for switching to formal controls. In this way, institutional form and content are interconnected and interact over time.

The following success standards and conditions for success may be distinguished:
- *functional relations*: performing the function concerned to the best of one's ability. This standard plays an increasingly important role in the current thinking on quality (ISO standards);

---

[3] Both forms are known in insurance economics and are termed 'adverse selection' and 'moral hazard' respectively.

- *survival-oriented relations*: balancing input and output;
- *growth-oriented relations*: obtaining as much input as possible. This is taken for granted in private enterprise (economic position, economies of scale, etc.) but for public organisations it is seen as a potentially dangerous factor and for this reason nearly always needs justification;
- *self-regulation oriented relations*: on the one hand this function concerns the best possible management of resources (positive motivation), while on the other it lays emphasis on acquiring good facilities for private use (dramaturgy).

The various relations form as it were the norms for dealing with the outside world. The specific correspondence between the various relations determines the nature of the organisation. In modern systems the most important value is still attached to (economic) *growth*. According to Vickers (1965) an emphasis on one of the relations takes place at the cost of the others. I would go a step further by postulating that a prolonged emphasis on one of these relations by an organisation in a modern society will lead in the long term, to miscommunication with the environment and reduced learning capacity (declining innovativeness and responsiveness).

The Relationship Between Supraregulator and Regulator: the Form
The role of supraregulator can be interpreted in a formal as well as an informal way. The role of the formal supraregulator becomes clear from the way in which organisations acquire the resources for their continued existence and survival. Vickers (1965) distinguishes four different ideal organisational types according to their method of acquiring input (classified from production by the state to production by the market):

1. organisations which are funded with public money (**public supported**);
2. organisations which get income from sponsors (**donor supported**);
3. organisations which are funded by users of goods and services, whereby variations in legal status can occur (whether a payment has been received or not) and in the market structure in which the organisation operates (**user supported**);
4. organisations which are maintained by members (**member supported**).

The resource provider's formal control relationship has a reserve function which appears particularly important when expectations are not lived up to structurally. In that event, the formal supraregulator is able to discharge the regulator and appoint another. This actor is thus distinguished from other quasi-regulating actors who usually attempt to adjust the policy of an organisation by means of a *dialogue* (providing recommendations, criticism and support).

The *thrust* of the control is determined first of all by the way in which the organisation acquires its resources. This can occur in competition with other organisations (as with e.g. donor and market organisations) but this is not necessarily always the case (public and membership organisations). Furthermore, the impact of the control depends on the *degree of reciprocity* between those who provide the means for survival and those who benefit from those means. In a donor relationship this reciprocity is entirely absent, whereas in a membership organisation the most direct form of reciprocity exists between donors and recipients. In the absence of reciprocity between donors and recipients, the organisation itself will have to concentrate on external accountability. Finally, in cases where equal weight is given to several relations (as by definition with many government departments) external accountability will be more heavily accentuated. If great importance attaches to two of the four functions (e.g. functional success and growth), the internal control is often split into *subcultures* which need to be reassembled for the external accounts.

The following schematic illustrates this argument:

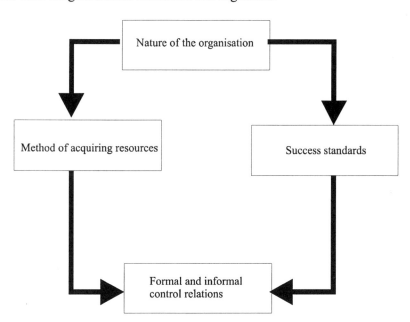

Figure 2.2 Theoretical Model of Control Relations at Organisational Level
(adapted from Vickers, 1965).

The above model and the terms used are rather abstract. Using the following examples, an attempt is made to clarify the line of reasoning and show how accountability is institutionalised by organisations. Consideration is also given to the question of where institutional efforts need to be made in order to safeguard external expectations (Vickers, 1965).

In the case of a *commercial market organisation*, for example, the capital for running the enterprise is obtained from the shareholders. It is they who have a direct interest in efficient operations and they call the management to account through profit expectancies (growth as a condition for success). As far as the management is concerned, however, this growth is not automatic: they must battle against competitors in order to win the consumers' favour. If the enterprise does not live up to expectations, the shareholders are free to sell their shares and reinvest the money as they see fit. The influence of formal control by shareholders and the board of commissioners is less relevant to the management if the formal auditors consider the commercial organisation to be a successful one. In that case, informal methods of control tend to be employed and other stakeholders gain relatively greater control. According to organisation ethicists, this is where the main difference between European and American enterprises lies. American enterprises give the 'shareholders' more control than the 'stakeholders' because they are aware that, due to the tough competition, they are directly dependent in the short term on the financial backers and not on the input of employees and other parties concerned.

Informal quality control also plays an important role if there is a relationship based on mutual trust (power relationship) and if no direct reciprocity exists between those who utilise the service and those who pay for or deliver it. This is the characteristic feature of the *donor supported organisation*. It is accountable for the substantive elements of its operations (functional success). The presence of reciprocity forces the organisation to be explicit about the charitable cause it is supporting. It may be expected, due to its philanthropic nature, that it will be as economical as possible in incurring costs for its own operations. Due to the competition for resources from other donor organisations, informal control will play a far more important role than independent accounts in the form of annual reports and information bulletins (Ben-Ner and Gui, 1993).

In a *membership organisation* (membership-based clubs, guilds and associations), on the other hand, the most direct reciprocity exists between those who contribute to the provision and those who receive services. This directness also applies to the relations between the elected officials and the membership. If a member so wishes, membership can be cancelled unless obligations are attached to the membership (what Coser, 1978, has termed *greedy institutions*), or if membership is compulsory (citizenship), or if no alternatives exist. A membership-based organisation is likely to consider functional success to be most important.

In a *public monopoly* with individual contributions the survival of the organisation is not at stake. It operates in a crystallised market in which growth is not a direct criterion of success. Reciprocity exists between income and expenditure, but the shareholder as beneficiary is not directly present. Financial performance is required, but this forms a condition for success rather than counting as an external success standard. The financial performance is designed in such a way that it is achievable for the individual organisation and takes precedence over the wishes of the supraregulator. The quality standard is thus transformed into a condition for success. In these cases, a form of *expert supervision* is often organised to gain more control over the supraregulator's expectations.

The quality standards set for public authorities depend on the type of department. There is less direct reciprocity than in the membership organisations due to the increase in scale. Government departments and private companies have to work within a budget drawn up in a political context; there is little scope for growth in the number of tasks and powers because these are laid down in legislation and ordinances. This will mean that the department tends to shift its attention to optimising its own survival (attractive offices, adequate staff numbers, etc) and there is an increased probability of policy introversion. For these organisations political control remains essential.

Table 2.2. Different Modes of Institutional Control (Vickers, 1965)

Organisation type	Reciprocity	Precondition for success (norm set by supraregulator)	Success standard (norm set by regulator)	Form of control Formal	Informal
Donor supported	absent	growth and functional success	survival and optimising self-enforcement (positive)	self-regulation	stingent assessment
Public supported	very indirect	functional success	optimisation self enforcement (negative)	budget discipline and political control	dialogue
User supported	direct	growth and survival	growth and functional success	price mechanism and shareholders	dialogue
Member supported	very direct	functional success	functional success	option to discharge board	dialogue

2.6.   Control of Activities

In the previous two sections we looked at how accountability is organised for economic regulation and various organisations respectively. Just as various organisations are found under economic regulation, so organisations can undertake various activities. In order to define the analytical framework more precisely, this section will examine the division of control among different sorts of activities. These concepts may be useful to the empirical analysis in describing the nature of concrete port activities.

Savas (1982 and 1987) elaborates his institutional theory by means of two principles to produce an ideal types classification for various activities. The *consumption* criterion makes it clear for whom the norm for use is set. The exclusion criterion demonstrates the consequences for enforcement. By cross-referencing the two criteria we get four ideal types of activities or goods and/or services.

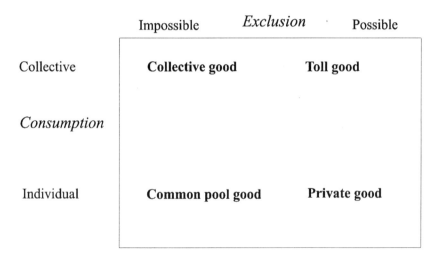

Figure 2.3. Classification of Goods Chracteristics (Savas, 1987)

Thus **private activities** can be distinguished which are characterised by individual consumption and the option of excluding consumption by others. In the best case, private goods are offered in competition so that the consumer has the opportunity to choose providers without incurring any penalties. It can also happen that there is reduced competition because the provider has a *niche market*. This does not affect the nature of the service, but the consumer will have to incur costs in order to find another provider.

A second group of activities are those with a **toll good character** which are consumed collectively and where exclusion is possible. The choice of whether or not to use such a facility lies once more with the user, but conditions are laid down for the consumer (the creation of a profit principle, having certain features, etc.). The exclusion of users is not entirely without cost (income is lost) and private auditors and enforcers need to be appointed to maintain the quality of the toll good. Numerous examples of such goods and services can be found in the use of infrastructure; the use of roads, railways, electricity, water, gas, ports, etc. all come with a price tag.

A third type of activities and goods/services are those with a **common pool character**. Their characteristics are that they are individually consumed and exclusion is not possible. Well-known examples are to be found in those sectors oriented towards the consumption of natural products (fisheries, water supply, etc.). Since by their nature common pool goods do not have a supply and demand coordination they tend to generate overconsumption so that the 'stock' becomes depleted. Due to its natural growth, this resource cannot adjust quickly enough to the tendency to overconsumption and it becomes threatened with 'extinction'. Common pool goods and activities thus have a tendency to be destroyed by their own success. In order to regulate the supply and to guard against extinction, collective action from directly interested parties is essential. Depending on the way in which the regulation is subsequently interpreted, a common pool good is then redefined as a toll or public good.

Public goods have a collective consumption from which no-one is excluded. The content of public goods is determined in democratic legal systems by elected representatives who endeavour to enforce this quality by means of a bureaucracy. When the state takes responsibility for the provision of goods and services, the regulation of policy implementation is formally guaranteed through statutory regulations. These include a legal procedure for assessment in light of the general principles of proper administration. It is possible ultimately to enlist the aid of the Ombudsman. The institutional control of public activities also covers the critical capacity of the civil service and the influence of interest and pressure groups or the media.

The distinction between various activities is an ideal type of division which provides criteria to determine the extent to which it is possible to speak of a *market situation*. According to Savas (1982 and 1987), a market situation exists when an activity is paid for by (potential) consumers. The strongest control mechanism is still the individual consumer's option to switch providers. This is the strongest expression of the principle of '*pay the piper and call the tune*'. In other words, the reciprocity (the relationship between the resource providers and the end users) mentioned in the previous section is at its strongest. If a market situation exists it may be expected, in an oriented market economy, that private activities will be created in competition with private providers.

Depending on the *market structure*, a legislator can deploy the sort of regulations that he will use to subject producers to conditions and to influence the operation of the market (e.g. through competition legislation). As with the choice of a particular regulatory political model, Savas (1982 and 1987) makes it clear that the defining of an activity is a *political choice*. Should enterprises in the port take care of their own security or will the police in the nearby town do it, possibly for a fee? Are the costs of port infrastructure recovered through tolls or are they paid from public funds? Political recognition of the two criteria has consequences for the way in which resources are obtained and determines the form and extent of influence which public authorities can exert on the quality outcomes.

By reasoning from the characteristics of an activity, the role of the public authorities can be further elaborated. Private activities can still be undertaken by public authorities in an oriented market economy, but the reasons for this become more transparent when the political discussion about the assessment of the characteristics is taken as point of departure. Such a consideration explains in advance the external expectancy criteria.

2.7.   The Regulation of Control Relations and Control Mechanisms

In sections 5 and 6 the institutional-theoretical effects of the control or accountability were shown at three separate but interconnected abstraction levels. These theoretical insights will be used in this section to establish a typology of *control relations* and *control mechanisms*. As we mentioned in the introduction to this book, central government reorganisations lead to new relations with previous government activities and social actors. Frequently, one cannot see the wood for the trees and it seems as if only organisation lawyers can tell us the difference between one relationship and another. By putting the control issue first and defining it in the light of state-market relations, it becomes clear that institutional differences definitely exist and these require other guarantees.

2.7.1.   REGULATORY PRINCIPLES FOR CONTROL RELATIONS BETWEEN STATE AND MARKET

The institutional control issue always boils down to two elements, i.e. to whom does the actor owe primary accountability and which role does a public authority fulfil in that relationship so that the nature and purpose of the relationship is safeguarded. These two elements are relevant to any system, be it statism or liberal states, and always involve the search for primary regulatory principles and how these are translated into concrete activities. In order to be able to define the complexity of concrete activities in terms of control relations and subsequently make an international comparison, it was necessary to trace the essence of the institutional control issue. By distinguishing a central principle, the essence could be laid bare and four unique control relationships obtained. This theoretical model of control relations appeared a suitable method for categorising the complexity of specific control relations found in the ports.

The distinction between the four control relations is based on the *principle of management* which can be defined as follows (Savas, 1982). A production process can be broken down into three different functions: a producer, a consumer and an arranger. A producer produces goods or services and/or maintains activities. A consumer uses and/or purchases goods and services. The arranger coordinates supply and demand and/or imposes specific quality requirements for the producer. The arranger's function is separate from that of the producer and the consumer since it is not always the producer or consumer who lays down specific requirements with regard to the production and/or the product. This might be due to specific market relationships or to the particular legal position of a producer.

The institutional analysis of a production process is aimed at the various relationships which can exist between these three functions. If we focus on the arranger's function, for example, we see that it is not always necessary for producer and consumer to have a one-to-one relationship but that another actor might be able to coordinate their behaviours more or less directly. Furthermore, this positional distinction can be used to denote the relationship between the arranger's and producer's function on the one hand and that of the consumer on the other, or alternatively that of the arranger and consumer vis-à-vis the producer. It is conceivable, however, that both producer and consumer perform the function of arranger concurrently. In short, the arranger's function helps to clarify the management function of a production process. It involves not only the production and purchase of goods and services but also the formulation and supervision of production and/or product requirements.

If we focus on the arranger's function, the dichotomy between state and market can be narrowed. The four control relations reflect the unique combinations between producer, consumer and arranger. If a public authority is exclusively responsible for the production and arranging of a good or service, this is termed *government monopoly*. If a government functions as exclusive arranger and the market is put in charge of production, this is termed *government regulation*. In cases where both the government and the market perform the arranger's and production roles these are termed *co-arrangements*. Finally, there is the concept of *market regulation*, which refers to the control relationship between state and market whereby the market exclusively takes on both the arranger's and the production function. The various control relations will now be examined individually and some examples discussed.

Government Monopoly

The control relation indicated by the term *government monopoly* stands for the relationship between state and social actors in which a government exclusively determines, by means of tax revenues, what the substantive service level will be and the quantities in which a good or service will be produced. The market structure can be described as a legal monopoly. It concerns the production of collective goods and services. The citizen as consumer has no direct influence on the decisions taken in the political sphere. The asymmetry between consumer and producer is at its most extreme in this relationship.

Examples of this control relationship are provided by those state agencies involved in implementing environmental policy, spatial planning policy, nautical management in ports, etc. These are activities with an undisputed public responsibility and entail the same restrictions and opportunities for everyone. Some other examples which can be considered as state monopolies but which also combine a number of characteristics of government conditioning are the former *state-owned industries* in the Netherlands. They came under the ministry and, despite their apparently independent and legally sanctioned behaviour, formed specific elements of the department. The staff (including the Board of Directors) consisted largely of civil servants. In addition to the respective Acts and Royal Decrees which gave concrete form to the tasks and organisation of the state-owned enterprises, the minister was empowered to provide further guidelines (e.g. on prices). Finally, the state-owned enterprises as a branch of the state conglomerate were not financially independent (did not have their own resources) but did have their own budget. Profits went to the treasury and losses were financed by it. Finally, an example from the field of government monopoly and government conditioning is conceivable where a private individual temporarily calls upon the services of a public authority (with payment if necessary). In this way, the collective good is temporarily transformed into a private good which does not affect the control over the quality of the services or the ownership of the production factors.

Government Conditioning

The state-market relationship which is referred to as *government conditioning* concerns the control relationship in which the government acts as exclusive arranger and in which the market undertakes the production. A government simply sets the standards and selects other public authorities or private enterprises for the implementation and these then offer their services to the consumer. The consumer has no direct control over the standards.

An example of government conditioning under competition is the option the state has to choose from a number of providers for the outsourcing. These are short-term contracts where outsourcing can take place under market conditions. If there is no competition between providers or if competition is undesirable, the freedom of choice disappears and the government then has to rely heavily on control methods. For the management of natural monopolies (unique concessions/franchises) Schmalensee (1979), for example, makes it clear that the control relationship takes on a character of mutual dependence and generates a deadlock between client and provider. Public authorities lack the required expertise or simply do not have the means to control the management themselves. Since restructuring is extremely costly, in the short term government can only resort to regulatory measures such as setting a maximum price or maximum revenue and other, material requirements linked with the performing of tasks and limiting powers.

In the electricity sector in the United States, for example, extensive regulation has been designed with regard to the bookkeeping (indicators) to enable comparisons between the operational performance of enterprises in different states. Furthermore, all documents are made public and third parties have the right to intervene so that certain costs passed on to the consumer can be questioned. Politically impractical and unrealistic control methods include such things as revoking a licence or issuing an instruction (compare the way in which the introduction of competition can generate strategic behaviour; Schmalensee, 1979).

Co-arrangements
The concept of co-arrangements indicates a state-market control relationship in which both the state and market act as arrangers and producers. This is fundamentally different from the 'government conditioning' control relationship in which the state alone performs the role of arranger.

The most well-known examples of this control relationship concern entering into *contracts* in which state and market are on an equal footing. Concrete examples of this control relationship can be found in the construction and maintenance of infrastructure, e.g. the cooperation between state and the business community in building and using new terminals, or the joint developing of new sites by the port authority and the private user.

In this context, state participation can also be seen as an example of a co-arrangement. By means of a block of shares and the appointment of a state commissioner, the state gains some control over the activities of the organisation which is usually operating in a competing market.

It can also happen that the role of arranger more or less disappears due to the fact that the producer is also the consumer; this is a characteristic feature of the various initiatives of public-private cooperation, for instance. When such co-arrangements are entered into for an extended period, very close dependency relations can arise which bypass both state and market controls. There is a good chance that in such situations the state will squeeze the market party out of existence because the latter is no longer directly held accountable for its own success standards. It is very important that the public authorities should indicate, via public law, the conditions under which they may have dealings with market parties. In co-arrangements in particular there is an increased likelihood that the distinction between imperium and dominium will not be adequately safeguarded. Market parties are often interested in cooperation with public authorities because what they really want is to have a share in the imperium.

Market Regulation
In the market regulation control relationship, the state itself does not participate in market activity nor does it attempt to steer it. The arranger's function and the producer's function are both performed exclusively by the market. The objective of market regulation by the state is to guarantee the consumer's freedom of choice and fair competition between providers. Market participants realise that in modern states they must keep to a set of minimum quality criteria which have elsewhere been defined as the 'limiting conditions of public regulation for the market sector'. These concern the conditions governing the market participants for the protection of public health, public safety, occupational safety, social security, etc. Market regulation by the state also includes all the legal facilities which enable commerce, such as the recognition of rights of ownership, the provision of procedures for civil law, the options for setting up organisations (legal persons law), etc.

An example of market regulation from the port world is the drawing up of a *port ordinance* by the port authority. This document lays down the general rules of the port which form the conditions for the business community's promoting its own interests. The legal document is not meant to be used for active intervention in market transactions but to be used for corrective purposes when necessary.

2.7.2. DEFINING CONTROL MECHANISMS

Accountability always involves two questions: 1. to whom and for what is one accountable? and 2. against which standards is the behaviour assessed? Regulation theory makes it clear that the way in which the enforcement of norms can take place depends on the way in which the norm has been developed. A specific, unique combination of regulation and enforcement is termed a *control mechanism*. In the case of government monopoly and market regulation, clear control relations exist for which the terms *budget mechanism* and *market or price mechanism* are used. It is likewise desirable in theory to delineate the control mechanisms for the intervening state-market relations using clear-cut concepts.

In the case of government conditioning, the consumer's freedom of choice is severely limited since the state is the sole arranger. The option of changing to another provider is not directly available or is costly (sometimes literally: moving house). The contribution which a citizen must pay for the goods and services is not an indicator to the provider that he has done a good job. The state as arranger fulfils an important regulatory role here in which it is aware that a high degree of mutual dependence exists. It can, for example, regulate the quality enforcement by acquiring a part of the enterprise's property. Furthermore, the state has numerous options for specifying additional formal stipulations (consultation and participation) and material matters (tariff amounts, maintenance requirements, training requirements, etc.).

The general control mechanism for the category of government conditioning I have termed the *mechanism of provider change*. A government as exclusive arranger can lay down additional provisions for the concession and may cancel it if necessary (even though this sometimes involves high transaction costs). It can then tender and award it to another provider and this is the sword of Damocles used to threaten the existing private provider in cases of serious dysfunction. The amount of opposition caused by cancellation will vary according to the remaining duration of the concession. There is often no immediate alternative available (due to the provider's exclusive knowledge and expertise, e.g.) and in many cases it is preferable to offer the present provider a second chance in order to avoid conflict (emotional ties). Nevertheless, this does not alter the fact that a government as exclusive arranger determines whether it wishes to continue its relationship with a particular provider.

No separate mechanism need be formulated for the category of co-arrangements since either the competition mechanism (several providers) or the provider change mechanism (long-term cooperation) is active.

2.8.   Summary

In this chapter the issue of accountability, also termed the control issue, has been explored using institutional theory. 'Institutions' were defined as the 'sustainable relationships between state and market'. By defining the phrase at regulation level, the academic position vis-à-vis other disciplinary insights into institutions could be specified. At the same time, regulation theory provided the normative starting point for the present analytical framework. This was expressed in the hypothesis that the learning capacity of organisations can only be activated when responsibilities are laid down in clear state-market relationships.

A regulatory principle for the institutional analysis of the control issue was distilled from the theory. Using the 'principle of management', the relationship between the arranger's function and the actual production was made clear. On the basis of this distinction, the state-market dichotomy could be contextualised and an analytical framework obtained with four unique state-market relationships. These were described as *government monopoly, government conditioning, co-arrangements and market regulation* respectively. The control relationship of 'government monopoly' indicates that the state is simultaneously exclusive producer and arranger. In 'government conditioning' the state is the exclusive arranger while a private actor functions as producer. In the case of 'co-arrangements' both the state and market act as (equal) arrangers and producers. Finally, in the 'market regulation' control relationship, the arranger's and producer's function is the exclusive responsibility.

CHAPTER 3

The Nature of Seaports

'Seaports are the mouths through which continents speak to each other'

3.1.   Introduction

The general introduction to this book briefly examined why it is interesting to study ports and more specifically port management. The main reason given was the dearth of literature on the nature of the various relationships between public and private actors. All the same, there is literature available on ports in a more general sense although it is not equally accessible for everyone. Three of the best known books are J.G. Baudelaire's *Port administration and management* (1986), A.E. Branch's *Elements of port operation and management* (1986) and E.G. Frankel's *Port planning and development* (1987) which supply useful concepts for port authorities and highlight various trends in port administration.

Since this study does not focus on the management level but wants to emphasise instead the constitutional elements of various port systems, this chapter will address the general and permanent issues which exist in every port. The second section explores those siamese twins, trade and shipping, and ports' dependence on international trade. The third section examines physical port issues and outlines a number of trends in port planning. Section four focuses on the various actors and activities which are so characteristic of the economic activity in the port. Finally, section five describes a classification of port types which is used internationally and concludes with a number of contextualising remarks.

3.2.   Trade: a Port's Rationale

The fundamental rationale underlying commercial ports is that of international trade. No less than 82% of the world trade measured in tons and 94% of the world trade measured in tons per kilometre is transferred by ship (Frankel, 1987). This is why it is always important for a port authority to know how transport flows change over time and how wars and other disasters influence the global economy. The larger ports often have their own staff who, together with customs officials and national research institutes, compile and analyse data on the various trade flows. This highly expensive method is also called the *indirect method*. The less wealthy ports employ a *direct method* and retrieve information directly from shippers and forwarding agents/shipping companies in order to deduce trends. A port authority likes to know how much employment the freight will generate in the port itself and how much cargo will flow straight through to the hinterland for processing.

Ports are often categorised according to the type of goods that are handled. These are variously termed '*dry* (coal, ores, grain) and *wet bulk* (crude oil, minerals, chemicals)' which are respectively 'dumped' or 'pumped' into the hold. *Non-bulk* refers to the traditional general cargo, *container handling* or *roll on roll off cargo*. The annual handling figures from the world's largest ports are always received with considerable interest although their lack of uniformity sometimes leads to miscommunication. Different criteria are employed across the world which makes standardisation difficult. In Western Europe, for example, the *metric ton* (a thousand kilos) is used whereas in Asia the '*freight* or *stevedore ton*' (forty cubic feet or $1.13m^3$) is the measuring unit employed. In South Africa the '*harbour ton*' (a thousand kilos or $1 m^3$, whichever is heavier) is used and in many ports in North America the standard unit used is the '*revenue ton*' (which cannot be expressed in either weight or volume) or the '*long ton*' (1.016 metric tons) (Institute of Shipping Economics and Logistics, 1996). The handling figures reflect the economic growth which the port has achieved in the previous year and form worldwide the most important success standards for ports in spite of all the philosophy about the transition 'from tonnage port to value port'.

Container handling in particular has comprised the worldwide innovative challenge for port authorities over the last 25 years. In modern ports, the container[1] has gradually completely replaced the traditional general cargo that used to be packed in boxes, crates and chests. Over the last decades every self-respecting port has invested heavily in container terminals which require huge, solidly-anchored cranes. Table 3.1. shows a multi-year overview of international trade grouped by commodity. It is striking that the rate of growth has decreased in recent years.

There has been an explosive growth in container transport, and in the transport of oil and ore, over the last decades. Ships with 6,000 TEU have become run of the mill and it may be expected that in the near future until 2005 there will be a further increase up to 9,000. TEU stands for *Twenty-foot Equivalent Unit* and is the standard measurement for specifying the amount of container cargo measured in twenty-foot containers.

---

[1] An interesting point about the container as cargo unit is that it was the invention of a truck driver. The construction of modern container ships is based on a concept which takes the truck and trailer as its starting point.

Table 3.1. World Trade via the Sea
(Source: International Association of Ports and Harbors, 1995:45)

Year	Liquid bulk goods	Dry bulkgoods	Breakbulk, container	Total (x 1 million tons
1989	1,692	965	1,234	3,891 +5.4%
1990	1,755	968	1,285	4,008 +3.0%
1991	1,790	1,005	1,325	4,120 +2.8%
1992	1,860	990	1,370	4,220 +2.4%
1993	1,945	985	1,382	4,312 +2.2%

A further expansion in container transport will mainly depend on the costs for the ship in terms of e.g. additional engine power. At the same time, market developments (e.g. mergers between large container shipping companies and the location of new companies) are highly important. Compare in this context for example the fact that due to falling oil prices the largest tanker ever built for Shell turned out to be no longer cost-effective. The recent expansion in container transport requires huge investments in the superstructure for container handling. Cranes must have a greater range which then has consequences for the construction of the quay due to the extra pressure exerted on it. The handling of huge numbers of containers to the hinterland also requires investments in e.g. *intermodal transport* to offer the terminal a choice of modes of transport (rail, water, road) depending on desired speed of delivery.

Another typical feature of container transport which has implications for the ports is the fact that container shipping companies arrange their transport in ever-larger alliances to ensure they only sail with fully-loaded ships. Initially, this operational expansion took place through joint capacity planning in '*conferences*' (alliances of cooperating shipping companies) but recently mergers between the different alliances (Nedlloyd and P&O; APL and Neptune) have occurred. For port authorities this means less ships paying port dues and tougher negotiating to keep the shipping companies staying in their port. At present there are six container alliances active: *Global Alliance* (Nedlloyd, Mitsui OSK Lines, American President Lines, Orient Overseas Container Lines and Malaysian International Shipping Company), *Grand Alliance* (NYK, P&OCL, Hapag Lloyd, Neptune Orient Lines), *Sea-Land/Maersk, Hanjin/Tricon, HMM/MSC/Norasia* and *K-Line/Yang Ming*. In Table 3.2. the number of ships and the capacity of each alliance is shown.

The major container shipping companies, such as e.g. the Danish Maersk and the American Sea-Land, operate worldwide using their own terminal or the '*dedicated terminal*' concept. The dedicated terminal is a terminal which is under the ownership of a private stevedore and which can be used in principle by any shipowner but where use by the larger shipping companies

Table 3.2. International Container Alliances and Their Transport Capacity
(Source: Rotterdam Municipal Port Management, 1996d:45)

Alliance	Number of Ships	TEU capacity
Grand Alliance	122	296,831
Global Alliance	113	210,886
Sea-Land/Maersk	100	244,056
Tricon/Hanjin	76	164,437
HMM/MSC/Norasia	60	118,557
K-Line/Yang Ming	49	123,102

can be given priority under individual contracts. This makes it possible for these shipping companies to plan their container transport within a very tight timeframe and prevent the ship spending unnecessary time in port. The consequence of such a handling concept is the oligopolisation of the container waterfront and the creation of '*container hubs*' (one or two ports per continent where the deep-sea container ships can put their containers ashore). At the moment Singapore and Hong Kong function as container hubs for the Asian market and from there other regional ports are served by smaller ships (coasters). In Western Europe the port of Rotterdam with its deep Eurochannel is vying for such a position but it has not achieved it yet. The port of New York functions as container hub for the east coast of the United States while Los Angeles and Long Beach perform the same function for the west coast. For the port authority, being designated container hub by the shipowners offers security due to the relatively long-term contracts (sometimes up to twenty years) which can be negotiated but it also means surrendering some control over the conditions under which the transport occurs. A concept such as intermodal transport can only succeed if a logistic link is actually established between the various modes of transport. In this area there is often a lack of effective communication and coordination since the shipowners are striving to minimise the number of handling manoeuvres because these generate the most costs. Furthermore, speed is usually a priority so that alternatives via the water are not always seriously considered in the political discussion on the use of infrastructure (De Jong and Stevens, 1995).

With the economic developments in Asia (particularly China), Latin America, the former Soviet Union and, lately, Africa it may be expected that the growth in world trade has not yet peaked. Although the current rise of the various trade blocs means it is still far too early to talk of a fully open world economy it is a fact that an increasing amount of freight is being towed around the globe and that production takes place where it is cheapest. Due to the low wages, this means mainly in Asia (initially in Singapore, Hong Kong and Taiwan and now increasingly in Korea, Vietnam and China) although it is becoming increasingly usual to assemble the components and construct the specific product nearer to the markets in Western Europe and the United States. Over the last 40 years international trade has increased fiftyfold and it remains in a port's interest that economies should be as open as possible. A continent such as Asia, for example, which produces for the world markets, has nine large multi-purpose ports whereas Europe has three and the United States with its large domestic market and other cheap modalities has only two.

Table 3.3. Container Handling in the World's Major Ports (to the nearest million TEU)
(Source: H. Stevens based on Rotterdam Municipal Port Management, 1997a)
* Anticipated amounts

Port	1995	1996	1997	2010*
Hong Kong	12,53	13,28	14,50	32,0
Singapore	11,83	12,95	14,10	36,0
Kaoshiung	5,05	5,06	5,69	
Rotterdam	4,78	4,94	5,45	8,5
Busan	4,50	4,68	5,23	
Hamburg	2,89	3,07	3,34	
Long Beach	2,74	3,07	3,50	
Los Angeles	2,56	2,68	2,96	
Yokohama	2,73	2,40	2,33	
Antwerp	2,33	2,65	2,97	4,0

Port authorities are always trying to tie as much freight as possible to the port ('*captured*' freight) while also attempting to influence conditions to make transport via their own port as attractive as possible. An attempt is made to achieve the first objective by attracting industry into the port area. This is not only beneficial for attracting freight and ships but also for employment. In the (petro)chemical industry, for example, cooperation is essential so that there is a clustering of companies which jointly attract and than transfer a substantial amount of freight. Transfer is not always by ship but can also occur via pipelines. The same approach applies in the automobile industry, for example, which needs its own suppliers.

The creation of favourable conditions is achieved by endeavouring to make the port area itself as accessible as possible (maintaining the draught of channels, removing locks etc.) and by minimising the length of a ship's stay in the port (communication technology, e.g.) but also by controlling external influencing factors as much as possible. There are a number of ways in which the port authority can achieve this. It can guarantee, in consultation with the authorities, highly favourable (tax) tariffs and (customs) procedures, it can safeguard effective infrastructural hinterland connections, and it can keep towing and pilotage tariffs as low as possible by introducing competition. In addition there are a number of unanticipated factors which a port authority can neither directly nor indirectly influence and which are linked e.g. to political and climatological circumstances.

3.3.   Physical Issues: Nautical Management and Port Planning

Ports come in all shapes and sizes. They can fulfil various functions, ranging from offering a safe haven in a storm, to providing a location for processing freight and passengers or offering support services to ships, to functioning as a basis for industrial development, to forming a central distribution point for various chains of transport. The major commercial ports in the world fulfil all these functions simultaneously and are therefore also termed *multi-purpose ports*: they possess various specialised *terminals*. Some writers maintain that the competition between ports will develop increasingly into competition between terminals (Heaver, 1995). Particularly in Western Europe, the United States and Canada this trend is becoming increasingly visible in the container segment which is linked with the expansion in this sector.

In addition to these elements, (implicit) political functions and aims are often attributed to ports such as e.g. providing direct and indirect employment, generating trade and economic activities or underpinning municipal funds. Most ports in Western Europe and some in Asia are trendsetters as far as the installation and introduction of economic success standards are concerned. The United States and Africa link the success of their ports far more to such standards as employment, the promotion of trade and economic activity. Such criteria play a role in the way port planning takes institutional shape and the framework used for establishing these criteria. The formation of ever larger trade blocs, the increasingly open nature of the international economy and the standardisation of international service levels is making it increasingly difficult for ports to be directly involved in employment effects and other political objectives. Far more important is the price shipowners and lessees have to pay for using a port.

This does not alter the fact that port authorities will do anything in their power to make their port as attractive as possible and to this end they have a number of instruments at their disposal. In the first place, with the aid of marketing and account managers, information is exchanged regarding the day-to-day running of the port. Management, often in consultation with the local business community and if necessary the unions, organise regular trips abroad to visit companies and ports and foreign delegations are frequently received in the home port. In this context, relationships with sister ports have sprung up over time. The port of Rotterdam maintains Sister Port relations for example with the ports of Kobe and Seattle. These relationships are intended to promote the exchange of know-how and experience pertaining to numerous port activities. On an international scale, the *International Association of Ports and Harbors* (IAPH) has been set up which links all the participating port authorities and organises a major bi-annual conference where various project groups give presentations on a wide range of issues. In the project groups various port authorities with specific knowledge about the problems encountered in each port are represented. The IAPH also functions as a lobby and consultancy group for global organisations such as the *International Maritime Organisation* (IMO) which draws up treaties for the United Nations to regulate international shipping.

In addition to the informative aspects of port planning, the physical aspects represent a particularly important task for the port authority. Physical port planning activities can be broken down into two parts. First of all, shipping lanes to and from the port have to be created and maintained. This falls under the *nautical management* or the wet component of port activities. It involves guaranteeing a safe and smooth passage for shipping traffic by e.g. constructing shipping lanes, locks and turning basins, dredging shipping lanes and harbour basins, erecting buoys and beacons, constructing, managing and maintaining a traffic guidance system, etc. All these physical activities are aimed at receiving ships in a safe and supervised way and are generally considered to be a public task. This is why this book defines nautical management as a separate category of port activities.

A second category of port planning activities concerns the landside or dry component of the port, such as the construction, management and maintenance of quays, sites, terminals, distribution centres, industry, dry hinterland connections, etc. This category of activities is aimed far more at the reception of freight and its handling and onward transport. This is often a private activity and the port authority can only control the physical conditions for attracting enterprises. Important instruments to this end are spatial planning and environmental standards. A port authority will need to devise a *location policy*; situating companies in the most advantageous position so the various companies can gain the most benefit from each other's location. Chemical companies in particular operate in clusters and often seek each other out in internationally operating chemopolies. At the same time a port authority must be abreast of the current environmental constraints, of the effects of dangerous substances and offer waste processing facilities. Particularly in view of the increasingly stringent environmental requirements, which are generally not imposed by the port authority itself, it is no easy task for the modern port authority to unite economics and ecology in concrete projects.

The construction of terminals has become increasingly expensive as ships have become larger and more technically advanced, so that in many ports the superstructure (terminal equipment) is often constructed by the port authority which then contracts out the actual operation of the terminal to a private operator. Another contemporary trend in the financing of the port infra- and superstructure is towards a reduction in the write-down period. In the first decades following World War II it was fairly normal for port infrastructure not to be written off for forty years or even longer. This has now been reduced to twenty-five years with a period of ten to fifteen years having become usual for the superstructure.

An important source of information for the port authority pertaining to both types of port planning activity is the type and number of ships which are currently sailing and which will be built in the near future. As we have already said, the world's major ports are multipurpose ports which have various specialist terminals for specialist ships. In recent decades there has been a steady increase in the number of specialist ships. There are tankers for various purposes (oil, chemicals), container ships, Roll on/roll off ships for rolling stock, refrigerated ships for perishable cargo, ships for liquid gas, dry bulk carriers for ore, coal and grain, car ships, passenger ships and finally ships which can transport different goods in turn. Two examples of this final category are the OBO carriers which are equipped to transport oil and iron ore in rotation and the general purpose ships which transport traditional general cargo and a number of containers at the same time.

In many modern ports increasing use is being made of modern communications technology, such as a traffic guidance system for conducting ships into the port and within the port itself. Another example is the *Global Positioning System* which can pinpoint a ship's exact position. For container transport in particular *Electronic Data Interchange* constitutes an important improvement in the dispatch of goods information. News regarding the arrival and whereabouts of containers can be found by means of electronic information via the computer. More experience is needed in this area in streamlining communications as the desired standardisation is often lacking and the importance of passing on messages is not yet universally acknowledged. The port authorities often do pioneering work in the use of new communication technologies but it is the companies themselves which must take up the opportunities and develop them further.

Due to the many functions which a port simultaneously performs, the port authority, which has primary responsibility for the orderly running of the port area, is faced with a variety of conflicting interests. This is why, in what appears at first sight to be the functional organisation of the larger global ports, a variety of political considerations also play a role.

3.4.　Port Services: Specific Port Activities

A port authority is not the only player in a port, of course, although it can perform a number of public and private activities at the same time. There now follows a brief review of a number of other faithful port actors.

This chapter began by asserting that international trade is a port's raison d'être. Trade is chiefly generated by the *shippers* who wish to transport goods (raw materials, semi-finished or end products) to a client. Shippers might be companies in the port's hinterland but can also be (industrial) companies which have chosen a waterside location at the port. In the present era of specialisation the shippers do not have to organise the transport themselves. To this end they can contact a *forwarding agent* who will manage transportation. The forwarding agent organises this in accordance with the wishes of the shipper such as e.g. the destination and the handling of the goods, which costs are to be charged forward, etc. Forwarding agents often not only manage transportation but also see to the storage of the goods. For sea transport they contact a *shipbroker or shipping company agent* who is paid to act as the (foreign) *shipowner*'s representative. The shipowner sees to the safe transportation of the goods by means of either his own or rented ships. The larger shipping companies often own a number of transport links which gives them the opportunity to draw up a complete transport plan and to reduce extra transaction costs paid to shipping agents and forwarding agents. Along with the shipbroker who tries to amass freight in the port hinterland, the job of forwarding agent has also become precarious.

When an ocean-going vessel approaches a port it is nearly always required to take a pilot on board who, in consultation with the *linesmen* (for the hawsers) and the *tugs*, guides the ship safely to the dockside. Then the discharging and loading of the ship by the *stevedore* can begin. This port actor knows exactly how to balance the loading of the ship so there is no chance of the cargo shifting or in any other way causing the ship to heel over during a voyage.

3.5.　An International Port Classification

As we have already mentioned there is little theoretical literature on the administration and operation of seaports. A famous international exploratory volume which is often referred to is Goss' 1979 work entitled *A comparative study of seaport management and administration*. He makes an initial inventory of the major ports and the formal lay-out of port administration in dozens of countries. Due to its exploratory nature, the study lacks a firm protocol for describing the ports. It is primarily an economically coloured study in which ample consideration is given to financial-economic matters such as e.g. the way in which tariffs come about and write-downs are carried out. The study also reads as a kind of travel journal since it records many of Goss' travel experiences.

During the same period, the European Community was conducting a study into the formal structure of European ports and the financial division of roles between public and private parties. The study served as a draft for the possible format of a European Seaport Policy (see chapter 6). A classification into port types had not yet been made in these preliminary studies by Goss and the EC. Shortly afterwards, however, in the eighties the present international standard divisions into *landlord port, toolport and service port* appeared.

In principle, public administrations can offer (combinations of) various services in ports.

- *infrastructural services*: construction of berths and parking zones for the relevant means of transport, loading zones for goods and so forth;
- *superstructural services*: offering opportunities for handling (cranes), for processing of goods (distribution centres), etc. (services of stevedores and warehousing companies);
- *additional services*: dealing with the administration of transportation and cargo, the organisation of the loading and unloading of goods, etc. (services of shipbrokers and forwarding agents).

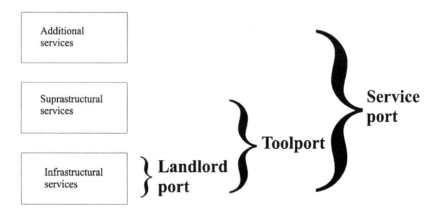

Figure 3.1. Standard International Classification of Port Types

The distinction is roughly based on the economic classification into various *production factors* (land, capital and labour). 'This comparison does not entirely hold water of course, because more than one production factor is usually necessary for the production of a particular port service. However, the comparison is useful as an indicator of the dominant production factor in a particular port service. It provides the opportunity to characterise the nature of the production of port services by the public authorities according to production factor' (Verhoeff, 1981:184). The present international port classification is based on this distinction. If a government only concerns itself with the construction of infrastructure, then the term 'landlord port' is used. If a government also provides the construction and operation of superstructure then the port is termed a 'toolport'. If, in addition, the actual work is also carried out by civil servants then this is termed a 'service port'.

3.5.1. CONTEXTUALISING THE DISTINCTION INTO LANDLORD PORT, TOOLPORT AND SERVICE PORT

One might wonder whether a distinction into institutional port structures on the basis of this classification is still useful (cf. Stevens, 1996). After all, many ports call themselves a landlord port whereas they themselves own superstructural items such as cranes and warehouses or other port services and sometimes even pay the wages of the personnel carrying out these activities. Furthermore, in an analytical sense it is often difficult to distinguish between land and capital because in some advanced terminals, for instance, new underground constructions have to be devised in order to give the cranes the desired range.

This preliminary distinction gives some idea of the role division between public and private actors but does not explain it nor does it show how the various port actors interrelate or to what extent they can form mutual expectations of each other's behaviour. The public vs private distinction, moreover, is very crude because port authorities are themselves often autonomous bodies at one remove from public authorities. In addition, linked to the land production factor, for instance, the division of ownership is not paramount since there can be a clear disparity between the intended use and the actual operational management of the water-linked property (Fleming, 1987). Compare in this context the distinction between construction and operation. Finally, private and public actors may cooperate in holding companies for investing in large projects or starting up small enterprises.

Not so long ago, Goss (1990d) came to the conclusion that in the future there will be three different types of ports, whereby the role division between public and private sector will be far more dependent on the scale size factor than on the distinction in ownership division between land, capital and labour. This criterion is even vaguer than the production factor criterion, however, because it is practically impossible to link the size of the theoretically correct scale to the size which the administration ought to have. The scale factor is thus too fluid a concept to use for classifying a role division between private and public parties. After all, if the scale size is taken as the starting point for the division of responsibilities and this changes as a result of external factors, then the role division will also need to be adjusted.

3.5.2. ANALYSIS OF PORT ACTIVITIES RATHER THAN OF PORTS AS GENERIC ENTITIES

Institutional relationships between public and private actors are of a sustainable nature. People have expectations of each others' behaviour. In my opinion, a classification of port relationships should begin by identifying the principles underlying mutual relationships. The question then becomes one of which role the state assigns itself in an economy. Once this political choice has been made then there should be consistent interventions on this basis. This imposes an institutional framework which determines the limits of a government's policy and provides a somewhat more stable definition of regulatory policy than the rather superficial criteria involving division of ownership into production factors or scale size. In this model, port authorities' proposals for privatisation will always have to be tested against existing regulatory principles.

Without directly making use of a new distinction, it might be interesting to consider the basis on which public and private actors make agreements pertaining to a role division and subsequently to look at the mechanisms which define the agreed norm. Such an approach allows more scope for cultural-historical relationships and norms and capitalises on other motivations in specific cases. There is also scope for arguments rather than just postulating the theory that a port only exists for shippers/carriers and for consumers at the beginning and end of the production chain (Goss, 1990a-d). In addition, one does not have to trust blindly to the hypothesis that increased competition between the ports is due to improved handling techniques and that this is why the management of seaports should follow the international trend towards privatisation.

The approach proposed here tries to analyse the *port community* (Fleming, 1987) with an eye to a wide range of activities which take place in and around ports. The larger international seaports are accorded a complex function; in addition to a handling centre in numerous transport chains for producers and consumers they are linked with specific industries or commercial sectors, they have strong regional or national ties and offer various services for ships, such as e.g. ship repairs and maintenance and the collection of waste or cleaning ships at port reception installations. By aiming the analysis at the issue of the normative foundation of various port activities, the specific structural elements of those port activities become apparent while such an approach also brings to light the nuances in the public-private division of responsibilities. The latter prove to be useful in the thinking on the possible privatisation of the port authority or port activities. Furthermore, by analysing the separate port activities it becomes clear how they can interact and how they relate to the general regulatory principles of the port. Finally, based on the above, expectations regarding the learning capacity of a port can be formulated.

A port harbours a multiplicity of interests which can all be traced back to specific relations between public and private actors. In the examples which were given in the empirical chapters, the relationships between public and private actors and their various institutional safeguards were explored.

3.6.   Summary

The descriptions which have been given in this general empirical chapter on the nature of seaports are intended to acquaint the reader with the specific problems with which port authorities and users find themselves faced. The preface advanced the idea that seaports exist by the grace of international trade and that this trade is often highly unpredictable, while sizeable investments are first needed before ships can be received in the port. The uncertainty regarding the period within which investments can be repaid is a crucial factor in the port world. This is why ports endeavour to tie as much freight as possible to the port by e.g. developing industrial zones or creating good access and hinterland connections. In the following chapters frequent reference will be made to the concepts that have been introduced and to the issues which have been outlined here.

Arguing from the theory, the present economic distinction between landlord port, toolport and service port was introduced. On closer inspection this classification turned out to be too broad for the purposes of the present study and a choice was made to take a close look at a number of different port activities. An initial division of port activities had already been made by distinguishing between the public tasks of nautical management, public and private responsibilities with regard to port planning and finally activities relating to the provision of port services which frequently have a private character. In the following chapters we will explore the type of control relationship between state and market which exists in each of these categories.

CHAPTER 4

The Port of Rotterdam

'Stronger through struggle' (heraldic device of the city of Rotterdam)

4.1.   Introduction

The industrial port complex that has been designated the *port of Rotterdam* stretches for 40 kilometres from the city of Rotterdam at the south-west point of the Netherlands to the North Sea. The total port area covers 10,500 ha, 3,500 of which are water and 4,081 ha. of which were being let out to rent or lease in 1996. 46 kilometres of wharf was being leased in 1996. 15,000 km of pipeline for the (petro)chemical industry runs through the port area. There are 1.1 million residents in the Rotterdam region. In 1996, there were 63,000 jobs directly linked with the port. This resulted indirectly in 276,000 jobs in other parts of the country. In 1998 the port handled about 30,000 ocean-going vessels and approx. 135,000 inland vessels which together transhipped 315 million tons of freight. With the total number of containers handled in 1997 amounting to 5.45 million TEU, Rotterdam retained its position as the largest container port in Western Europe (Centre for Research and Statistics, 1995; RMPM, 1997).

The present port of Rotterdam developed in a number of phases. In his thesis, Van Klink (1995) talks of its evolution from *seaport to port network*. In a spatial sense the port of Rotterdam has undergone explosive growth. This was made possible by sound business sense on the part of both private investors and the city. They acted then and now as the developers and managers of this area. Against the backdrop of the massive expansion which the port of Rotterdam has undergone and continues to undergo, it is interesting to look at how the division of responsibilities between public and private parties is given shape. Furthermore, it is worthwhile analysing what steering capacity the port authority in the largest port in the world has at its disposal in order to ensure that the port meets with the most up-to-date operational and political requirements.

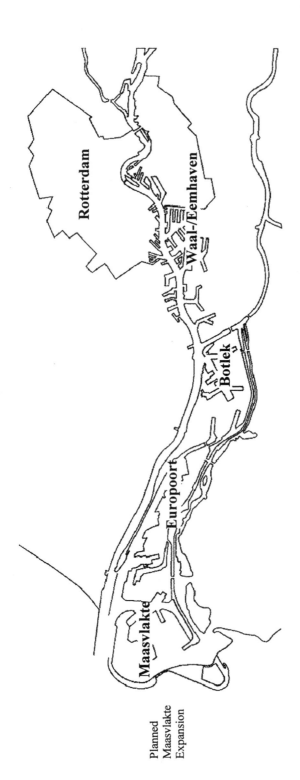

Port Area of Rotterdam

4.2.   Development and Change of Function for the Port of Rotterdam

The development of the port in the twentieth century is divided into two periods by the Second World War. Up until the Second World War the port still performed the transit function for the German Ruhr region whereas after the war an industrial port complex was developed. Early in the twentieth century the limits of port expansion on the right bank of the Maas were reached (border with Schiedam). On the left bank grain, coal, ore and petroleum were already being handled. In contrast to the older ports, the Rhine and Maashaven ports which were constructed early in the century were intended for larger ships with a greater draught. The annexation of Pernis made it possible for the city of Rotterdam to construct new petroleum ports and this occurred between 1934 and 1938. Whereas the older ports in the immediate vicinity of the city of Rotterdam principally handled mixed cargo, the ports in Rotterdam South were intended for bulk goods. A brilliant future was predicted for bulk goods and in Rotterdam South a new city grew around a fast-growing port area. After the war the largest dredged harbour basin in the world was constructed here: the Waalhaven.

The city of Rotterdam was badly hit in the Second World War, and it was many years before it was completely rebuilt. In the period after the war, the RMPM engineers were faced with a difficult task because the Germans had systematically destroyed the Rhinehaven and Maashaven docks in order to delay the allies' landing. The period after the Second World War displayed an unprecedented economic growth which did not slacken until the oil crises in the seventies. This growth can be broken down into four periods:

1) the period 1945-1950: rebuilding and modernising the port;
2) the period 1950-1960: constructing the Botlek plan and various facilities for the handling of bulk goods and mixed cargo in the old existing harbour area;
3) the period 1960-1970: the construction of the offshore development Europoort Maasvlakte, the dredging of a channel for the mouth of the Nieuwe Waterweg (the Eurogeul channel) and the advent of new mixed cargo ships, such as the container ships, roll-on/roll-off ships, LASH ships, etc. and
4) the period 1970 to date: two oil crises made the port of Rotterdam very aware of a number of facts: that it is entirely dependent on the developments in world trade; the almost total disappearance of general cargo in Rotterdam and the rapid growth of container transport; the steadily increasing need for space but also the hard confrontation between economy and ecology in this region. This was previously absent and implies that

further expansion plans must follow completely different decision-making processes than during the first three periods.

4.3.   Rotterdam Municipal Port Management (RMPM)

The RMPM is the municipal department responsible for nautical management and the socio-economic development of the port. It functions as a spider in the web between various local, regional and national tiers of government and the national and international port business community. On the one hand it must promote the public interest in an accessible, safe and clean port and on the other it must ensure that the costs of port investments are recovered in a sound and responsible way. Balancing all these competing interests at once is the major, exciting yet frequently stressful challenge facing the Rotterdam port authority. This section will explore how various relations between the Rotterdam port authority and layers of government, and with the port business community are formally shaped and how they work out in practice[1].

The RMPM was set up in 1932 and had sole responsibility for port management and port planning. The Public Works department was to be responsible for the actual construction and technical maintenance of the ports (Van Walsum, 1972). By integrating port management and port planning in one department, coordination was simplified and for the entrepreneurs it was clearer who they should tackle about port management (the modern 'single counter' idea). 'By establishing the Municipal Port Management department the Rotterdam city council acknowledged the port's great importance for the development of Rotterdam' (De Goey, 1990:253).

4.3.1. POSITION OF THE RMPM IN 1997

The RMPM is a municipal department whose Chief Executive Officer is appointed by the city council. The municipal executive at present has a

---

[1] The ex-director of the RMPM, Mr M. Molenaar, once told me that the role of the RMPM could be clarified with the aid of the following anecdote. A man is travelling through the desert on a camel when he comes to a Bedouin camp where everyone is in mourning for their leader who has just died. The deceased has bequeathed seventeen camels to three children: half for his elder son, a third for the younger son and two camels for his daughter. The problem is, how do you divide seventeen camels into a half and a third? The visitor offers his own camel and the division can begin. The elder son gets nine camels, the younger gets six and the daughter two, which leaves one camel for the visitor to continue his journey on. The moral of the story is that when things are functioning well, the RMPM adds something to the system to help it carry on working, but without intervening directly in any further developments. The RMPM is the traveller and ties itself now to this party and now to that.

separate port committee chairman. He and the port director meet formally once a fortnight at the *Port Coordination Meeting*. Within the political regulatory framework, the port committee chairman is assisted by the port commission from the municipal council. He may ask for recommendations from this committee and other municipal council committees concerning port affairs, such as spatial planning and environment, traffic and transport. Each political party has a representative on the committees. The committee for the port consists of seven people and is headed by the port committee chairman. It convenes once a month. The opinion of the committee carries a good deal of weight as far as the decision making within the municipal council is concerned and their advice on how to vote is in practice always followed.

The director of the RMPM has been given a mandate for various tasks and powers through the Municipal Port Management Mandate Decree. Thus he is limited in his actions by the RMPM budget, the recommended prices for the long let of the sites and the maximum size laid down for a site which is to be let (5 ha.). Furthermore he should take account of the limiting conditions which are laid down in consideration of spatial planning (local plan) and the environment. He is also responsible for the coordination of policy with the administrative department and other municipal agencies and companies. If the decisions or measures to be taken could involve political consequences, the director should provide feedback to the committee chairman or the committee at his own discretion. In an annual report he affirms accountability for the use of his powers. The previous mandate decree requires specific audits to be conducted by the municipal council but nowadays this is laid out in more general terms.

The RMPM is the only agency in Rotterdam with its own budget which is uncoupled from the municipal budget. The reason for this uncoupling, which took place in 1970, lay in the political concern that the explosive development of the port would block the normal municipal facilities (Van Walsum, 1972:151). Since 1964, the RMPM's revenue had no longer been sufficient to cover its investment costs and began to put too much pressure on the general municipal budget. Against the backdrop of the commercial operation of the port, the uncoupling also made the financial transactions more transparent for the port. It is formally expected of the RMPM in the long term that they run a closed commercial operation. In 1996 the RMPM had a loan of 2.6 billion outstanding. The RMPM's sources of income are the seaport dues, leaseholds, ground rents and wharfage, inner port monies and other revenue. In 1996 the total income came to 702 million guilders comprising just under 400 million in seaport revenue and 248 million in leaseholds, ground rents and wharf monies. The RMPM's operating profits were 82 million guilders (RMPM, 1997b). Formally, the RMPM hands over 8% of its turnover (in 1995: over 50 million guilders) to the city but it regularly occurs that the RMPM is also forced to contribute to other municipal expenditure. A part of this can be found in the 'contributions/subsidies' entry in the annual accounts. In 1995 this was a sum of about 20 million guilders (RMPM, 1996c).

In the last twenty-five years the turnover of the RMPM has more than quadrupled and its debt to the city has doubled. The net result varied over time and was only negative in the early seventies. This was due to the tremendous construction costs of the Europoort plan and the time lapse before the first ships could be accommodated.

Table 4.1. Some Commercial RMPM Indicators (in Guilders) Since 1970 (Source: H. Stevens based on RMPM Annual Reports and Annual Accounts)

Indicator	1970	1975	1980	1985	1990	1995
Turnover (in millions)	150.8	275.5	464.2	529	560	680.4
Net result (in millions)	+4.9	- 30.4	+28.2	+56	+ 45	+46
Net result (as percentage of turnover)	3.24	11.03	6.07	10.58	8.14	6.79
Long-term debts (in millions)	1284	1998	2105.3	2399.3	2402.7	2222

4.3.2. POLITICAL DIRECTION OF THE RMPM

The director of the RMPM is formally accountable to the port committee chairman. He must also be available at the port committee meetings to answer any questions. Rotterdam council has a number of formal instruments for regulating the RMPM and adjusting it via consultation. These comprise the annual report, the corporate plan which remains in force for four years, the budget and the committee meetings. The political steering of the RMPM has changed a number of times over the years. Initially, the port came under the mayor of Rotterdam. Between 1946 and 1956 a committee chairman was responsible for the port and from 1956 until September 1970 it was once again a mayor. In 1970 it became once and for all the responsibility of a committee chairman. In the democratisation period of the seventies, great importance was attached to the discussion on whether to place such a task under a politically elected person. There is not a lot of difference between the Port of Rotterdam being directed by a mayor or by a committee chairman, but for the function of mayor as an institution in Rotterdam it means a great deal. Within the framework of accountability, a committee chairman is a logical choice for this position since he can be held directly accountable by the city council whereas under the Dutch system of appointed mayors it is, formally speaking, far more awkward for a mayor to be held directly accountable[2].

From 1932 to 1970 the port committee consisted not only of elected representatives, as is now the case, but also of citizens from the business community (employers and employees). 'The idea was to involve experts from various sectors of the port in the preparation of decisions, by showing the city council or mayor and committee chairmen around the port. This was intended to counterbalance the objections of people who felt that those with direct interests in the port had insufficient control over its management' (Van Walsum, 1972:148). As a result of the wave of democratisation and the desire to separate political and technical expertise, since 1971 the port committee has comprised only elected representatives in order to reinforce the position of the city council, to allow more scope for the consideration of others than those with direct interests in the port and to reduce the risk that port administrators could become the puppets of the business community.

Political control by non-elected representatives over specific questions is now exercised far more by the unions and interest groups and lobbies. They draw up a chair at the policy preparation table which is now less formally organised that it used to be. The Port of Rotterdam has a separate interest group for industry (the Europoort/Botlek Foundation; EBB) and an interest group for the stevedores and shipbrokers (Shipping Union South; SVZ) which also develop a number of joint initiatives. In addition, the Chamber of Commerce plays a role in representing the interests of the multiplicity of shippers. On balance, it is the RMPM and Rotterdam city council who give most consideration to the SVZ and EBB. The latter argue the case for their proposals to the council in shifting coalitions and occasionally in cooperation with the RMPM.

The political control of the RMPM is formally carried out, as mentioned above, by the committee chairman, the committee and the city council, on the basis of the annual plan and the corporate plan, the budget and the question time during the committee meetings. The direct influence of this political control needs to be contextualised. Most council members lack specific knowledge and a committee chairman's term in office is also limited (committee chairman's lack of continuity in knowledge and experience compared to civil servants) and is not necessarily a port specialist. Furthermore, the political administration in Rotterdam lacks a politico-economic policy on the basis of which activities of financially autonomous

---

[2] Nevertheless, the mayor of Rotterdam still can and does intervene in laying down conditions for the port. In the framework of the regionalisation of the administration and the self-reliance of the RMPM, for example, the mayor plays an important political role. The mayor also steps in to fill the power vacuum during a period when there is no committee chairman.

agencies can be assessed (cf. sections 4.4.4 and 4.5). The general administrative department, consisting of financial and legal specialists who have to monitor the planning and control of various council departments, does not actually know what is happening in these various departments because it is dependent on the input from those same departments. The administrative department, for example, occasionally makes use of an external consultancy to monitor the efficiency of a municipal department. It even occasionally happens that the consultancy is subsequently paid by the municipal department concerned.

Formally speaking, the city council can of course implement emergency measures to discharge the chairman of the RMPM but a great deal of political clout is needed for this and it has never yet happened.

4.3.3. REPOSITIONING OF THE RMPM

By means of the 4-year corporate plan (RMPM, 1996a) the RMPM makes it clear to the city council what it plans to do over the next four years and what the city council may expect of the RMPM[3]. In the new corporate plan (1997-2000) the RMPM has brought up substantive arguments to make it clear that it cannot substantiate its future external expectations in the present institutional context. The present financial scope in particular is inadequate for this. Such issues and similar arguments have also played a role in the past. Van Walsum, ex-mayor of Rotterdam, wrote as long ago as 1972: 'The value of an administrative structure depends to a great extent on the financial opportunities which it offers'.

The new corporate plan can be seen first of all as a plea to the city council to allow the RMPM a more independent position whereby the plan increasingly fulfils a political function. The external expectations are not immediately defined by the politicians, however, but follow an initial recommendation by the RMPM. The corporate plan sets the broad course which the RMPM intends to steer and contains few concrete indicators on the basis of which it is held accountable. The new corporate plan brings up the position of the RMPM for discussion and this is by definition a political discussion. The corporate plan is thus transformed from a primarily substantive touchstone into an overture to the political discussion about its position which has yet to take place.

---

[3] The character and content of industrial plans can vary enormously. The previous RMPM industrial plan (1992-1996) focused on the developments in transport and distribution and explored the market options (chemicals, food and distribution) for the port as a whole. The present industrial plan focuses on the RMPM's own goals and is based on general international economic trends which are not port-specific.

Having specified the framework and the function of the corporate plan, it is interesting to look at the substance of the new corporate plan in order to clarify what the business community and provincial and local government may expect of the RMPM over the next five years. First of all a number of trends are distinguished which will influence the development options for the port in the future (RMPM, 1996a:5-7). These involve a globalisation of production and transport, increased flexibility regarding labour and capital, the sustainable development and use of land and energy, increasing difficulties in obtaining a wide support base in society, the increasing mutual dependence of organisations, the increasing economic expansion of the port, the growing arsenal of IT and the increasing competition between international governments and regions. The points on which the outside world may call the RMPM to account are also indicated. These comprise those factors relating to *the locational climate, logistics, distribution, industry and space.*

As a municipal branch of service, the RMPM considers itself inadequately equipped for its task in view of the planned developments and the other demands which the rapidly changing environment makes of it, and thus feels that it should evolve 'from port authority to director and facilitator of not only logistic but also industrial processes, in other words: the image of the RMPM will become more that of a business partner' (RMPM, 1996a:9). This is the central reasoning inderlying the Corporate Plan 1997-2000 and the details for the various policy areas are described in the following pages.

Returning to the statement made by ex-mayor Van Walsum, it is clear that financially speaking the RMPM has already been steering a different course in recent years and it is this course which is presented in the new corporate plan as strategically cohesive policy. A different course is already being steered within the existing institutional frameworks and this contextualises the deliberations on repositioning through the corporate plan itself. Examples of financial innovations within the present institutional setting include: the joint financing, together with central government, of the infrastructure-plus in the Delta 2000-8 project; acting as intermediary in the purchase and resale of terminals (piers 6 and 7); and the establishment of a participation company in cooperation with a private actor in order to help set up small-scale businesses using venture capital.

In addition, in the field of acquisition and leasing the RMPM intends to implement *'strategic acquisition'* (linking companies, customer ties and public-private cooperation), *'market-oriented facilitating'* (see the already existing financial constructions and the tailoring of infrastructure to the wishes of the client), the *developing of products and concepts* (such as the planning of the Waal-/Eemhaven area and Maasvlakte II), *'networking'* (maintaining social contacts and arranging cooperation between companies) and finally *'flanking'* (influencing the decision making on spatial planning, environment, hinterland connections and IT through its own studies) (RMPM, 1996a:10-11). In the field of nautical management, the RMPM wants control over the management of the nautical safety level, transport safety, environmental safety, crisis control and nautical service provision.

From the description a picture emerges of a port authority which wants to be in control as much as possible. There is something to be said for this on practical grounds because the RMPM is familiar with the day-to-day running of the port (expertise), but it remains to be seen whether the tight integration of public and private powers in a form other than the present one will be able to foster cross-fertilisation in the long term. The RMPM already has its own budget and delegated/mandated powers at arm's length from everyday politics which leaves the practical interpretation of the success standards up to the RMPM itself. To whom will an even more independent RMPM be accountable for its actions and how is that institutionally organised? Which safeguards are in place in the other position for maintaining the care which is so essential to the public tasks?

It remains to be seen whether the range of necessary RMPM activities described in the new corporate plan is essentially different from in the past. Plenty of examples could be provided of cases in which the port authority also took on the role of director and facilitator for unique issues which had arisen and for which a different institutional structure was not necessary. Van Walsum (1972) concludes that the municipal structure of the port authority has never stood in the way of port development. Why should that suddenly be the case now? In developing new initiatives it was always the city council's confidence in the RMPM that ensured that the latter never had the slightest obstacle put in its way. Furthermore, the city council functioned as a financial safety net for the RMPM in less economically prosperous times (early seventies) and, as a municipal department, the RMPM pays less interest on its loans that it would if it were a company. Finally the question is whether, in a different institutional structure with different relations and expectations, the RMPM would accomplish its objectives more effectively than in the present familiar intermediary situation. Would the increasing self-reliance of RMPM not lead rather to unclear relations and the exclusion of the outside world in the long term?

Up to now, no concrete proposals for repositioning have been submitted to the city council. The municipal executive will first conduct a study into the shortcomings of the present institutional structure. The results of that study will be published in the course of 1997. In addition to convincing the city council of the need for more room to manoeuvre, the RMPM is also attempting to interest central government in a self-reliant RMPM. Its current involvement in the context of the mainport policy should be a reason for central government to play a role in a repositioned RMPM. In this context comparisons are readily made with Schiphol airport which is managed by a public limited company comprising various layers of government. An extended comparison between the management of an airport and of a seaport does not form part of this study but may nevertheless provide interesting insights. From the perspective of institutional theory, however, there is the far more relevant question of whether central government's temporary financial involvement in and control over infrastructural projects pertaining to Rotterdam should also disaggregate into a permanent control construction.

4.4.   Nautical Management in Rotterdam

In Rotterdam the RMPM has sole responsibility for nautical management. The objective of nautical management is to attend to the safe and fast transfer of shipping traffic and can be broken down into a number of subsidiary tasks. First of all, it involves the supervision of shipping traffic by means of a traffic guidance system and regulations and/or announcements to shipping. In addition it concerns the safety and environmental aspects, e.g. the control of dangerous and harmful substances. Third, nautical management involves the depoloyment of buoys and beacons to aid navigation, e.g. traffic signs, lightlines (entrance to the Nieuwe Waterweg) and for providing anchorage. Finally, it entails 'round the clock' maintenance and crisis control in the port should the need arise. Given that there are over 30,000 ocean-going vessels and 100,000 inland vessels which call at Rotterdam each year, the nautical management organisation is a critical factor in the port's success. The nautical management tasks are by definition public tasks in which the implementation of public authority and functional care are the most important general success standards. It may therefore be expected that nautical management will be implemented by means of the *government monopoly* control relationship. The objective of guaranteeing the safety of shipping in a commercially operating port such as Rotterdam may conflict with the objective of speed. For this reason it is interesting to determine to what extent various quality standards are guaranteed in the institutional framework. This section explores the way in which the position of the RMPM and its relation with provincial and local government is regulated with regard to various nautical management activities. On the basis of the Port Ordinance and Port Regulations on Dangerous Substances the RMPM carries out a number of unique activities (section 4.1). In addition, it deals with the delegated powers and Inspectorates of the Ministry of Transport, Public Works and Water Management (section 4.2) and the involvement of DCMR Rijnmond Environmental Protection Agency concerning floating transhipment and the port reception installation (section 4.3). These diverse relations together provide an interesting picture of the way in which nautical management takes shape.

4.4.1. THE PORT ORDINANCE

The municipal instrument used for regulating issues concerning order and safety in the port is the *port ordinance*. As far as it is possible to ascertain, Rotterdam has been using the port ordinance, in which the position of the harbour master was created and regulated, since 1642. In the turbulent seventeenth century, it was essential to create the institution of *'harbour master'* in order to keep the port attractive to trade, since he served to guarantee a fast and safe passage for ships. Furthermore, his position could be financed from the harbour dues. One of the harbour master's most important tasks in the Port of Rotterdam has always been to act as a facilitator for shipping traffic. Nowadays the principles of *speed and safety* are still essential in order to provide the port with both workable and realistic limiting conditions. Concrete measures and initiatives for modifying the port ordinance are always assessed in the light of by these two principles.

The director of the RMPM's Shipping Department (DSV) is harbour master of the Port of Rotterdam and his primary task is to conduct the nautical management of the port on the basis of delegated powers under the port ordinance. The port ordinance is formally drawn up by the municipal executive and approved by the city council and it lays down the powers of the harbour master. The port users can also address complaints and objections regarding the harbour master's actions to the municipal executive. The port ordinance regulates a broad range of matters relating to the orderly and safe control of shipping traffic in the port itself, i.e it lays down the conditions for the activities which occur in the harbour basins and for which the city is the competent body (the harbour basins within the city limits). These comprise such matters as: determining the area of application and regulating powers for the control, coordination and admission of shipping traffic; allocating berths; instituting procedures for the transport and handling of dangerous substances; allocating port reception installations for ship cleaning; compulsory pilotage; and enforcing order and safety in the port. The port ordinance provides the city (here, the RMPM) with the legal instrument to enforce order and safety in its own port area.

At present in addition to the Port Ordinance there are also a number of other ordinances with regard to sector specific matters in the port, such as the linesmen's ordinance and the towing ordinance which regulate the conditions for these activities (see also section 4.6.2). These rules will shortly fall under the port ordinance which means they will take on the function of a convenient general heading for any port affairs that are still to be regulated. The present Port Regulations for Dangerous Substances is an example of this. By classifying the regulation of various port matters increasingly under the port ordinance they will gain more weight as an administrative instrument in the future. Furthermore, in this way the divisions of authority between the RMPM and the provincial and local authorities will be more clearly laid down; after all, the unique powers of the RMPM are being constantly refined.

The port ordinance can be modified on the basis of changing realities. At the *Coordination meeting* (RMPM top management and the committee chairman) proposals for an amendment are made by the RMPM to the committee chairman and this is subsequently discussed by the port commission (and, if necessary, the commission for legal affairs). The Municipal Executive then submits the proposal to the city council which takes the final decision. If the proposal is processed in the normal way this can be arranged in three months.

The Harbour Master's Staff

The concrete implementation of nautical management (including the operation of the Traffic Guidance System)[4] is carried out interactively by Traffic department staff at the radar posts and the harbour division of the RMPM's Shipping Department. The traffic division mans the diverse radar posts in the various areas of the port (Europoort, Botlek and Stad), oversees the shipping traffic and provides directions when necessary. The central radar post is located at the RMPM's main office and functions as Port Coordination Centre (PCC). 24 hours prior to their arrival in the port, sea-going vessels must report to this centre which then gives them permission to enter the harbour. They are then allocated a berth by the Regional Officer of the watch. Similarly, the PCC must be informed three hours prior to departure from the port or before removing to another berth. Also, the necessary information regarding the ship (name, owner, insurance agent, shipbroker and technical details) and any irregularities (e.g. fire having broken out on board, a malfunctioning propeller, etc.) must be reported to the PCC. In the case of dangerous and harmful substances the PCC plays an important coordinating role in guaranteeing clear-cut responsibilities (the single entrance and exit principle). By means of these aids and information the PCC can supervise the shipping traffic and guarantee safety although this does not mean that every eventuality is covered since a captain always remains responsible for his ship. The PCC also functions as a centre for passing on information to the various port service providers (pilots and tugboats) who can then gear their services to one another. It thus fulfils a clear, coordinating public function.

The harbour master of Rotterdam has staff with policing powers at his disposal. Depending on the particular location in the port (on the river or in the

---

[4] The Traffic Guidance System (VBS) is the successor to the previous shore radar and consists of a radar chain which can spot and guide ships right from the Eurogeul entrance (at Ostend).

port basins) the harbour master's staff have either combined supervisory powers and powers to investigate or solely powers to investigate. The staff have the combination of powers solely by virtue of the port ordinance (in the port basins). Powers to investigate are given them solely under the Pollution of Surface Waters Act (PSWA) and in such cases they cooperate with the Directorate General for Public Works and Water Management staff. The port ordinance regulates the position of the staff charged with supervisory and investigatory tasks ((assistant) shipping masters, dangerous substances inspectors, shipping inspectors and the bridge- and lockmasters). The RMPM (formally the municipal executive) has arranged, in agreement with the Minister of Justice, for these staff to have powers of investigation so that the staff can check licences and draw up an official report, if necessary with the assistance of the (river) police (responsible for the general enforcement of public order). From the water, the harbour division monitors the activities in the port and, when necessary, assists the city police or other inspectors who have powers of inspection landside. Port users are duty bound to offer all possible assistance to these staff; they may enter any space or place in the pursuance of their duty. The Port Ordinance and the Port Ordinance on Dangerous Substances have instructions attached in which the delineation of responsibilities between various hierarchic levels and between the regions and the PCC among others are regulated. The harbour division works round the clock every day of the year and has 18 patrol boats at its disposal.

River Police and Fire Services
The Rijnmond area river police are also active in the port. In 1986, a cooperation regulation was drawn up between the Ministry of Transport, Public Works and Water Management (Directorate General for Public Works and Water Management (RWS) and the Directorate General for Shipping and Maritime Affairs (DGSM)) and the river police in order to reach a more efficient division of tasks. The river police supervises the port on the basis of its task as enforcer of public order and providing assistance to those who need it. They assist the port department when necessary in its management of the fast and safe handling of shipping. In principle, the department which is first on the scene where a direct shipping procedure is called for will take the necessary steps. The division concerned, however, immediately warns the other. For the enforcement of public order, the river police have principal powers to take investigative action and draw up an official report. This power does not affect the fact that the harbour division has investigative competence in its own policy areas (enforcement of shipping regulations, taking a berth and preventing danger, damage or disturbance). The river police have principal powers to draw up an official report in the case of collisions. If it is first on the scene, the port division can start the investigation but in the event

of an accident the investigation is principally the task of the river police. Agreement regarding the method of enforcement and other operational matters is reached in the joint consultation between the divisions and with the Public Prosecutions Department.

Security in the port area is managed by the Port Security Service Department which acts for the members of the Shipping Association South. It is a private security service which supplements the city police within the port area by virtue of the Militia Act. The Department can be seen as an example of self-regulation. The aim of the Department is to police the area and it has its own emergency unit for this purpose. Its work includes carrying out gate porter's services, mobile surveillance services with police-approved dogs and observation duties. The staff are not armed and once an arrest has been made they call in the city police.

Since 1996, the regional fire department in the Rijnmond area (on the basis of a joint regulation between seventeen municipalities) has been formally integrated into one public body with the private fire services of the various industrial enterprises in the Botlek and Europoort areas. The chairman of this joint organisation is the mayor of Rotterdam. In the framework of an effective and efficient safety policy and industrial catastrophe planning, this direct coordination is a logical step which moreover has a positive influence on the business location climate. There is a joint private training centre (RISC Education and Training) with 30% participation from the RMPM. On behalf of the mayor of Rotterdam, the fire department has operational management of industrial catastrophe planning but is supported in this by various municipal departments (including police and harbour division). It is the duty of the RMPM to carry out fire-fighting operations from the water. On behalf of the RMPM, the harbour master of Rotterdam serves on the Public Order and Safety staff group of the municipal crisis staff. In their respective policy areas, the separate departments are responsible for drawing up subsidiary plans for industrial catastrophe planning. The harbour master, together with the directors of other services, advises the mayor on public order and safety issues.

4.4.2. RELATIONSHIP BETWEEN THE RMPM AND THE MINISTRY OF TRANSPORT, PUBLIC WORKS AND WATER MANAGEMENT

The Ministry of Transport, Public Works and Water Management (henceforth Ministry of Transport) is the national waterways manager for the rivers and for the Dutch section of the North Sea. Traffic movements on the water are regulated under the Shipping Act whose policy is dealt with by the Directorate-General for Shipping and Maritime Affairs (DGSM). The DGSM is also responsible for the shipping traffic facilities, such as the buoys and beacons (Navigational Markings Service) and the construction of a traffic guidance system (the traffic department in cooperation with the RMPM). Furthermore, it is the appointed body in matters of 'port state control': activities which are performed by the Shipping Inspectorate in the context of the Netherlands as flag state. Finally, the Ship Monitoring department is a part of the DGSM. The care for the water quality and the technical shipping lane management (including dredging the shipping lanes and recording water depths) comes under the other directorate-general of the Ministry of Transport which deals with ports: Public Works and Water Management (RWS). This is also the body responsible for actually clearing away wreckage in the harbour basins as well as on the river, on which subject it advises the RMPM.

Shipping Traffic Policy
For many years there were a variety of implementing departments at national and local level for the various port areas. Thus, the DGSM had its own regional office and its own National Harbour Division which oversaw the traffic movements on the river. In accordance with the division of ownership between Central government and the city, the municipal port service had the authority over the harbour basins. There was a similar dichotomy regarding the pilots (Civil pilots came under the Ministry of Defence) which sometimes led to a lack of clarity regarding responsibilities particularly in border regions. Furthermore, such a division did not promote a *cohesive maritime policy*, a principle used since 1987 to form the basis for the allocation of responsibility. Thus, in 1988, nautical management in Rotterdam was restructured. The separate pilots organisations were integrated into one organisation which would operate in the future in a self-reliant form. Moreover, the implementational tasks of the National Harbour Division in national waters were assigned to the harbour master of the city of Rotterdam, so that from then on the DGSM would only be involved at policy level with shipping traffic in the Port of Rotterdam.

The new relationship between the DGSM, the pilots and the RMPM was agreed in April 1987 and was laid down in a covenant on 31 October 1988. 'The respective responsibilities may be summarised as follows:

- the (national) harbour master is responsible for the daily supervision of the safe and fast transfer of the shipping traffic in his remit;
- the traffic division is the operational unit which the (national) harbour master has at his disposal; the tasks of the traffic division comprise among others the provision of information and, when necessary, traffic directions;
- the pilot service (pilots) advises the individual traffic participants for the purpose of safe navigation; with the permission of the captain, the pilot

himself can act as traffic participant; in certain circumstances the pilot can perform his advisory function from the shore;
- neither the (national) harbour master nor the traffic division will encroach on the responsibility of the individual pilot or vice versa.' (DGSM, 1994:11).

In the covenant between the city and the DGSM concerning the Rotterdam-Europoort shipping traffic, a number of matters are regulated with regard to the relationship between central government and the city of Rotterdam. First of all, the establishment of a 'Joint Policy Body for Shipping Traffic Rotterdam-Europoort' on which, in addition to the chairman, three representatives chosen by the Ministry and three chosen by the city serve. The primary task of the policy body is to advise the Minister and the municipal executive regarding:
- the shipping policy to be implemented (assessing the implementation regulations of the Pilotage Act and the Shipping Act);
- the tasks, size and service provision of the staff that henceforth come under the Rotterdam harbour master;
- the powers of the city harbour master and the national harbour master;
- making arrangements in the framework of the transfer of powers.

Since the transfer of powers and staff from central government to the city took place, the policy body mainly performs an advisory role towards central government and the city on new policy issues concerning shipping, such as was the case e.g. in the framework of regionalisation. The policy body also advises the Minister when the city nominates a new harbour master.

Furthermore, the covenant contains concrete agreements regarding which powers will shift from central government to the city (article 8, paragraph 3). These comprise powers in the framework of port entry policy, providing traffic directions, the ad hoc enforcement of compulsory pilotage, and issuing ad hoc orders for navigational markings. The tasks which will henceforth be performed by the RMPM for central government are clearly delineated and the DGSM has laid down general policy rules for the implementation of these tasks. In placing traffic signs, the RMPM must also adhere to the guidelines which the DGSM has established for this purpose. In the covenant, provision is also made that the Minister of Transport, Public Works and Water Management (henceforth Minister of Transport) can give instructions in the implementation of those tasks and powers, which virtually never happens since Rotterdam would see it as a breach of faith. The Minister has delegated the management tasks and powers primarily to the RMPM (and even attributed some of them) because on the one hand it can provide the desired quality level and on the other because it is a public department (public law character of the organisation and officer respectively). Despite the relationship based on trust, the DGSM would have preferred the powers to be mandated because public consultation is now no longer part of the decision making process: the RMPM bears the responsibility for these tasks on behalf of the delegation.

The DGSM would only be able to regain control over the decisions of the RMPM by such draconian measures as, e.g., revoking the delegation provisions or by cutting down on funding and/or by issuing directions or instructions. Such actions are not anticipated given the relationship based on trust between the DGSM and the RMPM. The DGSM has also become dependent on the RMPM for the arrangement of matters with a general character so that the principle of a cohesive maritime policy is not always realised. In order to have general application, official announcements to shipping on the *river*, for example, can only be made in accordance with a ministerial regulation. The RMPM still makes use of the port ordinance to make public its own official announcements and assumes that they apply generally (both to the harbour basins and to the river) which is not the case since they lack the necessary legal status. The regulation regarding the duty to report has as yet had no effect because the case regarding the official announcements has not yet been settled and the RMPM is in no hurry for this to happen.

In order to achieve a cohesive structure for nautical management in Amsterdam, therefore, the DGSM will no longer shift to the delegating of responsibilities but will switch to mandating in order to retain an escape hatch. Rotterdam no longer has control over collisions in the shipping lane, in particular, and despite the fact that in 99% of the cases nothing goes amiss, the Minister of Transport is immediately notified of any accidents in Rotterdam because he is politically responsible for maritime safety. From this general framework, the DGSM cannot stomach the fact that it no longer has any control over the Port of Rotterdam. Another example in this context is the DGSM's dependence on the RMPM in implementing legislation. Because the RMPM is lord and master in its own port, this connection produces little result and this may lead to minor flare-ups in the future. In the present situation the DGSM can only negotiate with the RMPM by means of *package deals*, for example when the RMPM needs funds (for the Traffic Guidance System) or when it finds itself faced with other problems. However, this occurs too sporadically to make policy necessary.

Inspectorates
The guaranteeing of the safe handling of traffic is carried out by one of the various inspectorates. Due to the steadily decreasing profits made from maritime transport, it is important on the one hand that necessary investments

for safety are made and monitored and on the other that ships do not remain in the port any longer than necessary. A well-coordinated inspection unit is therefore an important precondition for the success of a port.

In the Port of Rotterdam various inspection units operate which do not always function in a coordinated way. The *Customs* are responsible for the regulation of import and export and, as a division of the Ministry of Finance, it collects the seaport dues for the RMPM. The *Shipping Inspectorate* of the Ministry of Transport is charged with the technical inspection of foreign ships (also called Port State Control: checking equipment, safety and documents). The *Dangerous Substances Inspectors of the RMPM* are responsible for the supervision of dangerous and harmful substances. The National Traffic Inspectorate (RVI) of the Ministry of Transport's Directorate-General for Transport performs a similar role. Finally, in the framework of the enforcement of public order, the river police are involved in the inspection of shipping.

The inspection of ships by a number of different inspectorates tends to delay the turnaround time. For the sake of speed and also to aid cohesive implementation it is in the RMPM's interest that the inspections should be carried out by as few agencies as possible to minimise the chances of the business community playing them off one against the other. In recent years an attempt has been made by means of mutual consultation (declaration of intent) to achieve a reduction in overlap between inspections and in the number of inspections per ship. This is no easy task, since reorganisations in the past have not been carried out with sufficient care so that today a lot of animosity still exists between the various inspection agencies. An example of this is the reorganisation of the Dangerous Substances Inspectors Corps which for some obscure reason meant a large number of national inspectors were brought to the Port of Rotterdam. This led initially to a competency battle with the Dangerous Substances Inspectors of the RMPM because the RVI were accused of incompetence. At a later stage, working agreements were made (the RVI covers inland navigation, roads and explosives) and at present the work is divided on an ad hoc basis. As yet there is no clear regulation for the allocation of responsibilities.

By means of the *Inspection Services Rotterdam Project* that started in November 1994 and that continued for six months, an attempt was made in cooperation with all the inspection services to seek solutions for the complaint made by the Royal Association of Dutch Shipowners regarding the large number of uncoordinated inspections. The aim of this broad cooperation was to reduce the number of inspections on board within the framework of the existing international and national legislation and regulation and without affecting the tasks and powers of the inspectorates concerned. The final report contains eight concrete recommendations aimed at improving coordination between the inspectorates. Although these agencies are still determined not to alter their tasks and powers, it is nevertheless considered desirable to develop a formal cooperative contract in which binding agreements and procedures must be laid down and in which the management of information must be regulated. A desire has been expressed to channel all information via the RMPM system (here, the PCC) which would underline the central role of the RMPM. With a relatively small budget, technical solutions for improved coordination can be achieved (electronic diary, compiling handouts, questionnaires, consistent working methods, data on dangerous substances inspectorate, data on management systems, data on legislation). Although the project group concluded that the competency struggle between various inspectorates has abated and that action had been taken in various areas by the inspectorates themselves to improve cooperation and reduce the number of inspections, only time will tell how useful the presence of this conglomeration of inspectorates actually is. Perhaps it would be even more effective if only two of the seven inspectorates were to remain. The solutions have been sought in the operational sphere and it remains to be seen whether an overlap in tasks and powers is still formally necessary.

A relatively young organisation in the field of inspections is the Green Award organisation. This organisation was established by the RMPM in cooperation with the Ministry of Transport to certify product and oil tankers from 20,000 *Dead Weight Tonnage* which comply with modern quality standards (with regard to ship management and technical safety). Certified tankers then receive a discount on port dues and on various port services (harbour master, pilots, linesmen, tugboats, etc.). The initiative, that started in 1994, has certified thirty ships to date. The Green Award certificate fulfils a supplementary role in the existing arsenal of instruments that the *International Maritime Organisation* (IMO) has at its disposal. The certificate, for which the shipping companies and/or shipowners pay and which is reviewed after three years, has a dual advantage. Ownership of the certificate means in the first place that the quality of the crew, the management and the engineering is checked so that certified ships need to be inspected less often which represents time saving for the shipowner. Moreover, certified ships receive a discount on port dues and other port services, such as the pilotage and towing services.

The Green Award certificate is actually a combination of various inspectorates. The intention is to give the certificate international status via the activities of the organisation. At present, the certificate is accepted by the ports of Spain and South Africa and certified ships also get a discount in those ports. With its recognition by foreign ports and port service providers, the independent organisation seems to be gaining increasing international standing. However, it is as yet financially dependent on the RMPM so that it is mainly the RMPM which is promoting the certificate at international conferences and fairs. It is doubtful whether this will benefit its intended independent character. The organisation has an advisory committee comprising international experts who monitor the standards which the Green Award sets.

Investments

In principle, the investments for nautical management are paid for by the owner. Thus the Ministry of Transport (Navigational Markings Service) finances the installation of traffic signs and maintaining the appropriate draught in the shipping lanes (RWS), while the RMPM is responsible for maintaining the correct draught of the harbour basins. In the past, the financing of the maritime access lanes and the construction of a traffic guidance system took place on the basis of the unwritten rule that central government paid two-thirds of the investment costs and local (or regional) government had to raise the other third. These days, the financial division between central government and the city is project-based and the indicator for central government is whether the city will be able to pass on the costs to the end users. The deepening of the Eurogeul to 68 feet in order to be able to accommodate deep-draughted oil tankers could at the time be passed on to the users. However, due to the oil crisis and the falling energy needs as a result of economic stagnation this proved not to be the case for the next stage of deepening to 72 feet (approx. 24 metres). The RMPM initially paid the full hundred percent but was able to reclaim half the costs from central government due to the guarantee regulation (as safety net).

Another project on which there was cooperation on a project basis between the city of Rotterdam (RMPM and the Public Works Department) the province of South Holland and the Ministry of Transport (RWS) was the construction of the 'Slufter' which was completed in 1987 as a reservoir for contaminated dredging sludge (class 2 and 3). Central government and the city of Rotterdam split the costs for the construction (200 million NLG) and the commercial operation on a fifty-fifty basis. The province supported the project with manpower but is not further involved in its management. The construction of the 'Papegaaienbek' as a reservoir for highly contaminated dredging sludge (class 4) is an initiative from the city of Rotterdam and the Ministry of Transport. In this context, the unique private-law approach to the producers of the contaminated dredging sludge should be mentioned. The RMPM has entered into covenants with Germany and Switzerland in which it is agreed that the RMPM will refrain from further legal proceedings if those responsible for the discharges will employ cleaner techniques.

4.4.3. RELATIONSHIP BETWEEN NAUTICAL MANAGEMENT AND ENVIRONMENTAL REQUIREMENTS

Environmental aspects have also entered the field of nautical management. It is not only the RMPM, however, but also the DCMR Rijnmond Environmental Protection Agency, simply known as DCMR, which are charged with supervising compliance with the environmental rules. DCMR is responsible for the overall environmental quality in the area and provides environmental licences to companies or imposes fines in order to promote innovative behaviour. This section addresses only those DCMR responsibilities which relate directly or indirectly to the nautical management of the port. The relationship between the RMPM and DCMR can be illustrated by means of the following examples of floating transhipment and the creation of a port reception installation.

Floating Transhipment
The Port of Rotterdam has a number of locations where floating transhipment can take place. This forms a relatively cheap facility for the RMPM, supplementing the fixed quayside installations and increasing the flexibility and therefore the attraction of the port (Maritime Economic Research Centre, 1994). At present the floating transhipment handles about 11 million tons on an annual basis. At first sight it might be thought that the control over floating transhipment would be a matter for the RMPM, since it takes place at a specially equipped location in the port and the wet area of the port is the area over which the RMPM has responsibility and control. As a result of a judicial decision, however, it now looks as if floating transhipment in general will come under the Environmental Management Act which is enforced by DCMR[5]. This is intended to create consistency in the supervision and enforcement of environmental rules for the ship-to-shore and the ship-to-ship transhipment.

The impact of the decision will be that for a floating transhipment buoy subject to licensing procedures, the RMPM will lose its flexibility in allocating berths and that costs will rise for handling companies and industry. After all, a licence cannot be cancelled at the drop of a hat and the costs of issuing a licence vary from ten thousand to millions of guilders (Maritime Economic Research Centre, 1994:21).

However painful the decision may be for both the RMPM and the business community, argued from the distinction between imperium and dominium it is

---

[5] Under the Environmental Management Act, a buoy location is defined as "premises" if cargo is transhipped on it for more than 18 days per year.

perfectly tenable. In this example there is a good chance that the nautical objective of speed will be compromised by the guaranteeing of safety and environmental interests. Against a backdrop of checks and balances it would become dubious if the RMPM were to have sole control of the supervision and enforcement of floating transhipment since from a marketing viewpoint it has an interest in keeping the port attractive by fast transfer. This example illustrates that the distinction between *wet and dry infrastructure* is not a sound principle on which to base the division of responsibilities.

Floating transhipment is an example of a situation in which various layers of government with various interests affect the conditions for port activities. Pursuing environmental objectives is an activity which by definition must be carried out by means of the government monopoly control relation. Just as in spatial planning, however, various government interests intersect in the concrete implementation in the port. There is not always direct coordination but where this does occur it does not always have direct results because political issues are involved. For a number of years, a separate division of the RMPM has specialised in consultation with the various layers of government and the business community. It wants to contribute to the development of government policy and performs an intermediary function both internally and externally. An annual report of its activities is published each year. However, the RMPM can do little more than cautiously lobby, contribute to and possibly participate in formal consultation.

Port Reception Installation: Tank Cleaning Rotterdam (TCR)
Another example from the nautical field in which various layers of government were involved and where the RMPM had no direct control over the conditions for effective port management is the case of Tank Cleaning Rotterdam.

This clear example of public and private intertwinement has deliberately been addressed under the heading of the relationship between nautical management and environmental requirements because on the one hand it is about the port's duty to shipping under the Marine Pollution or MARPOL convention and on the other it sheds light on the relationship between the RMPM and DCMR. DCMR acted as coordinator for the enforcement agencies.

Under the MARPOL convention (dating from 1973 and amended in 1978) ports must have a port reception installation (HOI). Under this international convention, governments must ensure that ships have the opportunity to dispose of their waste substances in ports although it does not compel ships to do so. However, it does make provision that dumping at sea must be registered and that tankers must be certified. The international obligation to have an HOI in a port is understood to be a public task. This is also the reason for the public involvement with the HOIs.

An HOI collects ship waste substances, is able to clean ships' loading bays (cleaning) and may process the waste substances into usable fuel. The latter is not a formal requirement. The collector must pass on, at a charge, the waste substances which he cannot process to a processor who must have the necessary licence. The aim of the Netherlands government, however, is to designate as HOI only those companies which can also process the substances. The MARPOL obligation to have an HOI flowed on more or less logically from the existing practice. The market for cleaning tankers had existed since the first half of this century and with the rising oil prices it was worth recovering fuel oil from the waste substances. In the seventies a licensing system and duty to report existed in Rotterdam to regulate the proliferation of collectors.

In the context of the MARPOL obligation and in accordance with the spirit of the times, in 1984 the Minister of Transport elected to hand over the activities for collecting and processing ships' waste substances to a private operator despite the fact that it was common knowledge that this branch of industry comprised only 'grey' and 'black' companies. This private operator would be able to carry out its activities with the certainty that it would receive subsidies and thanks to a unique designation by the RMPM. In any case, entrance to the market was regulated. The Tank Cleaning Rotterdam (TRC) company was chosen as private operator by the Minister of Transport for 'financial reasons'. In view of the contribution from central government (Ministries of Housing, Spatial Planning and Environment, Economic Affairs and Transport) towards the construction of the HOI and its unique designation in the initial phase, the control relationship between various layers of government and TCR can be termed *government conditioning*. On the basis of its public function, TCR had to take all ship waste substances offered and it could only process these if it had a licence; otherwise it had to pass the waste on at a charge to another processor. Government behaviour ensured that the position of TCR was secure until the arrival of another large HOI in the port. The arrival of another HOI was made possible owing to a Council of State ruling rescinding the designatory powers of the RMPM and requiring that Botlek Waste Processing (Booy) should also be designated. This decision deprived the RMPM of a market regulating instrument.

From the start of TCR's operations in 1985 until its bankruptcy in 1994 the government got into one muddle after another over TCR's HOI. From the outset various layers of governments lacked confidence in TCR. As mentioned above, this branch of industry comprised nothing but grey and black companies. It was a very long time before TCR had the necessary licences and it was actually operating all that time in the temporary forbearance zone.

The Commission for the Enforcement of Environmental Rules Tank Cleaning Rotterdam blamed the lack of adequate governmental supervision of TCR among other things on the 'institutional errors' with which TCR's operations were ridden and which in the long term could not be guided by political-administrative controls. For example, an HOI generates income before any costs for processing are incurred. Moreover, the lion's share of the start-up costs for construction of the installation were reimbursed by central government which failed to adequately carry out the audit of the accounts. During operations it appeared that the information asymmetry between TCR and the various layers of government was so great that TCR's cover-up devices were successful for a long time. It also appeared that the granting of subsidies to TCR resulted in (political) dependence and 'policy entanglement' and it was not until 1993, after all the official reports had been drawn up and after considerable negative publicity about TCR, that a criminal investigation finally got underway. All in all, it took ten years for the huge environmental scandal to surface for which a number of successive governments must share the blame.

Various layers of government, through diverse legal frameworks, imposed requirements on the TCR's HOI premises and enforced these using their own agencies. The Ministry of Transport (DGSM and RWS) is responsible for implementation of the MARPOL convention (control and inspection of discharge equipment and storage tanks and availability of HOIs) and guaranteeing the water quality (monitoring and licence preparation). The Ministry of Transport is responsible for the permitting for the processing of chemical waste substances and ensures an adequate collection and processing structure for them. The province issues licences and, through the public body of the DCMR Rijnmond Environmental Protection Agency, both carries out preventive checks and imposes sanctions. DCMR also draws up licences. The RMPM has powers to investigate under the Pollution of Surface Waters Act (WVO); it undertakes the investigatory task on behalf of the Ministry of Transport (RWS) and has a group of DGSM-trained MARPOL inspectors[6] for this purpose. From various studies the conclusion has been drawn that although consultation between the various layers of government had taken place under the direction of the province/DCMR, there was no real coordinated action. Once political interest in TCR declined or even disappeared due to a change of government at provincial level, the coordinating role of DCMR also appeared to be played out. The political control method thus proved far too erratic for this institutional relationship and absolutely no match for the opportunistic behaviour of TCR. Even when

---

[6] The enforcement of contraventions of discharge provisions is carried out under criminal law.

considerable political interest focused on TCR it proved impossible to take a hard line against the company due to its (political) dependence.

For all government parties the whole TCR affair was a very frustrating business. The example demonstrates that the production of a public good by a private monopolist (later an oligopolist) - at any rate without stringent regulation of the concessions - leads to countless opportunities for passing the buck and shows that political control in such cases is not effective. Due to the large number of government parties involved it was not clear who actually had an interest in effective control. The port authority has an interest in a good service level at the port but in this system it was not able to take real corrective action. In the early stages, the offer made by the City of Rotterdam to participate in TCR was baldly rejected because national government's priority was the goal-rational consideration of privatisation regardless of the desired market regulation.

Fairly soon after the creation of TCR, the HOI designatory powers of the RMPM were overruled by the Council of State so that the RMPM could then only bring TCR into line by either taking water samples and drawing up official reports or by means of their lease. This latter option was never really considered and in 1991 the RMPM was requested by the Public Prosecutions Department to stop taking water samples. Finally, TCR pleaded guilty to the discharge violations in bulk and settled out of court and about a quarter of the charges in the official reports were finally dropped. The port authority's control of TCR was rather limited in view of the interests it was supposed to be promoting for the Ministry of Transport, i.e. to secure the presence of an HOI. DCMR enforced the rules primarily in the interests of the environment but could not prevent public resources being misused and public requirements not being fulfilled.

The TCR case is a good example of the impossibility of government steering when the institutional format does not fit the nature of the service to be provided, despite any number of controls and other enforcement instruments. Criminal law provides the emergency exit in the final analysis. The Enforcement of Environmental Rules Commission Tank Cleaning Rotterdam quite rightly concludes that there were *institutional errors* which in my opinion cover a broader area than simply the fact that 'the cart was put before the horse'. The abundance of enforcers and the distance between administration and private implementation places a great strain on the mutual coordination mechanism which apparently cannot be sustained in the long term.

In the meantime, the bankrupt TCR assets have been bought up by another private company which will plough sixty million guilders into the reconstruction of the installation. The RMPM has designated it as HOI and in this new situation the government control (DCMR) appears once more to be concentrating on laying down rules by means of licences, auditing the bookkeeping and taking samples. This type of monitoring represents a huge commitment for DCMR in terms of manpower and political interest weakens after a while. It remains to be seen whether in the long term the same risks will not arise again. After all, the state-market relation remains unchanged and the public service obligations remain the same. The new private operator may be less 'grey' than TCR but in the new situation an abundance of inspectors and enforcers will also be involved and it remains unclear who is ultimately held accountable for the control deficit.

More generally speaking, the market anticipates that due to technical improvements on board ships, the waste substances will in the future decrease in volume and be more concentrated and therefore no longer suitable for processing in the present standard installation. Transfer of ownership of the HOIs to a couple of internationally operating companies also means that the authorities will have increasingly less control over the methods of cleaning and processing. The Rotterdam approach demonstrates very little insight into such developments, however.

Little imagination is needed to transfer the conclusions drawn from the TCR case to other port activities. During the activities of the processing industry in the port, large amounts of waste substances are also released and dangerous/harmful substances are also used which must be inspected by various public enforcement agencies. Once again, these are complex installations and careful handling is not directly in the industry's interest. This book does not further explore the environmental enforcement in this port sector, but can provide useful insights. Particularly in those economic sectors in which multinationals chiefly operate, which can wield a considerable degree of national political control and which use highly complex technical systems, for practical reasons environmental objectives will be defined in consultation. This is also expressed in the way in which the enforcement is given concrete shape and it would be interesting to determine to what extent such co-arrangements safeguard the care for the public interest.

4.4.4. CONCLUSION: NAUTICAL MANAGEMENT

The activities which fall under nautical management have a public character and traditionally formed part of the exclusive domain of the RMPM and the Ministry of Transport. Until 1988 there was a strict division of responsibilities; Rotterdam was responsible for the harbour basins and the Ministry for the river. Since 1988 the Ministry has delegated its responsibilities to the RMPM. This was intended to promote the cohesive maritime policy and it fit in with the RMPM's existing expertise. It is debatable whether the relationship between the RMPM and the Ministry of

Transport as regards the shipping traffic policy is based on sustainable regulatory principles. At present, the aim of a 'cohesive maritime policy' is to provide justification for the high degree of RMPM autonomy with regard to nautical management in the port, but in the long term it could encourage technical dilettantism. After all, what mechanism is there to induce the RMPM to be accountable for these activities?

In the present situation, it is not surprising that environmental competencies come under other government agencies than the RMPM. In this way the RMPM's autonomy is relativised to some extent. DCMR supervises and enforces ship-to-shore transhipment and floating transhipment. This would seem to supersede the traditional regulatory principle of wet versus dry port management. The activities relating to the various ship inspections have also been given no sustainable regulatory principle to date. It may be anticipated that in the future old problems will reemerge. Finally, there appear to be serious objections to the government conditioning of related nautical management activities, as the TCR example clearly showed. The general character of the conditions was partly to blame for the fact that there was so much room for evasive behaviour.

4.5.   Port Planning in Rotterdam

In the second half of the twentieth century a tremendous industrial port complex has sprung up between the city of Rotterdam and the North Sea. Port planning and development are primarily the task of the RMPM as manager of the sites and also because the RMPM is judged by a growth in income (rents and ground lease) and freight (more wharves and sites mean more ships and yield more port dues). Opportunities for the port authority to interpret port planning and development at its own discretion have decreased over the decades as a result of spatial planning and environmental requirements. Moreover, the RMPM is less close to the market than to the commercial sector so that many initiatives for the improvement of its competitive position are given shape in concerted action with the business community. This is why it may be anticipated that in the various port planning activities *government conditioning and co-arrangements* will be encountered as control relations. The role of the RMPM in relation to other layers of government and the port's industries will be explored in more detail in the light of the port's development in this century.

General Principles
As already mentioned in the previous section on nautical management, in the past there were clear agreements between central government and the city regarding the financing of maritime approaches to the sea and hinterland

known as the '2/3-1/3 scheme'. Nowadays, the approach tends to be project based and the division of costs is determined separately for each project. A similar approach is used for the region-specific port projects, such as the proposed construction of a second Maasvlakte. The forthcoming railway connection, the Betuwe Line, which is seen as a hinterland link, will be financed entirely by central government (including the section of the line which is located in the port area). The relationship between the RMPM and the business community has also always been clear. The RMPM is responsible for the construction and operation of adequate infrastructure. It finances the construction of new bays, sites, quay walls and so forth. Maintenance is also funded by the city, such as repairing quay walls and maintaining the appropriate draught in the harbour basins. Central government does not contribute to these costs. The business community is responsible for the construction and maintenance of the superstructure (cranes, sheds, warehouses, office space, loading and unloading gantries, trucks, etc.). This is not to say that the RMPM has no direct or indirect involvement with the superstructure. On the one hand, efficiency considerations may provide a reason for it to become indirectly involved with the deployment of superstructure. At the RMPM's insistence, various stevedore companies have merged to form the ECT (Europe Combined Terminals) and various petrochemical companies have established the Maasvlakte Oil Terminal which has reduced the need for the RMPM to build quay walls. On the other hand, the RMPM plays an increasingly facilitating role (see examples from the corporate plan 1996) in order to make a contribution to companies in constructing superstructure, starting up activities or expansion.

The port sites are rented or leased and never sold. 'The reason why Rotterdam only rents or leases its sites lies in the fact that the public interest requires that the city should retain as much control as possible over the use of the sites now and in the future. This policy also has its disadvantages, of course, for example in attracting major industries which can sometimes buy sites very cheaply in foreign ports. (..) Rotterdam has certainly encountered difficulties in attracting industry due to its leasing policy, although I can't think of a single industry which has rejected Rotterdam solely for this reason. The port authority has certainly benefited from it because it has always endeavoured to create the best possible facilities for the port users' (Posthuma, 1972:34).

In order to explore the various relationships between the RMPM and the external environment, port planning and expansion in Rotterdam has been broken down into four periods. The first runs up to the Second World War during which the expansion of the general cargo docks was continued. The period following the Second World War can be subdivided into three periods: from 1945 to 1970, from 1970 to 1990 and from 1990 to the present day. In the period between 1945 and 1970, the main success standard was growth and the RMPM had no need to demonstrate the necessity for expansion and received no criticism when carrying out the construction. After 1970 the situation changed drastically. Entirely different quality standards were applied to port planning so that the powerful position of the RMPM was relativised by the policy of other layers of government. After a 'pure culture' period (De Goey, 1990:262) in the early eighties it appeared that industrial development would once more be permitted. However, it was not until 1989 that the RMPM formally integrated its attention to industry, transport and distribution (RMPM, 1989).

Until 1932, port planning and development mainly came under the former Municipal Commercial Premises Department (now Public Works) which had a predominantly civil engineering approach to port management. It constructed docks which were specifically intended for the transport and transhipment of bulk goods. In the period prior to the Second World War, the city tried to reduce the distribution's sensitivity to economic fluctuations by attracting industry. It got no further than the planning stages, however. Initially, the call for industrialisation appeared to be an arbitrary argument that vanished when trade picked up again. The period of economic decline turned out to be too short to initiate industrialisation. Moreover, the Dutch pre-war entrepreneurs were not interested at all. The policy of the municipal authorities was passive and compliant rather than enterprising. It did not construct new docks until the demand for berths was overwhelming. The city council was only compelled to action by the massive unemployment: minimum prices, loans and the construction of a petroleum port in Pernis. Finally, the industrialisation in Rotterdam itself was also slow getting underway due to a shortage of suitable industrial sites.

4.5.1.   THE IMPLEMENTATION OF THE BOTLEK, EUROPOORT AND MAASVLAKTE I PLANS

Immediately after the war, efforts concentrated on the reconstruction of the port. Within only five years the port was made fully functional once more. In the early post-war years the city and the business community collaborated closely. The 'Reconstruction Commission' was set up to 'ensure the necessary contact between the city and the business community in order to make the best possible use of the overall port layout'. This commission possessed considerable expertise, particularly in the technical field, but funds were lacking. These had to come from central government since the reconstruction of the port was in the national interest. The cost of the port reconstruction was estimated at around 40 million guilders. This government agency, the Rotterdam Reconstruction Agency, worked in productive cooperation with the

RMPM and Public Works. Not only was the infrastructure restored, but a number of modernisations were introduced such as deepening the draught at the quays and increasing their permissible load plus a more efficient layout of sites.

During the Second World War, directors and civil servants from various municipal agencies gave talks and presentations on what should be done with the port after the war. The intention was to branch out in a different direction from in the thirties by making it less dependent on economic fluctuations which had been the case with the transit trade to Germany. Rotterdam itself had to find a way to hold onto industry and the work flowing from it. The *Botlek* plan which appeared in 1947 was a product of the war years and was wholly based on this vision. The RMPM was supported in this plan by the business community which insisted on upholding the old motto of the port: the ship must not wait at the quay, but the quay must wait on the ship. This turned out to be a good maxim but most entrepreneurs in the early fifties were biding their time. The RMPM was thus the lone driving force behind the development of the port. Although after the war there was a rush to repair the destruction wrought by the Germans and implement modernisations, it was another eight years before the first lessee was assigned his place in the Botlek.

There were a number of reasons for this. For a long time there had been an ongoing discussion on the issue of which ships this area should be able to serve. The compulsory purchase of the land was also a very slow business because the farmers held out for a good price for their land. Moreover, they had the backing of the Province of South Holland and the Ministry of Agriculture who approved the plan in 1950 and 1953 respectively. This was a new phenomenon for the RMPM because the expropriation of land had always been very easy in the past. It will be seen that this clash between the RMPM and the province was by no means the last. Posthuma, the RMPM's ex-Chief Executive Officer notes that: 'The Province of South Holland provided little cooperation in this matter of compulsory purchase, which in fact was the case in nearly all matters relating to the development of the port between 1945 and 1970.' Furthermore, the Botlek plan was being attacked from Dordrecht: it was opposing the construction of a Botlek bridge because this would mean less ships would be able to enter the port of Dordrecht. A third obstacle that was linked to the first was the requirement imposed by agriculture that in the plan use should be made of locks. This frightened off the business community (transhipment companies and petroleum companies) because the construction of locks would mean delays and an increased risk of collisions. It was only by remaining obstinate that RMPM were able to remove this point from the agenda. In 1955 the American Dow Chemical company was the first enterprise to be located in the Botlek area.

The second major plan to be developed after the war and which moved the port of Rotterdam a little nearer the North Sea was the *Europoort* plan. During the decision making pertaining to this plan, the Suez crisis (1956) flared up which ultimately led to the Suez Canal being closed for twenty years. Because ships could no longer use the Suez Canal they had to reach Europe via the Cape of Good Hope. This meant an unprecedented scale increase in shipping. Within a very short time, the tonnage of oil tankers shot up from 60,000 to 100,000 and 250,000 and there was even a ship built for Shell (though it was never put into service) with a 500,000 DWT. The scale expansion also entailed consequences for the lay-out of new ports (turning circles and depths) and by anticipating this (in so far as that was possible) Rotterdam ensured that this industry would choose this location.

In terms of area, the Europoort plan was three times larger than the Botlek (3940 as opposed to 1250 ha. gross area) and was implemented more quickly but preparations were entirely different than they had been for the Botlek plan. 'Whereas these (previous Rotterdam port plans, HS) had been developed in a relatively small circle, in which only the Rotterdam agencies (...) and RWS had been involved, Europoort was the work of an extensive commission. Not extensive in the sense of size but rather as regards composition. It comprised civil servants from various departments, not only RWS. In the Europoort Plan, considerable account was also taken of planning aspects rather than just port construction' (De Goey, 1990:92). The working party, entitled the Rijnmond Harbours Development Group, comprised 19 people of whom 7 were from Rotterdam. Rotterdam was thus in the minority. One point of contention between Rotterdam port opinion and national planning opinion was the issue of whether or not to relocate the village of Rozenburg: Rotterdam was in favour but it was refused for planning reasons. Ten years later Posthuma, the RMPM's Chief Executive Officer, was proved right by those same planners but by then it was too late. It was nonetheless the first successful attempt by central government to influence port development via formal channels, so in that sense the Europoort plan was not Rotterdam's plan.

The RMPM had learnt its lesson from the clashes that had occurred in implementing the Botlek plan and in 1958, after the foundation stone had been laid in Europoort, the first ship could be received in the port. In contrast to the approach used in the Botlek plan, this time the RMPM organised a comprehensive information campaign for the local residents and other parties involved. After all, the villages of Nieuwsluis and Blankenburg, and ultimately the De Beer nature conservation area, would have to be cleared to satisfy Rotterdam's hunger for land. It was also the spirit of the times in economic thinking, both at local and national level, whose flexibility ensured that the construction of Europoort entailed relatively few problems. The Port of Rotterdam was seen as a prime example of a macroeconomic instrument.

Due to economic growth and the increasing demand for industrial sites, political decision making regarding the first *Maasvlakte* was more or less bundled in with the Europoort plan. While the construction of Europoort progressed, plans had already been developed for the further expansion to the west. Financially, the two port plans were linked together because the first funding for the Maasvlakte came from the Europoort project budget. The Europoort plan that covered 3940 ha. gross area (1800 ha. net) ultimately cost 712 million guilders. In comparison with the Botlek project which involved 1250 ha. gross area (750 ha. net) and a cost price of 165 million guilders, this was a huge cost increase. It had already been foreseen in the early seventies that the first Maasvlakte would be an expensive project, but it was calculated that the costs could be recovered from the oil terminal and the power plan. The first Maasvlakte has a gross area of 2690 ha. (1270 net) and its construction ultimately cost the city 689 million guilders. In this context it is interesting to hear that for the projected construction of the Second Maasvlakte cost prices of 8 billion are being mentioned.

Closely related to the decision regarding the construction of the first Maasvlakte is the construction of the Eurogeul which will enable accommodation of the largest tankers. In close consultation with nautical experts from Shell a decision has been taken to commence construction. In 1996, the estimated construction costs amounted to seventy million guilders with the annual maintenance costs being estimated at three million guilders.

Locational Policy

The supply of companies was so large that the RMPM management had to observe a strict location policy. 'Rotterdam had no complaints about a lack of interest from the business community. Between 1955, the first leasing in the Botlek, and 1969, 54 lease contracts were approved by the council. Of the 54 new lessees in Botlek and Europoort, about half (28 companies) were industries. The others, 26 in all, were storage firms and handling companies (bulk goods)' (De Goey and Van Driel, 1993:119). In the planning process a choice had been made for (petro)chemical industries, iron and steel industries, food companies and car industries so that new companies had to meet these criteria. New companies primarily had to generate considerable 'tonnage' from which Rotterdam could obtain its income in order to recover the costs on its investments. Moreover, in the early sixties there were to be no labour-intensive companies because the labour market was too strained. For example, Ford wanted to open a tractor and truck factory in the Rijnmond area but this was considered inadvisable due to the shortage of labour. Finally, companies were excluded from Rotterdam on environmental grounds. This criterion was to gain increasing emphasis in later years.

A first taste of the fact that account would gradually need to be taken of other quality standards was the dispute between the RMPM and the province/central government in the first half of the sixties on the subject of the '*demarcation line*'. The demarcation line (laid down by central government in 1964) indicates the southern limits of the Maasvlakte and sets the temporary boundaries of any further expansion of the port area. This decision put an end to the spatial growth thinking and the RMPM began to calculate how much land it could still grant. From that moment it also became clear to the RMPM that it would have to take account of other forces, forces which could at times be far more unpredictable than the input of central government in the Rijnmond Harbours Development Group.

But even after setting the demarcation line, the Rotterdam central council did not simply resign itself to the situation. Contracts with the province of Zeeland were negotiated and the provinces of North Brabant, South Holland and Zeeland were in contact with each other regarding the 'utilisation of space'. In 1967, this resulted in the creation of the Consultative Body on Seaport Development South-West Netherlands of which expectations were high. The body was given too little clout in terms of financial resources and authority, however, to be able to achieve any real coordination between the various partners. The setting up of an acquisitions agency was a complete farce because there was no overview of site types and no price lists. Furthermore, no trips were made abroad to attract companies. 'The activities are aimed at the exchange of information regarding the notification of industries locating in the area and effective two-way communication in the hope of allaying the existing differences of opinion between Zeeland and Rotterdam' (Hazewinkel, 1978b:6). Criticism in the media focused on the fact that the consultation was not effective as real cooperation because as soon as the arrival of a new industry was announced everyone once again actively promoted their own interests. 'Not a bad attitude in itself, but it didn't work, it was a bubble. There was too much rivalry between the partners. The law of the jungle always applied; in concrete terms it achieved very little. It was a non-starter' (Hazewinkel, 1978b:6). In the present period of port planning and development we see an entirely different set-up at the RMPM: more concrete projects are being created which should prove more viable.

4.5.2.   PORT PLANNING 1970-1990

The early seventies were a turbulent time for the RMPM. In the 1970 elections, older city councillors were replaced by young ones who no longer worked on the exclusive assumption of port interests. Moreover, the portfolio for the port passed from the mayor to a separate committee chairman who immediately converted the opinions of the young councillors into policy. In the third place, the port commission was organised on which henceforth only elected representatives would serve. And at the town hall a new secretariat was formed for Port and Economic Affairs. All this indicated that after the previous decade in which the RMPM had reigned supreme, the town hall was now looking for more status. A number of initiatives had previously been taken even before the municipal executive or the city council had approved them. Moreover, the distance between the city and the business community was deliberately increased. 'The old method of arranging matters within a small, closed circle was strongly criticised. The carefully-erected post-war network of formal and informal relations between local government and the business community was broken up. The result was that an increasingly large group of people began to get involved in the location of companies. In other words: there was a sharp increase in the number of "counters"' (De Goey, 1990:259).

From the early seventies, central government adopted a different stance towards the port of Rotterdam. It was felt that the environmental pollution in the area should not be increased and it was deemed necessary in the framework of regional-economic policy to develop other areas of the Netherlands. This was termed the 'dispersion policy' and had been practiced by central government since the mid-sixties (policy document on Seaports 1966) and was to cost Rotterdam and its immediate environs a number of interesting companies[7]. The initial reaction from Rotterdam was a complete rejection of this policy but gradually the Rotterdam city council went along with it and in the early seventies it was disaggregated to the location provisions. 'The most important change involved the extraordinarily reticent stance of the central council with regard to industrial locations. New locations were vetoed, and expansion of existing companies discouraged' (De Goey, 1990:259). During those years, the central council made a conscious decision to focus on the transport and distribution sector which was felt to have been neglected. This is surprising considering the earlier experiences with the sector's sensitivity to economic fluctuations in the pre-war years. The further development of Europoort and the Maasvlakte would be subjected to the highly critical scrutiny of the city council. In particular there were heated and protracted local discussions on the projected use of the Maasvlakte. Agreement was reached fairly quickly with regard to three projects: a municipal power station, a related coal and ore storage firm and a joint oil terminal for the largest petrochemical companies in the port. In 1971, a

---

[7] Thus Rotterdam missed out on a number of companies. BasF, Bayer, Esso, Degussa and Union Carbide went to Antwerp; Ford went to Genk in Belgium; Volvo and Texaco chose Ghent; Amoca settled in Geel (Belgium) and Dupont de Nemours chose a location in Ireland (De Goey, 1990:160).

contract was signed with the Maasvlakte Oil Terminal and in 1972 with the present EMO.

As for the rest of the site, there were two options: a steel company or a container handling company which could not both operate on the Maasvlakte due to the space they required and the spatial planning criteria. It looked as if the container handling company would win hands down. After all, the criteria on which the decision would be based (employment, added value, income for the RMPM, taxes from the traffic infrastructure, investments and pollution) were all in favour of the clean container handling company for which a bright future was forecast. Furthermore, central government had thoroughly scuppered the plans for a steel factory by its Selective Investment Regulation (comprising an investment levy, a licensing system and a duty of notification) and various environmental acts (Pollution of Surface Waters Act, Air Pollution Act, Nuisance Act). Finally, the RMPM's contacts in the steel world turned out not to be as effective as those in the oil world. Since before the Second World War, Rotterdam had been wanting a steel factory within its civic borders. Many years on it had still not succeeded in attaining this desire. The competition in the steel world was far greater than in the world of oil and furthermore the steel world proved to be a far older world with other values and standards than those of the relatively young oil world. It took a long time for these facts to sink in at the RMPM.

From the early seventies, the RMPM was also increasingly restricted in its port development and planning by the behaviour of the province which is charged with enforcing the various environmental Acts. As a result of the anti-growth attitude of central government and the province, the relationship between the RMPM and the business community deteriorated. The RMPM became tired of having to reject companies and started to refer them to the town hall.

4.5.3. 'LATEST' DEVELOPMENTS (1990 TO DATE)

1991 saw the publication of the concept version of the Port plan 2010 which was to prepare Rotterdam for the next century. The objectives of the plan are as follows:

- The promotion of activities through:
  - increasing added value and employment by creating favourable conditions;
  - devoting more attention to the industrial and trade function of the port by improving hinterland and pipeline connections;
  - devoting extra attention to goods flows in growth sectors which have a high social yield and endeavouring to create a mainport for these.

- Optimising space through:
  - clustering functions: grouping similar activities together;
  - optimising the use of space by redeveloping older port areas;
  - possible cooperation with Vlissingen, Moerdijk and Arnhem/Nijmegen.
- Improving the environment by:
  - developing the port on the basis of sustainable development;
  - laying down strict, clear-cut environmental rules;
  - endeavouring to establish environmental policy internationally.

The Central Planning Agency's goods flow model formed the quantitative foundation for the plan. Although the plan does not specify the various alternatives, it remains flexible due to the prognosticative growth figures for trade (low, middle and high scenario). If reality does not come up to the prognoses, the timescale for achieving the concrete objectives should then be adjusted.

The plan was designed after identifying a number of trends in world trade and shipping which affect the layout and operation of the port of Rotterdam. In the first place, the '*globalisation*' of the world has become an increasingly familiar concept. The competition between various parts of the world to provide the location for assembly plants has become increasingly cutthroat. This concerns the locating of factories in regions which have a large market or which generate substantial cost savings. In particular, the integration of the former Comecon countries and the great strides made by the Asian Tigers (Singapore, Hong Kong and Taiwan) are important in this regard. The search for scale advantages by the major companies means they only have to maintain a small number of locations worldwide. This results in relations at ever greater distances but this is perfectly viable given modern telecommunications. In the eighties, the container shipping companies cooperated in 'conferences' but following a number of megamergers that period now seems to have made way for competition between various major shipping companies[8]. The 6600 TEU ships belonging to these companies only put in at ports which can handle them fast and effectively.

In Rotterdam, the ECT terminal on the Maasvlakte was fully equipped to accommodate these ships under the Delta plan 2000-8. This type of ship needs to be kept permanently running for reasons of cost and has neither the time nor the physical ability to stop at every port on a continent. This is where the

---

[8] In the world of container transport 'conferences' (international collaboration between shipping companies/sea carriers play a major role. The economic concentration of power continued in 1996 with a merger (between Nedlloyd and P&O) and a collaboration (Maersk and Sea-Land).

'*mainport*' concept comes in which designates the most important port for a shipping company on a continent. Rotterdam had been the mainport for wet and dry (ore) bulk since the sixties but not yet for containers. The major container shipping companies have a strong negotiating position in Western Europe due to the presence of a number of large container ports (Antwerp, Rotterdam, Hamburg).

A second trend which has a huge influence on port planning and port development is the increased attention devoted to quality. In this context the maxim 'from tonnage port to value port' is telling. This is expressed on the one hand in the criteria used in decision making and on the other in the nature of the work that the port wants to undertake. As far as the latter is concerned, Rotterdam no longer wants to bring into port as many tons as possible and then ship them straight out again but would prefer to do something with them, such as stripping and stuffing containers, and also wants activities derived from the distribution function (packaging, assembling, processing). This concept is not essentially new since the transformation from a transit to an industrial function had already been made after the war, but the awareness has once more arisen of the transit function's sensitivity to economic fluctuation. It remains to be seen whether the RMPM will be able to use the port to pursue employment policies. The market finds its own level which in concrete terms means that many container transport activities which yield added value take place in the port's hinterland.

A project initiated in 1993 that fits within the framework of quality improvement in the harbours is the redeveloping of the relatively old harbour area in the Waal- and Eemhaven. This area produced good money for the RMPM but with the trend towards the west and the scale expansion it had been rather overlooked. Renewed attention was needed, however, since the area was slowly decaying (poor infrastructure, congested roads, inconveniently arranged businesses, etc.). Although the RMPM had plans for the area, it was faced with a number of limiting conditions. In the first place, because the area includes one residential district (Heijplaat) and is adjacent to another (Charlois), the plans had to be watered down both on the grounds of environmental considerations and from the rationality of port development. In the second place, the RMPM had no actual experience with the redeveloping of sites because until now new areas had always been constructed and built on. In the third place, it was not the RMPM which was directly involved in the development but a different municipal department which did not immediately grasp the rationality of port development. This meant that the drawing up of plans was a slow process and the RMPM first had to make its position clear. And finally, it was a project in which 600 companies would be directly involved. By organising various sessions with the business community, the

RMPM ascertained the complaints and desires of the business community and a number of plans were launched.

In May 1996, the definitive plan for the redevelopment was approved by the city council and the businesses were given the opportunity to register for various projects thus demonstrating their commitment to the implementation of the plan. At present, many sub-projects are still in the research phase while others have already been given the go-ahead. One of the first sub-projects which was started up on the initiative of the companies themselves and which boosted interest and motivation for the plan as a whole was the Shortsea 2000 project which was also accorded high priority in the European transport policy. The shortsea operators (e.g Geest Lines) cooperated on this project and they can now all be found at one location in the port so that the quay among other things can be used far more efficiently. In contrast to the preparations for Europoort and Maasvlakte, this entire plan was accomplished with a proactive input from both the business community and environmental agencies so that a wider support base for the implementation was created in advance. The RMPM will continue to act as overall project coordinator in order to maintain the motivation of those involved which is necessary partly because some sub-projects do not produce immediate results.

The attention to quality also appears in the regional development area (ROM) covenant, entered into in 1993 between the Ministries of Housing, Spatial Planning and Environment, Transport, Economic Affairs, the Province of South Holland, the city of Rotterdam, fourteen Rijnmond municipalities and the business community (including the Chamber of Commerce). The aim of the covenant is to promote the integration of living, working and welfare. This affirms the aims of the Port plan 2010. By correlating the aims of the Port plan 2010 with those of the ROM plan, difficulties and indications for solutions were tracked down.

The Problem of Space
The shortage of space in the port is becoming serious. The strategic reserve will soon have to be called on, though with reluctance. Moreover, turning away business is the worst possible advertisement for the largest port in the world. In 1995, through the 'Extensification'/Intensifying the utilisation of space project, the RMPM has gained a picture of the empty sites in the port area and the limited options at its disposal to gain control of those sites. 'extensification' is optimising the use of space on undeveloped sites, intensification is the efficient clustering of activities on occupied sites' (RMPM, 1996b:1) On 1 January 1995, the RMPM situation regarding the availability of grantable sites was as follows.

Table 4.2. Grantable Sites in the Port of Rotterdam per 1 January 1995
(Source: RMPM, 1996b:2)

Category	Total in ha	Total in %	Not in use, % of total
Granted	4247	80	
of which internal reserve	362	8.5	6.8
Option	580	11	11
Immediately available	45	1	0.8
Not immediately available	424	8	8
Total	5296	100	26.6

The conclusion to be drawn from the Table is that a quarter of the sites, which are scattered over the whole port area, are not being used. The non-utilised sites can be broken down into a number of categories and the category determines which options the RMPM has to regain control over these sites.

Internal reserves are sections of leased sites which a company has left unused for at least two years. Among other things, a company maintains a reserve:

- with a view to future expansion (the petrochemical industry in particular opts for long-term expansion and since the site lease forms a relatively small part of the total cost they lease the reserved site too);
- in order to forestall competition (any space which a company leases cannot be allocated to a competitor);
- in order to avoid soil cleanup (leasing is cheaper than giving back a site to the RMPM and running the risk of claims for damages and cleanup costs).

The RMPM has four options for regaining control of the internal reserve which is located mainly on sites leased by the chemical industry and the oil refineries. In the first place, the RMPM makes enquiries to trace which companies are prepared to voluntarily give back their reserved land. It is anticipated that in this way only 22.6 ha of the 362 ha can be reclaimed, because even if companies have no immediate plans for the land, they are seriously concerned about the threat of soil cleanup. A second option is to reclaim the land via legal channels. Of the eighteen companies, eleven have a provision in their contract which states 'If, during the lease, the lessee does not make use of or makes insufficient use of the whole site and/or a part of the site for more than two successive years, the lease may be terminated.' Four companies with an internal reserve have an article in their contract which states that this provision does not apply and three contracts have no provision on this subject. However, the option of reclaiming sites by taking legal steps on the basis of the contract has not yet been used by the city. There are strong fears about damaging commercial relationships and the image of the port.

Such a provision in a contract can be used at most as a *coercive measure* in negotiations with the company and must therefore be discarded as an option for achieving results in the short term.

A third option is to allow the owners to sublet the empty sites to temporary subtenants, for example for the storage of empty containers or aluminium rolls. The RMPM would have to play a strong intermediary role and it all depends on the willingness of the main lessee. Thus it is important that the activities of the lessee are not threatened by the subtenant and that the subtenant pays a reasonable price. Finally, the RMPM may act as intermediary by seeking options whereby existing and new companies work in complement. A combined heat and power plant or a recycling company are examples in this regard. They play a key role for a number of companies. Such an approach requires an active, intuitive approach from the RMPM and needs detailed knowledge of the existing industry. Until now that knowledge has seldom been encountered within the RMPM. For example, there is little idea of the projects undertaken by the EBB (Europoort/Botlek Foundation) to encourage companies to cooperate more. Together with the temporary subletting idea it is anticipated that the accommodating of allied businesses will yield 232 ha. The final report also contains a number of reservations regarding RMPM's difficulties in reclaiming internal reserve sites: 'the implementation of the extensification measures requires an active role from the RMPM and a cooperative stance on the part of the companies involved. Many companies will not be prepared to cooperate on extensification because this will restrict their own potential expansion' (RMPM, 1996b:7).

A second form of space utilisation which is not directly economically utilised is that of the *option*. Options are used by the RMPM as a marketing instrument in order to attract new enterprise. In the present case, 44 companies had an option on sites with a total area of 580 ha. The possibilities for the RMPM to reclaim opted sites depend on the duration of the option and the type of option. Normally, options are issued for two years but it can happen that options on more sizeable pieces of land, on which potentially greater investments will be made, are issued for a longer term. As per 1 April 1995 the breakdown of the duration period of the various options was as follows.

Table 4.3. Expiry Date of Options (Source: RMPM 1996b:8)

Expiry date	No. of companies (%)		in ha. (%)	
1995	19	(43)	135.7	(23)
1996	4	(9)	101.2	(17)
1997	6	(14)	14.7	(3)
1999	1	(2)	110	(19)
2003	1	(2)	58	(10)
not determined	13	(30)	161.8	(28)
Total	44	(100)	581.4	(100)

The RMPM uses four different types of option: the paid and non-paid options, the billiton arrangement, under which the RMPM only has a duty of notification if another interested party appears, and the reservation for planned activities by the RMPM itself. The paid and non-paid options are valid for a particular company and legally speaking cannot be retracted by the RMPM. The reserve option can also be used by the RMPM for giving companies a few weeks to consider. The breakdown of the various options is shown in Table 4.4.

Table 4.4. Various Types of Option (Source: RMPM 1996b:9)

Type of option	No. of companies		in ha. (%)	
Paid option	12	(27)	303.5	(52)
Non-paid option	19	(43)	203.3	(35)
Reserving	7	(16)	21.7	(4)
Billiton arrangement	6	(14)	52.9	(9)
Total	44	(100)	581.4	(100)

The chances of RMPM reclaiming control over the opted sites is small when we look at the table and see that 31 of the 44 companies, with 87% of the opted sites, have the legally stronger type of option, i.e. the paid and non-paid options. It is anticipated that 29% of the outstanding options (168 ha.) will not actually be issued. Over the last fifteen years, the size of the area of opted sites has increased due to economic growth, the RMPM's marketing and the availability of new sites (Maasvlakte). At present, the RMPM is carefully considering its options policy and striving for a more stringent use of the method (a maximum of two years and maximising use of the billiton arrangement).

Finally, the RMPM has a number of 'temporary woodland' zones in its area, planted by the jobless and subsidised by the Forestry Commission. The production woodland zones need to stand for a minimum of 15 and a maximum of 25 years. The woods cannot start to be cleared until the year 2000 when 14.5 ha. will become available for small-scale businesses. Earlier clearing could only take place in consultation with the Forestry Commission and the Public Works department. Space may become available through the clearing of storage sites (e.g. the Papegaaienbek which is used for dredging sludge), but costs will then play an important role.

The picture which emerges from the final report on intensifying the utilisation of available space demonstrates the limited options RMPM has to regain control of sites. It is not surprising that the solution to the shortage of space is mainly being sought in the second Maasvlakte. The costs of this project are estimated at 9 billion guilders and could never be furnished by the RMPM alone. The financial involvement of central government in the project is essential and this, in contrast to the construction of Botlek, Europoort and the first Maasvlakte, puts the RMPM in a dependent position. Furthermore, the economic and ecological advantages and disadvantages of the project and its social desirability were gauged in the national discussion on 'usefulness and necessity'. The results of this discussion led, in July 1997, to a cabinet decision in principle for the construction of 1000 ha. of industrial sites and 750 ha. of nature and recreation areas. Partly in response to calculations by the Central Planning Office and misgivings on the part of the Ministry of Finance regarding the economic feasibility of the project, a definitive decision has been deferred until 1999.

A striking feature of the institutional pattern of decision making regarding this expansion is that the various public organisations have emerged as the most important initiators. In the 'usefulness and necessity' discussion a number of social organisations were heard but the further decision making lies with the RMPM and primarily with the Ministry of Transport. The exact commitment of the market remains unclear. From the perspective of the normative framework used in this study, the main question first of all is to what extent have the market parties made clear their willingness to invest? Central government can then step in (by setting its own conditions and if necessary offering a financial helping hand). Public organisations themselves, however, are not in a good position to assess the opportunities and risks of investments in the market.

The shortage of grantable land has been clearly shown, but the need for expansion assumes commitment on the part of the port business community and the port authority. With the construction of the second Maasvlakte, the RMPM aims at attracting investors from the chemical industry and container handling. It is difficult for the government to judge for itself how fierce this demand is. Expansion of course offers substantial opportunities but also involves costs and risks. The developments within the various market segments are uncertain. For example, Rotterdam's market share in the Western European import of oil and oil products is dropping. This is mainly due to the consolidation of this industry in Western Europe by more energy-efficient cars and the replacement of fuel oil by coal and gas. The petrochemical industries are under permanent pressure to reduce costs. Solutions are sought e.g. in the reduction of the storage capacity by streamlining logistics and reducing diversity. As a result of the permanent over-capacity of the OPEC and the economic liberalisation of the former Soviet Union, the storage of oil is no longer viewed as the core business of the petrochemical industry. In other words, oil tanks are being scrapped on a large scale (said to be about 60 x 10,000 m^3 per year) to release space for new industrial purposes.

The design of the plan for the chemical (actually: petrochemical) industry also needs to be contextualised. Investment expectations regarding new chemical parks are doubtful. Producers are increasingly achieving their required capacity by gradually stretching their existing capacity. Shell Chemical in Moerdijk, for example, has stretched its ethylene factory in Moerdijk, which produced 450,000 tons of ethylene per year when it started in 1975, to 600,000 tons in its current state. It has now reached capacity, and Shell has announced a larger increase to 900,000 tons. But there is ample room for all of this within the existing sites.

A final reservation regarding the port expansion concerns the choice for container handling. On the present Maasvlakte there is still 200 ha. available for these activities. To what extent this can be considered adequate depends on the expected growth. In making a decision about expansion, account will need to be taken of the construction and operation costs. In view of the increasing competition between Western European ports, the assumption of overcapacity and its related costs is not inconceivable.

These examples are not intended to suggest that a further expansion of the Maasvlakte should not take place. They merely serve to demonstrate that there are market risks attached if it is decided to expand. Weighing the costs and potential recurrent benefits can therefore only take place in a market context. This has important implications for the decision-making process. The first responsibility devolves upon the port authority and the port business community, not upon central government. It is highly probable that the first parties in the market will not be able to carry this type of expansion alone (certainly not if the RMPM emerges as the sole interested market party) and so will make an appeal to central government. Central government can then offer a helping hand on its own terms (e.g an investment subsidy including risk and profit sharing, or on the condition that spatial and environmental interests are compensated). Central government must not rule the roost, however.

It is surprising that concern about the granting and location policy has only started to materialise now that the Port of Rotterdam has nearly run out of space. After all, the demarcation line was laid down in 1964. This extremely precious possession, land, ought to be RMPM's main directing instrument but in fact it is not. In a recent interview with the *Trouw* newspaper, the head of the RMPM's allocations department admitted that the system of land allocation in the past had been far too liberal. The Port of Rotterdam apparently lacks a mechanism (political control or competition) to safeguard the careful management of land in the long term. Another surprising point concerns the apparent absence of coordination with the market with regard to the economic viability of the second Maasvlakte plan. Why has Rotterdam opted for more of the same while the existing industry apparently has sufficient space? In other words, to what extent is the plan also supported by the market?

Reacquisition of Sites: Voluntary Return and Relocation of Companies
Despite the impossibility of expansion or extensification of sites in the short term, situations are also conceivable in which the RMPM can regain control over the land. This is the case when a site is voluntarily handed back (termination of a lease) or when a company relocates to another site within the port area. In these situations, however, the decision making needs to go through other channels in addition to the RMPM. An example of each situation is described below.

On the Nieuwesluisweg site on the Hartel Canal, Texaco had a small location with a number of storage tanks which it wanted to return to the RMPM. All the tanks had a spillage trough to catch any spills or leaks from the tanks. In 1992 during the dismantling, it was discovered that one of the connecting pipes between the tanks had been leaking with the result that a substantial soil cleanup was now necessary. In consultation with the Rotterdam Environmental Agency (MR) and the DCMR, Texaco agreed via a temporary exemption order that it would carry out the cleanup operations itself. The South Holland Regional Environmental Hygiene Inspectorate (RIMH) did not go along with this decision, however, because they felt that deliberations on the various cleanup methods had not been sufficiently meticulous. This lack of unity finally led to a conflict between the city of Rotterdam (and Texaco) versus the RIMH in which the RIMH threatened to go to the Public Prosecutions Department and publish a press release and Texaco responded by threatening court action. This was finally prevented by the Rotterdam Environmental Agency which had conducted an investigation into the decision-making processes surrounding the situation so that the cleanup operations which had already been initiated could proceed.

Another way in which the port authority can regain control over land is by moving companies within the port area. This does not happen often since it is an expensive operation and a new location for the company does not necessarily have surplus value. Appendix 2 to the *Enforcement of Environmental Legislation Handbook,* published in 1995 by the Environmental Policy section of the Rotterdam Public Works department, contains an example of the effective relocating of a storage and container handling company in the port of Rotterdam. This company handled scrap in the Waalhaven and since 1986 this had been causing a good deal of noise nuisance. DCMR received complaints from residents about the site and, from 1992, complaints about dust nuisance were added to these. Meanwhile, in 1991, the municipal executive had imposed a pecuniary penalty on the company which finally led, after the civil proceedings in 1994, to a guilty plea and an out of court settlement. During this period the company's lawyer attempted to drive a wedge between the political administration and the public enforcement organisations which was ultimately unsuccessful because the chairman of the Rotterdam environment committee stuck to his guns.

The actual impetus for the relocating of the company, however, only came when the city decided to tackle the dust pollution problem. The position and actions of the RMPM then also become clearer. 'It only started to get interesting once the RMPM came very emphatically to the fore. The RMPM felt that it was inappropriate to impose a pecuniary penalty because it was already involved in negotiations with x (the steel container company, HS) regarding the relocating of the company. Taking legal action would, in the RMPM's opinion, adversely affect the climate of those negotiations. This illustrated the need within the city of Rotterdam for effective consultation with RMPM to achieve a clear policy line. It is of course a time-consuming business to demonstrate the different points of view and convince each other. Meanwhile, complaints continued to come in. Once again and quite correctly, DCMR proposed to the Rotterdam Environmental Agency (MR) that it should advise the board to take legal steps. In late 1993, the MR's recommendation reached the municipal executive and contrary to the RMPM's view, in the MR's opinion there was no question of agreeing a "code of conduct" with x as this would mean forgoing the option of enforcement action. This would damage the credibility of the administrative enforcement process, which had been so laboriously and patiently built up in Rotterdam' (Veerman, 1995).

The municipal executive agreed to a pecuniary penalty order with the provision that the RMPM and MR had to enter into negotiations with the company. RMPM and MR devised a strategy and both initially reached the conclusion that the pecuniary penalty and the negotiations conflicted with each other but on further consideration decided that the pecuniary penalty could play a useful part in the negotiations. After all, it could be used to put pressure on the company to expedite its relocation. This strategy was also successful because coordination took place in advance with the Public Prosecutions Department which meant the authorities were able to display a united front. In principle, the Public Prosecutions Department dealing with environmental offences can adopt a position independent of that of the administrative enforcer. In this case, which involves the care for public welfare on the one hand and business interests on the other, it would probably have had an adverse effect. Due to the private law relationship between the RMPM and the lessee, the RMPM would probably have chosen to defend the lessee's position, thus damaging the goodwill basis of the joint search for creative solutions.

Both examples illustrate that balancing public and business interests in the port is possible and that by placing the enforcement of environmental policy under a different municipal or regional agency, relationships between municipal government and the business community are kept clear.

Cooperation with Other Ports and Major Transport Centres

Another option for reducing the problem of space which is already being used is the cooperation with the port authorities of Moerdijk and Vlissingen/Terneuzen and the second-line major transport centre Arnhem/Nijmegen. According to the spatial economist Van Klink (1995), the cooperation with other ports and urban centres forms the major challenge for the RMPM for the coming decade. In his view, the current cooperation with Moerdijk, Vlissingen and the Arnhem/Nijmegen major transport centre is merely the forerunner of a more comprehensive cooperation in the Delta of north-west Europe (the area between Rotterdam and Antwerp). Although the town and country planners Kreukels and Wever (1996) endorse the functional and territorial widening of the mainports, in their view it is a fallacy that such a development can be strategically planned by a port authority. I tend to agree with Kreukels and Wever. Enquiries revealed that the alliances had not been as strategically planned by the RMPM as had at first been suggested. It was more a case of 'opportunity windows' which were taken advantage of on an ad hoc basis. The RMPM for example has still not entered into an alliance with the successful Rail Service Centre in Veendam.

Kreukels and Wever (1996) see more potential in the flexible planning of capacities and facilities so that a port authority can react adequately to developing trends in the evolution of transport and distribution. This does not undermine the traditional role of the port authority nor that of the business community and mutual expectations remain clear. Examples of a flexible approach can be found in the development of various projects linked to the transport of containers to the hinterland, such as the projects in the framework of Rotterdam's Internal Logistics and Combi-Road. The RMPM or the Ministry of Transport are the chief advocates of these projects in which numerous parties participate and which often have a high innovative content (particularly in their use of information technology).

4.5.4   CONCLUSION: PORT PLANNING

Port planning in Rotterdam during this century has developed both in a variety of institutional contexts and in various social environments. Port planning in Rotterdam before the war was a municipal affair. After the war the involvement of the province and central government steadily increased and the port authority's options for setting conditions became increasingly limited. The Europoort plan (in the fifties) was thus no longer called a Rotterdam plan because the government's town and country planners, to the great annoyance of the RMPM, were directly involved in it. This plan was able to be developed remarkably quickly, however, because there was complete agreement at national and local levels regarding the economic need to construct this industrial port area. After the realisation of this plan and the ensuing wave of democratisation, a change occurred both in the local political climate and in national involvement in the port area. The economic dispersion policy became an important national economic objective. In the early seventies, in addition to the spatial aspects, environmental interests arose and more stringent requirements were gradually imposed on (the location of) companies and the port's further development. There was to be no steel industry on the Maasvlakte and the demarcation line was a sign of the political limits to growth at that time. The relationship between politics and the business community made a u-turn and it was not until the eighties (high unemployment) that thinking was once more allowed to turn to commercial enterprise. Since the early nineties, the port of Rotterdam has again become an economic spearhead (mainport policy, seaport policy, major urban centres policy) and the local level appears once more to be able to interest the national level in the port (Betuwe Line and proposed second Maasvlakte), as was the case in the fifties and sixties. The considerable political attention devoted at present to national infrastructure is only temporary, however, as once the projects are completed Rotterdam will have to rely on itself again. Furthermore, Rotterdam remains dependent on the political climate at national level for the completion of these projects.

With regard to some sectors of industry (small-scale businesses, storage and container handling companies, stevedores) the RMPM has increasingly taken on the role of facilitator, creating a fusion of the RMPM with those companies. For the (petro)chemical industry, the situation remains largely unchanged in the sense that facilitation can occur to bring in such a company (e.g. by laying a pipeline) but that once the contract has been signed, the RMPM knows little about further developments. Joint projects with industry in the framework of environmental improvements have not yet been established.

In this context, the very limited options of the RMPM for regaining control over sites may also be mentioned. Port planning hardly exists in this sense; the RMPM is too dependent on the goals of the business community and there is no mechanism to ensure that land is managed carefully. The alliances which the RMPM has set up with other ports and transport centres appear to be the result of ad hoc opportunities and are not (as yet) underpinned by policy. Furthermore, the cooperation seems to have been a victim of its own success, such as the client who wanted to lease the entire lot belonging to the "Exploitatiemaatschappij Schelde Maas" (ESM; joint venture between Rotterdam and Vlissingen) which would have meant the cooperation would have achieved its aims in one fell swoop.

The redevelopment of old port areas requires a different approach from the RMPM than it is used to, since this does not immediately concern the expansion of the port area, as was the case with the Botlek, Europoort and the Maasvlakte. The immediate proximity of the city and the large number of small-scale businesses that were already located in the area also meant RMPM had to move very cautiously and tentatively in its search for options. The determining of its position vis-à-vis other municipal agencies also needed to be addressed. In direct contact with the business community a comprehensive vision was drafted and various projects were initiated. Just as in the construction of new sites, the initiative remains with the RMPM and the success of the project once again depends on the risks which the business community is willing to take.

4.6.   Industry and Port Services in Rotterdam

In the port of Rotterdam the principle traditionally applies that the business community must stand on its own two feet and that government parties only set the market conditions so that they can pass the costs on to the port users. Traditionally, therefore, it is not customary for the port authority to meddle in private activities. It lacks the specific market knowledge for this but above all, by acting as a market party it will lose its original independent position whereby all its public powers will be at stake. After all, its present position offers the RMPM the opportunity to tie itself short-term to private parties in order to get processes underway, for instance. A permanent role as market party increases the risk that public and private responsibilities will ultimately become lumped together and the independent position of trust will be damaged. On the basis of the market regulation principle for private activities in the port, it may be anticipated that the port authority will dissociate itself from participation in market activity. This section looks at how the port world manages to relate this principle to industry and the port services.

The port business community in Rotterdam breaks down into three categories. In the first place there are the internationally operating (petro)chemical industries (Shell, Esso, Texaco, Arco, Eastmann, etc) which are represented by the Europoort/Botlek Foundation (EBB) to which ninety companies to the west of the Benelux tunnel are affiliated (donor status). The EBB is actually an industrial sites association or industrial circle which even has its own representative serving on the board of the Chamber of Commerce. It is aimed at initiating and implementing projects for the joint advancement of the various competing industries. Examples of this include running a joint fire service, negotiating on port dues, but also developing initiatives for environmental life cycle management. The RMPM maintains relatively little contact with this clientele and it knows little about these companies' internal affairs apart from the odd staff member who happens to have worked in the industry. For instance, when the EBB initiated its own plan regarding the image building of the Second Maasvlakte (the Hollandse Poort plan) this was initially accepted only by the port committee chairman. But once the chairman moved to another job there was no-one left at the RMPM who wanted to use the plan for further discussion. For many projects the EBB does not need the RMPM either since these directly benefit the EBB. But it would help communication and new development initiatives (consider the shortage of space) if the RMPM were to increase its knowledge of this commercial enterprise. At present the companies are still regarded by the RMPM as separate entities so that a concept such as environmental life cycle management cannot actually take shape but with some help from the RMPM this might well prove possible. The EBB is also a member of the executive committee of the ROM-Rijnmond organisation.

In the second place there are the companies which make a living from the fact that freight enters and leaves the port of Rotterdam. These include stevedores, shipbrokers, transporters, container companies, ro/ro companies, etc. They are represented by the Shipping Association South (SVZ). This association actually consists of two separate organisations, the Association of Port Operators and the Association of Employers. The first represents the professional groups listed above and its task is to maintain the lobby circuit around the various layers of government, while the Association of Employers is concerned with socio-economic circumstances: collective labour agreement negotiations and working conditions. The SVZ is particularly keen to achieve the uniform application of European regulations. On this point the situation in the port of Rotterdam leaves much to be desired according to the SVZ (see the discussion on floating transhipment in subsection 4.4.3).

With regard to port development, the EBB and the SVZ work very much on the 'mutual backscratching' principal. In other words: government cannot expect the business community to develop initiatives with an interorganisational interest if it does not participate or create the limiting conditions for them. In various consultation forums and other joint initiatives (Port Interests Organisation, Rotterdam Port Promotion Council) the port business community and the RMPM attempt to clarify joint aims, keep to them and put them into effect. For example, the initiative to open up the Rijnmond region more effectively (the Rotterdam Internal Logistics project) has a large number of participants in addition to the RMPM and both the SVZ and the EBB take part in it. Both interest groups also undertake joint activities: these include publishing a magazine and participating in the Educational Information Centre which was set up in collaboration with the RMPM (50%) to provide about 25,000 schoolchildren and students per year with an introduction to the commercial enterprise in the port.

The Chamber of Commerce concentrates on the third category of the port business community, i.e. the interests of the shippers. It will offer advice on request to politicians or the business community. They are co-signatories of the ROM covenant and also take part in management consultations between the RMPM, SVZ and EBB. Finally, there is the trade union movement with which the RMPM formally has little contact because the RMPM itself does not carry out such things as stevedoring operations and because employers and employees are perfectly able to make agreements together. Both parties see it as in their joint interest to ensure the port remains attractive by preventing labour unrest. The unions in the port only use their right to strike in extreme circumstances (when fundamental rights and agreements are jeopardised) and try to enter into discussion with employers through proactive idea building (e.g. for improving productivity and throughput). The RMPM may mediate, as it did in the early eighties, in the event of all-out and prolonged strike actions.

The following subsections focus on the port services. This is the category of the port business community with which the RMPM maintains regular contact and which it can influence to some extent.

4.6.1. CONTAINER HANDLING AND THE ROLE OF THE RMPM

The port of Rotterdam has a number of container handling companies for various market segments. Europe Combined Terminal (ECT), with a market share of about seventy percent, is by far the largest container stevedore in the port partly due to its monopoly position for the deep-sea segment. It is located on the Maasvlakte where the largest container ships can be accommodated. Hanno and Uniport, both located in the Waal-/Eemhaven area, handle the market segment related to this which has a partial overlap with that of ECT. They will only be competing with ECT if the scale expansion of the container ships does not persist, because then their present minor function on the overlapping portion could turn into real competition. This is not yet the case. A third segment in container handling is formed by the coasters (Geest Lines, among others) which use totally different handling techniques and which at present are undergoing tremendous competition from the Channel Tunnel between France and Great Britain (which is able to be highly competitive partly as a result of debt restructuring).

Due to ECT's monopoly position in the deep-sea segment, the recent opening of a 'state of the art' container terminal on the Maasvlakte with financial aid from the city of Rotterdam, and the historical relations between ECT and the RMPM, it is interesting to explore the relationship between the port authority and ECT in more detail.

From General Cargo to Container: Cooperative Incentives
In his book *Cooperation in port and transport in the container age* Van Driel (1990) describes the concentration in the general cargo and container sector in

Rotterdam. The *stevedore sector* in Rotterdam has traditionally been characterised by the tension between a strong competitive spirit and a collective approach to *capacity problems* by, e.g., establishing a labour pool or exchanging ships. Managing capacity was always a crucial problem in the general cargo sector. The construction of new cranes meant that the fixed costs formed a substantial part of the total costs: price wars were soon looming in an attempt to try and secure at least some profit. Each stevedore understood that this might ultimately cost him his job. Moreover, although there was no way to plan when a ship would enter the port, the stevedore had to have enough manpower ready to load or discharge the ship as fast as possible. In the old days no-one could afford to give the labourers a permanent contract so that on this point there was cooperation.

Then in the mid-sixties, when the first containers from Sea-Land arrived in the Netherlands and it looked as if these iron cargo containers would become the transport of the future, a number of motives suddenly reemerged for directing the strongly-rooted Rotterdam competition and developing joint initiatives. Thus, in the late sixties, two container handling companies emerged: *Europe Container Terminus* (the forerunner of the present ECT) and *Unitcentre*. Both sprung from cooperation between various stevedores which also continued to maintain their own general cargo activities. After all, this was a totally new transport concept and there was great uncertainty regarding its success and the quantities which could be anticipated. Furthermore, in the early years most of the cargo continued to be shipped in sacks, boxes, crates and chests.

The former Europe Container Terminus was set up in 1966 by the five major Rotterdam stevedores. Their motives varied, however. For Quick Dispatch and Thompsen the cooperation was a means to growth, while Pakhuismeesteren, Müller & Co., and C. Swarttow mainly participated in order to enter, without too much risk, a new and possibly important growth area. Unitcentre was set up in early 1968 by Heijplaat (SHV) and Furness who had different motives again. 'When it became clear after a number of years that the container would occupy an important position in the general cargo sector, the latecomers tried to expand their interests in container handling. Some enterprises attempted to reinforce their position in ECT through mergers, such as Müller which, by its takeover of Thompsen, raised its share from 18.5% to 41%' (Van Driel, 1990:317).

During the seventies a few changes and 'buyouts' occurred but they did little to alter the market balance between ECT and the later Unitcentre. Finally, in 1983, Pakhoed gained complete control of Unitcentre which until then it had had to share with SHV. The competition between ECT and Unitcentre took place primarily in the field of acquiring new clients since the large container clients tended to be strongly tied to a particular location and/or stevedore.

Unitcentre and ECT continued to be competitors until ECT's merger in 1989 with Quick Dispatch and Müller-Thompsen. The latter two companies handled not only containers but also other cargo units (former general cargo, cars, etc.). In 1989, these two multi-purpose stevedores merged with ECT to form the new ECT which then changed its name to *Europe Combined Terminals*. The common feature of these three companies was that they had joint owners (i.e. Nedlloyd and Internatio-Müller). From 1989 to 1993, Unitcentre and ECT were formally competitors although ECT, with a market share of 60%, was by far the largest container handling company. Finally, Unitcentre and ECT also merged in 1993 and Pakhoed became a shareholder in ECT.

In the deep-sea container handling segment ECT holds a monopoly position. Due to the scale expansion in shipping, the larger container ships (5,000 - 6,000 20-foot containers) can only use ECT. In addition to ECT, the port of Rotterdam has a number of other container handling companies which principally work in other market segments but sometimes have a partial overlap with the ECT market segment, e.g. Hanno in the Maashaven which can handle ships with a maximum of 4000 20-foot containers. Uniport and Deka specialise in container handling in the market segment for smaller ships, such as those used for feeder traffic among others.

This very cursory description comprises the supply concentration of container handling for deep-sea containers over a little more than the last twenty years. The only company which now remains invented the 'state of the art' container handling concept for the present Maasvlakte which has impacted on the format and requirements of port infrastructure and other transport systems.

RMPM's Interests and its Position on Industrial Concentration and further Expansion
Although in the sixties the RMPM was delighted with internal competition, it considered it wise to limit the competition in the container sector to some extent. It was not in its interest to have to construct separate quays for each container stevedore. It would have cost a fortune. This is why in the sixties Posthuma, RMPM's Chief Executive Officer, more or less forced the five largest stevedores to jointly construct a container terminal. The creation of ECT thus became a fact and the steering role of the RMPM was accepted.

In March 1974, an RMPM project group explored the options for handling containers on the Maasvlakte. This was a risky undertaking for ECT, however, since a substantial overcapacity would be achieved in one fell swoop. Furthermore, the location was thirty kilometres west of the Eemhaven site and, finally, ECT had already made heavy investments in the previous years. In 1975, the RMPM came up with the idea of getting the existing stevedores to cooperate in developing a container terminal on the Maasvlakte, just as in the past. Both ECT and Unitcentre had their own reasons for not being interested in this and the RMPM could not force them.

Finally, in 1978, ECT decided to take the plunge: it would not move all its operations to the Maasvlakte but would build an 'overflow terminal' there. Since no other private partner could be found for this initiative, in 1981 the city decided to offer ECT a favourable financial arrangement. 'Of the 310 million which ECT would have to invest in the layout of the terminal, the city of Rotterdam provided 149 million in the framework of 'prefinancing' at a rate of interest below the going market level. In addition, ECT received a postponement of payment on their site lease and wharfage which would have to be paid back with interest. And finally, under a risk cover arrangement, ECT could also receive a reduction, postponement or cancellation of prefinancing and (postponed) rent repayments. A reduction would be given if the growth over the three-year period fell more than 5% below the ECT prognosis (30,000 additional containers every year) due to "calamitous external circumstances", and postponement and possible partial cancellation would be provided if the company nevertheless met continuity problems. In 1984 ECT's Maasvlakte terminal, called the Delta terminal, became operational' (Van Driel, 1990:247-248).

Almost simultaneously with the ECT initiative to move to the Maasvlakte, Unitcentre was also planning to expand because its terminal had reached capacity years earlier than anticipated. For the RMPM, which had been unable to engineer the joint operation of the terminal on the Maasvlakte by these two companies, it was essential at that time to prevent the potentially adverse effects of a double overcapacity. As a condition of the construction of the ECT terminal on the Maasvlakte, therefore, it stipulated that both companies should allow the RMPM to scrutinise their integrated market position and market strategy in order to prevent them opposing the city's policy (preventing price wars and maintaining industrial peace). The request from the city was favourably received with the result that in the eighties the market relations remained as intact as possible. The RMPM guided as it were the market process in the eighties by also ensuring that the shareholders of ECT and Unitcentre, whose general cargo companies still competed, were committed to maintaining the Maasvlakte terminal. The RMPM had provided all the favourable starting conditions and ECT now had to substantiate them.

Until 1985, the RMPM continued to control the market conditions but after that it was no longer possible because the container market boomed and prognoses were readjusted upwards. This was why very little fuss was in fact made when the new and much smaller container stevedores such as Deka,

Transstorage and Uniport lowered their prices to quickly fill up their capacity. In late 1986 the competitive position of ECT, which had always charged high prices for high quality service, began to weaken when they lost four major clients in a short space of time to the newcomers.

Until the early nineties the RMPM did not play an intervening role in the container sector. It only began to assert itself once more when plans were made for the further development of that section of the Maasvlakte reserved for ECT. The increasingly large container ships forced ECT to invest not only in cranes with a far greater range but also in technology for the fast transfer of containers to their destination. Under the leadership of ECT chairman Wormmeester, an ingenious and technologically highly advanced terminal concept was devised which would be able to handle the anticipated container growth. The plan was called *Delta 2000-8* which stands for the construction of eight container terminals which will be ready in the year 2000.

The plan can be broken down into various aspects which each entail consequences for ECT's competitive position (relationship with shipowners and shippers) (see Stevens, 1997). In the first place, in fleshing out this plan ECT developed the third generation container handling facilities. The first generation were found in the Eemhaven, where the employees still had to do a relatively large amount of the work. The second generation was developed at ECT's overflow terminal on the Maasvlakte and in the early nineties ECT, together with Sea-Land, introduced the third generation. This is characterised by an almost total robotisation of container activities which is unique in the world. The operator on the familiar *straddle carrier* is replaced by an *Automatic Guided Vehicle* which, by means of sensors implanted in the surface of the quay, knows exactly where containers need to be taken and deposited. Instead of operators manning the cranes, ECT needed computer scientists and control engineers who were able to further automate the container handling and everything to do with it (planning, checking cargo in and out, etc.). The result of this automation was that fewer people were carrying out the actual work and more people were being employed in staff and research departments.

Another structural aspect of the terminal concept on the Maasvlakte is the principle of *intermodality*. The success of ECT depends on the fast transfer of containers to and from the terminals. With the congested roads around Rotterdam and the limited rail facilities in comparison to ports such as Hamburg and Antwerp, it was absolutely essential for ECT to have a link with other transport modalities (rail and barge). Without good hinterland connections the plan was doomed to failure: this was the unmistakable message. In the meantime, after effective lobbying from Rotterdam and the Ministry of Transport, the Cabinet has approved the construction of the Betuwe Line, so that condition has in any event been met. Before that time,

ECT had already started making its own contacts in the hinterland, such as the second-line major transport centre of Venlo. Moreover the RMPM, in collaboration with the shortsea shipping companies (incl. Geest Lines) and with the aid of EU funds, has regrouped the coastal trade in the port area. It is still not clear whether this branch of transport will have the desired logistic connection with the ECT terminal on the Maasvlakte.

The plan was finally designed on the *dedicated terminal* principle. This principle is a cross between the exclusive use of a terminal (literally the terminal as private good) and the public use of a terminal (anyone can load or unload there; first come first served). The art of the commercial organisation of a container terminal, just as with general goods handling in the past, is to ensure that as many ships as possible are berthed at the quays and at the same time to ensure that the ships do not have to stand in line. It is thus a question of achieving a balance between avoiding overcapacity and not falling below a certain capacity utilisation, trying to prevent queues forming even though in principle you don't know exactly when a ship will arrive. Automation and scheduled services mean that shipping traffic is becoming more regular but uncertainties regarding dispatch in other ports and the journey by sea remain. The 'dedicated terminal principle' ensures that large global lines do not have to cooperate fully in their use of terminals but can do so partially. Moreover, ECT gives them the opportunity to shape their own identity through the Delta 2000-8 plan (use of own logo on cranes and own IT). Sea-land, as one of ECT's faithful clients, was the first to sign a long-term contract.

In order to implement such an ambitious plan, ECT, the RMPM and other layers of government (Ministry of Transport, Economic Affairs) felt that cooperation was essential. The motives for cooperation lay mainly in the terminal's *technological integration* with the hinterland, the *national economic impact* that the plan would have (estimated 1000 guilders per container), the *distorted competitive relationships* between container terminals in Europe (Antwerp and Hamburg either do not or only partly pass on the government investments in infrastructure), and the direct involvement between ECT and RMPM in the past in the construction of the first Delta terminal. In 1992, under the leadership of ex-Minister Winsemius, a consultative committee was formed which made the initial inventories of projects and role divisions. Then in 1993 the *Incomaas* project group was set up in which employees from ECT and RMPM could cooperate in solving problems under the leadership of an external chairman. At the same time, various academic institutes (Delft University of Technology, Erasmus University, TNO) became closely involved in the project with a view to gathering additional information and solving technical problems. The outcomes and the individual monographs, which were financed by the National Economic Structure Fund of the Ministry of Economic Affairs among others, were ultimately compiled into one volume and published by the Centre for Transport Technology in 1996.

The great interest which RMPM had in cooperating with ECT lay in the '*package deal*' which it could arrange using this plan. The construction of this ambitious plan exceeded ECT's financing capacity so it was desirable that the RMPM should help out. A different input was expected, however, than had been the case in the construction of the first terminal on the Maasvlakte. It was decided not to lend ECT money but to take over ownership of part of the terminal and to pass on the costs to the user, so that a part of the terminals would become an investment item. Instead of the previous distinction between infra- and superstructure a distinction was now introduced between infrastructure, *infrastructure plus* and superstructure. Infrastructure plus refers to site surfacing, crane rails, electronic facilities, buildings, connections to the utilities and so forth. These items had previously come under superstructure. In total a sum of 600 million guilders was involved, to be recovered over a period of 25 years.

By creatively stretching the responsibility of its role, the RMPM was also able to achieve its own objectives. For example, it insisted on the merger between ECT and Unitcentre, and ECT's concentration on container handling. This meant it was finally able to refurbish the Waal-/Eemhaven area and fit it out for the relatively small stevedores (general cargo and feeder traffic). ECT had now relocated in its entirety to the Maasvlakte which meant that the stevedoring companies Hanno and Uniport could move to the former ECT site and their old locations became available for other useful purposes. Once Hanno was relocated, for example, the Kop van Zuid plan could be completed. It also became possible to concentrate the feeder services of, e.g., Geest Lines in the Waal-/Eemhaven area and thus provide them with the opportunity to develop joint initiatives.

The Delta 2000-8 plan in fact not only involves the development of the terminal itself but also the realising of various other projects and conditions which proved essential to the success of the plan. For the relocating of ECT, for example, RMPM bought the cranes (176 million guilders) which they later sold to Uniport and Hanno. The RMPM also decided to take out a loan with the European Investment Bank for 450 million guilders at a favourable rate of interest to finance the infrastructure plus and central government helped out with 150 million guilders for the construction of the harbour and the quay walls. In addition, the capitalising on anticipated growth trends in container transport will involve the construction of the Betuwe Line (eight billion guilders) and the conversion of the Caland bridge into a tunnel (an estimated 800 million guilders). Both projects will be financed from the national budget.

The Delta 2000-8 plan alone will cost more than 2 billion guilders, one billion of which will be raised by ECT itself.

The most important conditions which the city specified for participation were that the larger shipowners must conform to ECT's handling concept and that the city should have an escape clause that they could use in 1996 if so desired. This did not occur. The American shipping company Sea-Land was the first shipowner to go along with the concept but to what extent other shipowners would do so remained unclear. The situation became critical in late 1996 when Maersk announced that it intended to operate its own container terminal on the Maasvlakte without utilising the services of ECT. This jeopardised the whole philosophy of scale expansion to a single container stevedore. The RMPM could not and would not lose this container giant.

Finally, in July 1997 it was decided that ECT and Maersk would jointly operate a terminal but that Maersk would make no use of ECT's modern technology. At the time of writing, it was not clear exactly what form the cooperation would take. It is therefore difficult to predict what the effects will be. The arrival of Maersk, however, put pressure on the carefully erected, close relationship between the RMPM and ECT[9].

As regards the deep-sea container handling market segment, it must be concluded that neither the Rotterdam central council nor the RMPM have as yet pursued a consistent economic policy, i.e. consistent with the market regulation principle that 'the business community must stand on its own two feet'. It looks as if the choice of a cooperative construction with Maersk leaves ECT's monopoly position unthreatened and it seems that a good deal of credence is given to the corrective effect of the competition from the container terminals in Antwerp and Hamburg.

4.6.2. PILOTS, TUGBOATS AND LINESMEN

Other important services for a port are the pilotage, assisting/towing and tying up of ships. Although they do not form the largest part of a ship's initial costs, they partly determine the service level and the turnaround time of a port.

Until 1988, pilotage of ships was the direct concern of government and was carried out by a state monopoly (municipal and national pilots); obligations such as compulsory pilotage could be imposed on pilots and captains and pilotage tariffs could be laid down by the government. In 1988 political pressure was brought to bear on pilots to privatise and since then they have

---

[9] It is extremely rare for companies in the port to file a complaint with the municipal executive and the Council but when the RMPM opened negotiations with Maersk, ECT went to the town hall.

come under a privatised Pilots Corporation so that a private economic power position was created. The privatisation built on the former basic principles but because the service level has improved, the pilots have awarded themselves a salary increase: this has meant an increase in the transport costs for ships which put in at Dutch ports. Although the pilotage of ships in the Netherlands is still seen as part of the nautical safety policy, since the privatisation the Ministry of Transport and the RMPM, who are responsible for nautical management now have very little actual control over pilotage matters. They have become sharply aware of this recently on the competition-sensitive issue of pilotage tariffs.

Formally the pilotage tariffs are specified by order in council. The role of the Ministry of Transport is to assess and specify. In specifying the tariffs it must comply with two conditions: the tariffs must be nationally uniform and for politico-economic reasons, under the Scheldt Treaty of 1863 between the Dutch government and Belgium, the pilotage tariffs on the Maas near Rotterdam must be coordinated with the tariffs on the Scheldt. Due to this political objective there is no differentiation between the pilotage tariffs on the Scheldt and on the Maas which handicaps the competitive position of the port of Rotterdam vis-à-vis that of Antwerp. As an emergency measure, Rotterdam has lodged an appeal with the European Commission in which it explains that the provisions of the Scheldt Treaty have led to a price cartel and directly contravene the European competition regulations. The result of the appeal is not yet known but it is gaining increasing priority now that the Netherlands has undertaken a commitment to Belgium to dredge the Scheldt. Central government is attempting to reopen the discussion on the position of the Pilots' organisation via the *Frissen Committee*. In July 1997, the committee advised the Minister of Transport to end the monopoly position of the pilots by introducing regional variations in tariffs and safety conditions and by discontinuing the cross-subsidies between the larger and smaller ports.

Another option which the RMPM could use to undermine the pilots' monopoly is to influence the scope of compulsory pilotage which is laid down by the Ministry in the Compulsory Pilotage Decree (1995). Institutionally speaking, the existence of compulsory pilotage does not combine well with the (private) monopoly character of the Pilots' Corporation. At present only a small number of ships are exempted from compulsory pilotage, e.g. seaships of up to 60 metres not carrying a dangerous load, warcraft, fishing vessels, the port service boats, dredging vessels, tugboats, barges in so far as there are no extraordinary circumstances and seaships for which the captain has an exemption. In granting an exemption, the main thing is always the guaranteeing of safety. Ships are only exempted if, for example, they put in at the port more than a certain number of times per year and/or if the captain has completed a course in handling the Traffic Guidance Systems (VBS). The

VBS can be used increasingly by the RMPM as a technical aid to pilot ships in and out of the port. The RMPM is part-owner (50%) of the Marine Safety Rotterdam training centre which provides simulator training among other things. For the time being the Compulsory Pilotage Decree is fixed and the RMPM cannot directly change it. This is particularly hard on shortsea ships, e.g., which sail back and forth in a day between the Netherlands and England and which are nevertheless subject to compulsory pilotage.

Recently, the Ministry of Transport has introduced a separate VBS tariff which for many seaships comes on top of the pilotage tariff. On the principle that government investment in the port should be recovered in a commercially responsible way, it is logical that a VBS tariff has been imposed. However, as yet no direct link has been established between granting exemption from compulsory pilotage and a more active use of the VBS since the pilots and the VBS are not interchangeable and will continue to remain separate but necessary elements of the safety policy. The VBS has improved the service level in the port but has also pushed up the costs of maritime transport.

Tugboats
The tugboat community of the port of Rotterdam forms its own little world. In the sixties, Rotterdam had an Association of Tugboat Operators in which many smaller tugboat companies were represented and which voluntarily assisted each other despite being competitors. After a wave of mergers the New Rotterdam Tugboat Association was created of which less companies were members. And finally the association comprised only one member: Smit Port Towing Services. In 1979 there was a seven-week strike in the port of Rotterdam. The management of Smit was stuck between its own employees and the German consul and ambassador who turned up on its doorstep on the second day of the strike because the German steel industry had ground to a halt. Finally, the Chamber of Commerce paid the Smit employees two million guilders out of its own pocket on condition that they stopped the strike. Moreover, the pilots got one million guilders because they had taken up a cooperative position. During that hectic time, Smit acquired a competitor: the firm of Jan Kooren which had started its own business with about ten boats. Until 1987, there had always been two tugboat services in the port. Between 1979 and 1987 the tugboat tariffs for the urban section doubled. The tariffs in Europoort, which were already 50% higher, were increased by another 50% during that period. In 1987 a third tugboat company came onto the towing market operating under the name Kotug. Kotug's tugboats were extremely powerful which meant that less boats were needed to assist each ship. The arrival of a third provider ensured that competition was renewed between the towing companies. By negotiating directly with shipping companies (Sea-Land was a major client) Kotug was able to secure a position in the Rotterdam

tugboat market. This was greatly to the annoyance of Smit Port Towing Services which, with its inferior equipment, had to deploy more boats per assistance and ultimately had to offer below cost price. In February 1994, the Smit employees barricaded the Nieuwe Waterweg out of frustration with a number of drastic internal reorganisations. This effectively closed the port to all shipping traffic. The RMPM applied for a temporary injunction and immediately prior to the court case the tugboats withdrew. This incident formed the catalyst for the regulation of the tugboat profession which was included in the Port Ordinance in 1995.

In order to achieve a form of regulation, the RMPM and other layers of government took a softly-softly approach. After all, it concerned the regulation of a world which had previously been perfectly capable of sorting out its own problems. The only reason for the RMPM to resort to regulation was the long-term guaranteeing of tugboat services to ships. Not only must it be made possible for tugboat services to assist each other if one of them was short of boats, but the RMPM also felt that the continuity of service provision at an adequate technological level needed to be guaranteed. This referred to the financial scope which the towing service companies needed to have for the depreciation of, and creating reserves for, the purchase of new equipment. The Economic Affairs and Transport Ministries, however, prohibited the RMPM from exerting direct influence on the tariffs employed in the towing branch. Theoretically, there were three options remaining for regulation: a licence system, a book of standards or a stand-by regulation making it compulsory to provide assistance in the event of tugboat shortage.

Right from the start, the licence system was not considered a real option by any of the tugboat companies. It was considered rather offensive in view of how highly experienced the tugboat companies were and the equipment they used. The book of standards suggestion was not badly received but was dropped in the end because an attempt was made at the same time to harmonise the tugboats and the linesmen, which proved to be too great a challenge. The stand-by regulation also produced varied reactions because it provided more opportunities for companies that had more work than boats and vice versa. Finally, in consultation with the RMPM, a choice was made for a system of quality criteria which must be complied with and compulsory participation in the Central Coordination Unit for Port Towing Services. The RMPM has subsequently included a provision in the Port Ordinance to which an implementation procedure is attached which regulates the required quality criteria (number of tugboats based on capacity and size of the ship) and the relationship between the RMPM and the coordination unit. The procedure provides the RMPM with the opportunity to assess in good time the risks of unassisted, inadequately assisted or wrongly assisted seaships and to take the necessary measures (licence suspension or issuing instructions).

Linesmen

The linesmen are responsible for the tying up and untying of ships. In the last century, there was a whole range of small linesmen's companies between which cutthroat competition existed. In 1895, the harbour master provided the impetus for the Linesmen's Association 'Eendracht' to be set up which to this very day remains the focus for these activities in the port. Due to the phasing out of the fragmentation in linesmen's companies it became easier for the port authority to achieve mutual coordination. Moreover, the setting up of an association for the shipping agents also made it more manageable. During the activities, the linesmen work in close cooperation with the pilots and the tugboats. They maintain constant radio contact with each other. Eighteen months ago the linesmen, together with the pilots, the tugs and the RMPM port service staff, moved to the Nautical Service Centre in the Botlek and this has facilitated coordination between the maritime services. The linesmen now provide the 'communication sailing' service (for the pilots among others) which refers to the unscheduled transporting of people to and from ships at a charge. Nautical centres will also be created for Europoort and the Maasvlakte in which the linesmen's association will be represented.

The linesmen have no competition in the port and must meet the requirements which are laid down by the RMPM in the Linesmen's Ordinance (1951) (incl. training, recognition by the municipal executive, etc). The ordinance is the city's instrument for regulating the linesmen's market by means of quantitative and qualitative criteria. The number of licences issued (to individuals within the association) does not exceed the number of linesmen required. Moreover, unwritten quality criteria are laid down for the profession of linesman on the basis of which linesmen's organisations can be recognised by the municipal executive. The only recognised organisation of this type in the port of Rotterdam is the Eenus Organisation which is a collaboration between the linesmen's association Eendracht and the cooperative association Neptunus. At present the Eenus Organisation, in close consultation with the RMPM, is involved in turning the unwritten quality criteria into written recognition standards (with respect to equipment, training, availability, etc.) and incorporating them into the Linemen's Ordinance. Finally, at present the RMPM is still exercising price control on the linesmen's tariffs but this will disappear under the new regulation. The RMPM would appear to be opting for competition in the linesmen's market.

4.7.   Final Analysis: Port Management in Rotterdam

The port of Rotterdam plays an economic role which has (inter)national significance. In the national and local political arena, employment and added value are the most important assessment criteria for the success of the port, but an awareness exists that in other parts of the country unemployment is increasing. This is why local political parties are so insistent that the port should produce as much direct income as possible so that the RMPM's revenues can be used to provide 'extras' for the residents of Rotterdam. Due to the current national seaport policy and the port of Rotterdam achieving mainport status, over the next five years a relatively large amount of (political) attention will be devoted to the port and its infrastructural needs. After this national infrastructural stimulus, the city of Rotterdam and the business community will have to prove themselves.

As a municipal company that is responsible for the management and operation of the port, the RMPM has both public and private powers and enjoys a large degree of autonomy. There is back seat political control while a direct market mechanism is absent. After all, investments by companies are made for the longer term and companies cannot just pick up their superstructure and set it down elsewhere. Competition with other ports is mainly perceived when contracts are being renegotiated. All this means that the impact of strategic choices only becomes visible in the long term and that RMPM is not immediately confronted with its actions. In times of economic decline the city budget can serve as a safety net.

The RMPM, set up in the thirties as port authority to occupy an 'independent position' between the authorities and the business community, considers it extremely important that the port should retain its competitive position vis-à-vis the surrounding ports. It judges the success of the port on the basis of price level, cargo tonnage handled and revenue acquired. The RMPM considers it unproductive to use the port as a direct employment instrument because, as a government company, it is too dependent on the enterprising spirit of the business community which in turn is dependent on trends in world trade. In the Corporate Plan 1997-2000 both the city council and the RMPM have conformed to the general objective of optimising the port in a European context and not formally using it as an employment instrument.

Against the backdrop of the theoretical assumption that hybrid organisations cease to learn in the long term and become introverted, as far as the double role of the RMPM is concerned a number of reservations are in order. In the past, for pragmatic reasons, (expertise, single counter principle, cohesive maritime policy, etc.) both public and private powers were integrated in the RMPM. As the various analyses show, in the long term this hybrid position lays a heavy claim on the regulatory capacity of local politicians. They cannot actually live up to this because the distance between them and the RMPM is too great. This also holds for the national involvement in nautical management. 'Checks and balances' will have to come more from other local and national public services, as increasingly seems to be the current practice (environment and spatial planning). If these divisions of responsibility take a more structural shape (clearer regulatory principles) there is much to be said for the RMPM's expertise motif.

However, if in the future the RMPM demands an even more autonomous position for itself, grave doubts arise regarding this fruitful cooperation between public and private activities in the long term. There are a number of arguments for this. Rotterdam mainly implements financial controls (annual report, budget). Since there is no national politico-economic vision with regard to the port, political involvement in port affairs is very limited. Moreover, an attempt is made to activate the learning capacity of the RMPM by maintaining two separate departments (Shipping and Operations and Acquisitions). Another question is whether the consultation mechanism at management level is sufficient to justify the expertise motif. After all, nautical management and port planning concern entirely different activities which have each developed their own, entirely different cultures. Finally, the RMPM's behaviour is aimed at a larger private role in the port. This has now become the current norm in view of its facilitating role in the development of the existing Maasvlakte and the objectives outlined in the new corporate plan. These considerations raise the question of whether a linking of public and private tasks in the RMPM will be advantageous in the long term. After all, it is unclear which mechanism is meant to regulate the commercial risks taken using public resources in the long term. So if the RMPM demands even greater autonomy for itself, serious consideration should be given to splitting nautical management and port planning.

Two forms of repositioning are currently being contemplated. One possible option is privatisation of the RMPM into a public company in which central government and the city of Rotterdam participate. Another option is to accommodate the RMPM in the forthcoming regional government. For the first alternative, a minimum institutional condition is that a division should be made between nautical management and port planning because against the backdrop of the private law objective of the limited liability company it is highly likely that in the long term public and private tasks will become intertwined and there will be no controls. In that case there might even be something to be said for two separate private law port authorities for separate sections of the port (the sections to the east and to the west of the Benelux tunnel respectively). Moreover, the financial responsibility to the city must be made more explicit. For the second alternative a minimum condition that needs to be met is the clear delineation of public and private (financial) responsibilities and of the political and executive port committee. At the same time, a politico-economic view of the port would need to be defined; relationship with the city, function for the region, weighing against other interests, etc.

In the description of the sundry port activities, the various control relations between the RMPM, other layers of government and the business community were made clear. Each time, expectations were first stated regarding the type of control relationship which would be found based on the type of activity. Then various concrete activities in that category were described and it was discovered how the control was actually organised. The next section analyses the various port activities by contrasting the expected norm with the empirical facts found and questions are posed.

Analysis of Nautical Management
As a general guideline it was stated that the nautical management activities involve public tasks and public powers which are expected to be undertaken by various public services by means of the government monopoly control relationship. In studying the empiricism, there are a number of activities which by nature should come under nautical management yet where different control relations have been found which need to be questioned.

The description of nautical management began with an outline of the options which the RMPM has for regulating the port. Following on from this the relationship between the RMPM and the Directorate-General for Shipping and Maritime Affairs (DGSM) was sketched. The picture that emerged was that as far as the harbour master's activities are concerned, the RMPM in fact has a monopoly on the control of shipping traffic. For reasons of expertise, central government powers were delegated in the past to the harbour master of Rotterdam thus giving him his own responsibilities. Nevertheless, in the event of a crisis or in turning policy formation into policy implementation the RMPM and DGSM come into direct contact with each other although DGSM in fact has virtually no control over any of it. This is surprising given the national interest that DGSM has in national safety policy and central government's co-financing of the construction of the Traffic Guidance System in the past and of its present maintenance costs. Within the government monopoly control relationship, the 'checks and balances' between the various public bodies have been removed. The question is, how will such a relationship develop in the long term and how can the RMPM be stimulated to furnish accountability for its nautical management?

An entirely different relationship is that between the RMPM's dangerous substances inspectors and those of the National Traffic Inspectorate (DGV). For years, inspectors from these two bodies have been trailing around after each other on board ships to the great irritation of the shipowners. After the reorganisation of DGV, it appeared that with regard to this activity no clear relationship between the central and local level had been devised which was a source of irritation and frustration on both sides. The Rotterdam Inspectorates Project aims to reduce the number of inspecting bodies per ship. As yet, it is only considering the input of (technical) resources to enhance coordination so that the work ad hoc can be improved. But who can guarantee that the current improved relations will last in the long term if no clear regulatory principle exists between the RVI and the RMPM and while the present positional interest is allowed to continue?

The Green Award Organisation is trying to unite the various inspection interests for tanker traffic in a creative, institutional way. It was set up as an independent body to promote the interests of the Ministry of Transport and of the RMPM. It will have to substantiate its independent status by the sale of certificates which the shipowners can 'recover the costs on' thanks to reductions in port dues and port services in various ports, but potentially also by means of the shorter stay in port due to the reduced number of inspections. The certificate is aimed at the international tanker market and it is very important that it should be internationally accepted. This will work if as many ports as possible offer a reduction to certified ships. In their turn, ports will be prepared to give reductions to an independent institute that can guarantee control quality. How will the Green Award fare, however, if it remains dependent on the Dutch government and the RMPM?

Nautical management also covers those activities pertaining to the control of floating transhipment and the port reception installation. The careful balancing of the often conflicting interests of speed and safety in floating transhipment is guaranteed in an institutional way by designating the regional environmental body as sole enforcer. The clarity of this regulation of responsibilities was sadly lacking in the confusing and ultimately disastrous division of interests in the regulation of TCR's port reception installation. By making it compulsory for TCR to take on all the ship waste substances offered and by allowing the profits to precede the costs it proved impossible to produce the public good without problems in the framework of government conditioning. The abundance of enforcers and supervisors muddied the issue of who the primary interested parties in this government control were. The question then is how else could it have been done? Or in a broader context: is the distinction between wet and dry infrastructure a sound regulatory principle on which to base enforcement and supervision?

Analysis of Port Planning
The regulatory principles for port planning have changed over the years. Until the Second World War the city was expected to take the lead in the construction of sites. The business community in particular adopted a 'wait and see' stance and did not develop an overall concept. After the war, port planning was no longer only a municipal concern but had now become a national interest as was made clear by the development of the Europoort plan

and the acquisition of companies. In the sixties, the RMPM occupied a position of supremacy as the implementer of ambitious plans and due to the great interest shown by international companies. This position suddenly changed in the seventies when the national economic interest in the port waned and other quality standards were adopted by the Rotterdam city council. Due to the economic crises no further expansion plans were made and the RMPM concentrated instead on developing the existing area. They had to take increasing account of spatial planning criteria, environmental requirements and global locational criteria. In contrast to the sixties, they were now forced to develop the port area together with other layers of government and the business community. It is expected that contemporary port planning will take shape through the control relations of government conditioning and co-arrangements.

The current political attention devoted to infrastructure at a national level has ensured that the port of Rotterdam will be able to strengthen and possibly even expand its international competitive position. The construction of the Betuwe Line, to be financed by central government, will mean that the poor rail link to the hinterland will be phased out and Rotterdam could develop into a container hub. The spatial expansion of the port depends on the political discussion on a second Maasvlakte, the costs of which have been estimated at eight billion guilders. The question is, who will foot the bill in due course? Until now, port expansions have been financed by the city but the amounts were not particularly high. The current mainport policy has already accorded political goodwill to Rotterdam but the question remains of whether the construction costs will also be borne by the market and which financial role central government will play (full funding or partial hooking up with city initiatives). The main question which needs to be considered is how the land can be managed in a more innovative way in the future. What options does central government have to guarantee a balanced and innovative use of the land if it were to finance a second Maasvlakte?

In this context it also seemed useful to discuss two examples which illustrate two different methods for the releasing of sites, i.e. voluntary returning after soil cleanup and relocating a company within the port. In both cases, close consultation between the port authority and the local and regional environmental bodies finally had the desired effect. In the second example it was clear that the RMPM has a tendency to be too deferential in its dealings with the business community, fearful of disturbing a relationship based on trust. Finally it proved possible, in cooperation with the environmental agencies, to effectively deploy a highly vertical means such as the pecuniary penalty as a motivation for the company to relocate. It may well be possible to derive options from this for interaction with international industry so that the port authority can regain far more control over the use of the land than it has at present.

In adapting old port sites to modern requirements, co-arrangements also appear to be the appropriate method for getting new initiatives off the ground. The redevelopment of the Waal-/Eemhaven area is now starting but there is no sign of any actual financial participation from RMPM in this small-scale enterprise. The cooperation appears to extend only to joint plan formation in which the RMPM performs the role of general project coordinator and motivator.

Another example of a co-arrangement was found in the developing of the present Maasvlakte for container handling (Delta 2000-8 project). To this end numerous conditioning projects were implemented by various combinations of parties. The Delta 2000-8 project required an entirely different effort from public parties in which the old regulatory principle was abandoned on the grounds of expensive technical innovations and international competition. It remains to be seen, however, whether the competition with Antwerp and Hamburg will keep ECT sufficiently on its toes now that the city has agreed to a form of cooperation between ECT and Maersk which will put paid to internal competition for the present.

Analysis of Port Services
The port services comprise the private activities pertaining to the service provision to ships and for goods in the port. It may be anticipated that such activities in a port the size of Rotterdam will be provided by private actors. In other words, this involves the market regulation control relationship in which the public authorities and the port authority assume the regulatory principle of competition and lay down supplementary conditions on the basis of, e.g., safety objectives. In a number of cases the empiricism turned out to deviate from the norm.

The description began with a detailed analysis of the container sector and the relations between the RMPM and ECT. In order to secure its own interests (constructing less quays, securing income from the Maasvlakte terminal, integrating expansion plans), the RMPM engineered a supply concentration for the deep-sea container handling market segment which left ECT as monopolist. As was clear from the Delta 2000-8 project, the cooperation between ECT and the RMPM has become increasingly close over the years despite the fact that the RMPM had no influence on the layout of the terminal, the container handling concept or ECT's executive committee. Now that Maersk has checked in as the latest stevedore, the question is how should the RMPM deal with both ECT and Maersk? Do the same conditions for building a terminal apply to Maersk as were devised for ECT?

Another remarkable institutional design is the pilotage system. Compulsory pilotage means that piloting ships is a public activity but in Rotterdam it is carried out by a private monopolist over whom no public authority has any control whatsoever. On the other hand, the assisting of ships is provided under competition which regulates the market by means of the Port Ordinance in order to give the RMPM an escape clause. Nowadays, quality standards are also set for the linesmen and included in the Port Ordinance. Since the linesmen occupy a monopoly position, however, it is highly likely that the port authority will get all its information from the incumbent organisation. This means there is a chance that the quality standards will be laid down in such a way as to make it extremely difficult for other potential providers to enter the market.

4.8. Summary

This chapter discussed the institutional map of the port of Rotterdam. Via a brief description of the city's history and a more detailed exploration of the position of the RMPM, the relations between the RMPM, various national authorities and the business community were examined using the familiar three-way split into nautical management, port planning and port services. Since the chapter was of a highly descriptive nature, a number of interesting glimpses of the Rotterdam port authority were provided. Thus, it turned out that in fact the RMPM has exclusive rights in the field of nautical management, while for the increasingly important environmental enforcement activities it is dependent on the DCMR Rijnmond Environmental Protection Agency.

During the first three decades after the Second World War, the RMPM realised tremendous achievements in the field of port planning. Over the last twenty years, on the other hand, the growth of the port has been influenced by a decline in the anticipated growth of the world economy, but principally by the emergence of other quality norms (spatial planning and environment). Now that the port wants to expand further into the sea, it must be concluded that in the past the system of land allocation was too liberal and that there appears to be no mechanism which safeguards against this. Finally, the port services in the port of Rotterdam are characterised by the existence of various monopolies over which the port authority has no control. In the separate analyses of the individual port activities, questions were raised from the viewpoint of learning capacity and these will be returned to in the final chapter.

CHAPTER 5

The Ports of Antwerp and Hamburg

'The word "seaport" is not even mentioned in the basic Treaties of Rome or Maastricht' (Suykens, 1996:156).

5.1.   Introduction

Rotterdam's two largest competitors in Western Europe in the field of conventional general cargo and containers are the ports of Antwerp and Hamburg. All three ports fight for as large a share as possible of this freight while jointly attempting to convince the European Commission of the impossibility of a European seaport policy. Since 1993, the transport policy has been high on the Brussels political agenda: the role of ports as crucial components in the transport chain has been recognised, old institutional controversies have once more arisen and options and criteria for fair competition are being sought. It is interesting to see how Antwerp and Hamburg are preparing themselves institutionally for the approaching European unification. In contrast to Hamburg, where the port ordinance of 1970 still appears to be functional, Antwerp has recently decided to privatise the port authority and steer a new course.

This chapter is constructed as follows. Section 5.2 describes and analyses the institutional structure of the port of Antwerp. In the next section Hamburg is dealt with in the same way. Section 5.4 then describes the intentions and progress of the European seaport policy and the relationship between Brussels and the port managers. The final section contains concluding remarks.

Table 5.1 Comparison of a Number of Indicators from the Ports of Antwerp and Hamburg

	Antwerp	Hamburg
Total port area	14,055 ha.	8,700 ha.
Employment 1995	65,000 (direct)	140,000 (direct + indirect)
Total transhipment 1998	112 million metric tons	76 million metric tons
Container handling 1998	3.266 million TEUs	3.546 million TEUs

5.2.   Port Management in Antwerp

The port of Antwerp lies on the Scheldt river which runs from the North Sea about seventy kilometres inland and forms the physical boundary between the Netherlands and Belgium. Antwerp is the largest port in Belgium and the second largest in Europe. Over the years the port has expanded to the north of the city along both sides of the Scheldt and these areas are referred to as the Left Bank and the Right Bank. The port and industrial area comprise 14,055 ha. of which 7,657 ha. are on the right bank. In 1994, over 2,000 ha. of land was still available on the left bank.

In Western Europe, the port of Antwerp is known mainly for the high level of general cargo which it handles, its industrious workers and its pragmatic approach. In 1996, it handled 106.5 million tons of goods comprising 27 million tons of liquid bulk (crude oil and chemicals), 27.2 million tons of dry bulk (iron, coal) and 52.2 million in general cargo (iron and steel, wood, paper, cars, fruit, sugar). More than half (29.4 million tons) of the general cargo was transported in the 2.65 million containers (TEU as standard). In the same year, over fifteen thousand ocean-going vessels and over fifty thousand inland vessels put in at the port.

Port management in Belgium is traditionally a decentral affair. Van Hooydonk (1996) in his interesting and detailed legal-historical study traces the roots of the decentral port management in Belgium back to the early Middle Ages when the towns came into being as individual entities with their own powers. Van Hooydonk concludes that it was this *Hanseatic* tradition (named after the well-known trade network of towns) in combination with the commercial operations that in recent reforms have determined the autonomy of the port's management. It remains to be seen whether an interaction pattern derived from history will be successful in the long term as a norm for the regulatory principle between the central and local level.

The administration of the port is a municipal concern; the municipality needs no specific statutory authorisation or order to construct and operate a port. The port administration is thus held accountable /judged not so much for /by its national-economic impact but rather for /by the development of the actual port area: 'The local port legislation assumes that a port administration should actively **not** serve higher interests' (Van Hooydonk, 1996:115). It is only when a financial contribution from regional government is made to concrete, individual projects that the project's contribution to the national economy is taken into account. The domestic competition between the various Belgian seaports (Antwerp, Ghent, Ostend and Zeebrugge) has always been found to be highly productive, thus avoiding the need for a regional port policy.

5.2.1. POSITION OF THE PORT AUTHORITY

Until 1 January 1988, the day-to-day operations in the port of Antwerp were managed by a municipal department under the leadership of the local authority. After that time, it was assigned a more independent status. This gave it the opportunity to set up its own bookkeeping and to steer a more autonomous course. The targets of the port authority did not change, it continued to be responsible for the maintenance, renovation, schedules and operations of the port with an eye to harmonious development. Since the port authority is part of the municipality, *independent* action is expected from it which will serve the general interest. Making a profit is not seen as an end in itself, but a profit is of course welcomed. In Belgium, too, it appears difficult to pin down a port authority's specific tasks and functions. Belgian law makes it clear that port authorities are not expected to aim for profit. This is actually impossible to sustain because creating financial reserves is often stipulated and port authorities employ commercial methods and business standards (Van Hooydonk, 1996).

When it achieved a more independent status, the port authority did not yet have an individual legal status and it was still directly controlled by the municipal council and the locally-elected port committee chairman. Moreover, the port management felt that this transformation did not sufficiently stimulate the transition to a more businesslike culture, which was seen in the later change proposals as an essential objective. The organisation of the port authority consisted of a General Management covering two departments (the Port Captain's department and the Technical Unit) which employed a total of about 2,000 people. Those involved who desired a more dynamic port policy, felt that this step towards a more independent port status was not going far enough. For investment sums over 100,000 francs, for example, the approval of the municipal executive and the municipal council was still needed. Such a procedure took at least 121 days.

Since March 1995, it has become possible, on certain conditions, for the Belgian municipalities to set up autonomous municipal companies with legal status. In concrete terms this means that 'the municipal council takes the decision to set it up; participation in this department by other companies is not possible nor are any shares available; the majority of seats on the board are held by of council members; two out of three financial officers are also council members; participation by the department in other companies is only possible if the department has a majority of votes and acts as chairman; in general, the policy of the municipal company is under the constant control of the municipal council' (Van Hooydonk, 1996:105). On 28 September 1995, the municipal executive of Antwerp took the decision to expedite the establishment of a municipal port authority, which during the course of 1996 was developed in increasingly concrete proposals. In September and October of that year, the names of the first of those to fill the eighteen administrative posts were published in the media. The administrative council of the Port of Antwerp plc will consist mainly of members of the Antwerp municipal council plus a number of representatives from the port business community (industrial shippers and professional associations) and environmental groups (1 seat). It is forbidden by law for an individual private shipping or port authority to have a seat on the executive board. The role of the port authority management is to observe and advise the executive board and the management meetings are chaired by a delegated administrator.

The nature of the administrative council is already being questioned because the original intention was that the port authority should be able to take up a more independent and professional position in the political arena. It remains to be seen whether it will be better able to do that in the new configuration. After all, the involvement of the municipal council remains considerable. Moreover, the statutes of the privatised Port of Antwerp plc explicitly state that the public company retains the authority over the public management of the port, and also may develop private activities. A relevant question here is how the traditional norm of independence is safeguarded in the new design.

In the discussions on an autonomous port authority, various standpoints were presented. In the first place, a highly autonomous position did justice to the Hanseatic tradition so characteristic of Belgian port administration. A legal status would generate more financial resources and funding options: this has become increasingly urgent now that the Flanders Region is redefining its position on financing. In concrete terms, this means that the region, which is plagued by debts due to budget restrictions, will /would concentrate on maritime access so that in the future the ports will /would have to furnish a larger share of the financing in their region. Moreover, it would be /will become possible for the port authority to implement its own staffing policy and it would /will be able to enter into joint ventures and participation with private actors in order to set up initiatives (e.g. for tugboats, pilotage and waste disposal). By means of this institutional method (legal status and municipal representation) expression could be given to the involvement of the community and the ownership ties between city and port. And finally, the need for contact and ongoing dialogue with the users was cited to underpin the local embedding. Since 1 January 1997, the Antwerp port authority has had legal status amidst the port political arena and the port business community, which constitutes a clear choice on the part of Belgium for local autonomy and scale reduction. This is in complete contrast to, e.g., the port of Rotterdam which is managed on the principle of scale increase by the local administration.

In the past, the port authority made a profit of 1 billion Belgian francs on a turnover of 7 billion. This profit was distorted by the fact that the port authority foots the bill for the municipal pension budget. At present, a sum of 25 to 30 billion francs pension monies is outstanding which, with the creation of the autonomous port authority, will from now on formally count as part of the port authority's assets. In 1994, the port authority ultimately made a net profit of 'merely' 14 million francs.

Relationship with Central Government
As a result of the reforms in the Belgian political system, the port of Antwerp has come under the Flanders Region since 1988. Together with the other Flemish ports (Ghent, Zeebrugge and Ostend), and representatives of the employers and employees from the various modes of transport (road, barge and rail), Antwerp also has a representative in the Flemish Port Commission. This commission advises the Flemish government on investment projects of more than 300 million Belgian francs and on questions concerning the organisation of the port administration (e.g. the privatisation of the Antwerp port authority). It also appears to function primarily as a buffer against centralising powers in the Flanders Region. The Flemish Port Commission itself has no powers of decision and is only charged by the Flemish government with the preparation of important port decisions. The Flemish ports, which are competitors at home, are thus forced to cooperate in the advisory council.

The Flanders Region has been declared to be expressly qualified /authorised /given authorisation for the industrial development of the port areas and for laying down quality conditions for fringe matters such as pilotage, the rescue and towing of ships, aids to navigation such as beacons and dredging the maritime access lanes. In concrete terms, this means that the Antwerp port administration is responsible for the day-to-day running of the port and that it appeals to the Region for a financial and/or diplomatic contribution to port planning activities (see subsection 5.2.3). In addition to its financial supporting role in port development, the Flanders Region has special powers in a number of port administrative matters aimed at nautical management, such as laying down the police and shipping regulations for tidal ports and appointing the port captains. In practice, the latter concerns are of minor importance to the Region (Van Hooydonk, 1996).

The relationship between the Antwerp municipal port administration and the Flanders Region is mainly characterised by financial cooperation on major investment projects. This dates back to the 1860s when Antwerp appeared no longer able to pay its own way and the national government had to come to its financial aid. For national economic reasons, the Belgian government bought up the *Scheldt toll* (payment for passage over the Scheldt to the Netherlands) for the ports of Antwerp and Ghent, without demanding formal local control for it, although it did insist on a tariff reduction. This national contribution was the start of a national seaport policy. In complete contrast to France, however, direct state involvement in Belgium between 1870 and 1988 did not arise through centralistic interference but because port development was considered to be of national importance. All that time, national government viewed the running of the port as a municipal matter.

In 1895, the first national port investment plan was developed, aimed at expansion and improvement operations. In the period up to the Second World War, the foundation was laid for the relationship between the national government and the municipal port administration. Prior to the introduction of the national financial contribution, the development of the port was an entirely municipal affair and it was logical that ownership and management should fall to the municipality. In those projects which were developed through national funding, formal ownership remained in the hands of national (and after regionalisation: Regional) government, while the actual management was transferred to the municipal port administration by means of *management transfer and concession arrangements*. This institutional method of port development still applies today and characterises the (*contractual*) relationship between central and decentral port administration; various layers of Belgian government acted as partners in developing the port and a clear division of tasks (financial contribution versus operation) was created over time. The contractual relationship also means that centrally-imposed decisions do not work and that the port administrations themselves determine the tasks they wish to undertake.

After the Second World War, a welfare state was also established in Belgium and international trade underwent explosive growth which in turn led to the need for large scale facilities. The construction of these was financed largely from the treasury so that national port politics became increasingly dominated by infighting between the various Belgian ports to gain the support of national investments (Van Hooydonk, 1996). Sometimes it just seemed to be a matter of spatial and economic growth, because in the pragmatic dynamics essential, related matters tended at times to be forgotten. A second typical characteristic of Belgian port administration relationships can be derived from this which Van Hooydonk (1996) describes as follows: 'An even more extraordinary practice is that of failing to regulate the decentralised administrations of new ports constructed with government money, either by act or decree, or by contract, so that the local administration takes on the *de facto* operation of the port sector with no legal underpinning whatsoever' (p. 82).

In the past, various infrastructural port facilities have been constructed with the help of national funding (extension of the Scheldt quays, construction of container quays (Europaterminal and Noordzee terminal), various locks (Zandvliet lock and Berendrecht lock) and the Left Bank section. After regionalisation in 1988, all national port properties came under the ownership of the Flanders Region and the national investment policy was continued by the Flemish government. In principle, it can use two methods for this. In the first place, the Flemish government can construct new port installations with itself as provider. Although these then become the property of the Flemish Region, by means of a special statutory or decree provision or by convention they are entrusted to the stewardship of the Antwerp port administration. A second method is to subsidise local port works, which for the port of Antwerp means that *subsidies* can be allocated:

- 100%: for basic port infrastructural works (sea locks, moles, earthworks for docks, etc.);
- 60%: for the construction of equipment infrastructure (mainly mooring places such as quays and gantries) and
- 80%: for the renovation of existing infrastructure (Van Hooydonk, 1996:79).

Under municipal law, the Flanders Region is empowered to supervise port administration. A surprising element in this is that the Minister for Home Affairs is in charge of administrative supervision and not the Minister responsible for port policy. There are specific supervision procedures for laying down port regulations and port dues. The distinction between the domains of these two Ministers leads in practice to the administrative procedure being of a predominantly legal nature and taking less account of economic considerations in the port.

Another institutional element that is very important to the control relationship between the Flanders Region and the Antwerp port administration concerns a recent authorisation given by the Flemish Council to the Flemish government. This authorisation makes provision that the Flemish government (among others) may provide infrastructure in the port of Antwerp and to this end may enter into agreements with the Antwerp port administration which relate to the port administration's obligations regarding maintenance, repair and renovation of infrastructure and the regulation of strict liability and costs. Van Hooydonk (1996) quite correctly notes that there are a number of snags to this authorisation. In the first place, entering into the above agreements has no direct connection with budgetary matters. Moreover, the Flemish government is given carte blanche to enter into agreements which cannot apparently be regulated by the Flemish Council. And finally, it must be said that the assignments for port works by the port administrations are determined to a considerable extent by precisely this type of agreement, so that the principle of *'every supply creates its own demand'* comes into effect (Van Hooydonk, 1996:84).

The central (and after 1988: the Regional) Belgian government has never had pretensions to set itself up as administrator of the major Belgian ports. Its involvement in the port of Antwerp's development has always been for national economic reasons, taking place on a project basis and on the grounds of the local financial need. Due to the project-specific approach, no national seaport policy has been formulated. In the past, the desire to increase the complementarity of the Belgian seaports was more important than at present. In that sense the political reforms and/or regionalisation of 1988 certainly seem to have contributed to the scale reduction of the Antwerp port administration and the desire to liberalise the port authority.

5.2.2. NAUTICAL MANAGEMENT

Nautical management in Antwerp is the responsibility of the port captain who is appointed by the Flanders Region on the recommendation of the Antwerp port administration. The Port Captain has unique policing powers within the port area. Although he is appointed by the Flanders Region, he is employed by the port administration because the port policing power is defined as an essential element of port administration powers. The Port Captain does not form a general policing power but is there exclusively for the port users. In that sense, nautical management in the Belgian ports is not a public service. Moreover, this construction is based on the fact that in the port no division can be made between policing and non-policing user conditions. The wet and dry areas of the port are seen as a special area; it is a public area but at the same time it contains numerous specific risks. One of the Port Captain's main tasks is to prevent or settle day-to-day conflicts between port users.

The relationship between the Port Captain and the municipal police force is clear-cut: 'Anything which concerns the special port police is removed from the domain of municipal powers' (Van Hooydonk, 1996:331). Moreover, the chance of competency disputes is small now that the port and the city have gradually become physically more separate. The Port Captain's department is also responsible for manning the bridges and locks and for the maintenance of the traffic guidance system.

The port of Antwerp has never had a port reception installation for cleaning ships and recycling ship's waste. Although the Marpol convention has been in force in Belgium since 1984, the (partial) completion of such an installation may not be expected until late 1997. Its operation and management will be carried out by a private company once an Environmental Impact Assessment has been carried out and a licence issued. This company first carried out a thorough market research study. Many waste materials are still being disposed of at sea or are being sent to the Netherlands in smaller shipments.

5.2.3. PORT PLANNING

Section 5.2.1 dealt in some detail with the national government's input with regard to port planning. In addition, it is interesting to look at the construction which has been devised for the industrial development of the Left Bank, how the allocation of sites occurs and what may be expected now that the Scheldt is being deepened.

Until the early seventies, the port of Antwerp had always been able to expand along the Right Bank. The rapid economic growth which followed the Second World War and a steady scale increase produced a need for an additional site on the Left Bank which belonged to the province of Eastern Flanders and the municipalities of Beveren and Zwijndrecht in which it was located. The city of Antwerp would have liked to annexe these sites for the further development of the port. This idea was a non-starter, however, because the province of Eastern Flanders, the municipalities and those involved from the area known as the Land van Waas demanded public consultation procedures in the port and industrial policy. The Belgian national government had commenced work on the port in 1971, but it was not until 1978 that the famous Chabert Act, also called the Left Bank Act, was adopted. In fact, due to continuing dissension, the Act did not enter into force until 1987 (Van Hooydonk, 1996:57).

This Act regulated the management and operation of the Left Bank section. 'The special statute for the Waasland port is unique in Belgian administrative law. The port lies entirely outside the Antwerp provincial and municipal boundaries and yet is managed by the City of Antwerp' (Van Hooydonk, 1996:57). The area has been specially designated as a port and industrial zone and these 'must be functionally linked' (ibid.). In order to clarify the control relations in this area, its exclusive management has been assigned to a functional organisation in the form of an intercommunal group bearing the name 'Society for Site and Industrialisation policy on the Left Bank of the Scheldt'. Serving on this group are representatives from the Land van Waas, the municipalities of Beveren and Zwijndrecht, the Flanders Region and the City of Antwerp. The Society is responsible for the port and industrial zones and for the industrialisation policy of the industrial zone. It acquires the sites concerned, develops them for building and makes them available to industrial investors, or to the City of Antwerp if these are sites essential to the management and operation of the port. Thus the City of Antwerp does not have exclusive rights to the sites intended for industry but must convince the adjacent actors. This has not yet led to an integrated spatial-economic concept for the entire Right and Left Bank area, however, although there is increasing need for this now that the Waasland port is being increasingly developed.

The granting of allocated sites normally takes place under competition from various applicants. In the past, however, sites were also sold, e.g. to a number of large multinationals (Bayer, BASF, Ford, General Motors). These days this rarely occurs due to of the loss of planning potential. In all other cases *concessions* are used which articulate a customised approach tailored to the company. Long-term concessions used to be issued (sometimes up to 99 years) but since the Second World War they have become increasingly short-term to enable a more effective port policy to be pursued.

To ensure the port of Antwerp will have prospects for development into the next century, since the sixties the Belgian government (first federal and later regional) has conducted extensive consultations and/or negotiations with the Netherlands regarding the deepening of the Scheldt by the Belgians. The scale increase in container trade and the desire for intermodal transport meant that to keep Antwerp attractive the deepening of the Scheldt was absolutely essential. The port will then be able to accommodate 5,000 TEU container ships and it will be permanently accessible to ships with a draught of 40 foot (12.2 metres).

The Netherlands obstructed this operation for a long time and for various reasons. In the first place, it provided a splendid opportunity to once more put a spoke in Antwerp's wheel after the buying up of the Scheldt toll in the previous century and to indirectly promote its own ports (Flushing and Rotterdam). This informal motive was strategically concealed behind a number of other arguments. For example, environmental objections were made to dredging which were linked at a later stage to the cleanup of the Maas by the Belgians. Since 1993, the Dutch desire to determine the route of the TGV high speed train had also been a factor in the negotiations. Finally, however, on 17 January 1995 the treaty between the Netherlands and the Flanders Region was signed and the (Dutch) dredgers could start work. The project, that will cost six billion Belgian francs, is entirely financed by the Flanders Region and will not be offset against seaport monies. It is expected that the project will be completed in the year 2000 when Antwerp will once again have strengthened its competitive position for the container market segment vis-à-vis not only Rotterdam but also Zeebrugge.

5.2.4. PORT SERVICES

In principle, the port services in Antwerp are a private affair. The port administration has always worked as much as possible on the economic principle that the success and effectiveness of the port are best served by competition between providers within the port. For example, the port authority did not support a possible merger between the port's three container stevedores. Nevertheless, Hessenatie, in particular, thinks that competition no longer takes place within ports but between ports. For the time being the merger is off and competition exists within container handling. In the current situation, the port authority is in a position to divide the concessions for container terminals in such a way that competition is safeguarded. Thus Hessenatie, for example, failed to get the concession for the operation of the North Sea terminal despite being the largest container stevedore in the port with a market share of over sixty percent. When the concession was granted to the joint initiative of Noord Natie and Belgian Railways, plans for the merger were immediately shelved.

Another characteristic feature of container handling and other specialist handling which is mainly carried out by major companies (ro/ro, cars, fruit, fertilisers and other neo-bulk cargo such as iron and steel, wood products, etc.) is that private companies bear the costs of the cranes and other required superstructure. The port authority decided that the smaller companies would probably have to remain on the conventional quays so that they could lease port equipment short or long term from the port authority. In the past it had bought new equipment for this purpose (quay cranes, floating cranes and depots). The port authority now has over 85 electric quay cranes and three floating cranes and derricks. This cooperation has so far worked to everyone's satisfaction and is probably a contributory factor in the port of Antwerp's relatively large share of the general cargo market.

In contrast to the services for goods handling, the competition principle has not yet carried over to the service to shipping. On the Scheldt, national pilots, with regulated tariffs, operate and there is only one private tugboat service. Recently, this monopolist wanted to raise its tariffs by fifteen percent when the arrival of Kotug created real competition between towing services, first in Rotterdam and later in Hamburg, too. In the port itself, towing is carried out exclusively by the Port Captain's department which runs at a profit (21 tugboats). Within the port 'dock pilots' also operate who, although private, enjoy a monopoly position; use of the pilots is not compulsory, but you do have to pay for them (seventy percent of the regular tariff). The linesmen's work is also performed by a private monopolist who is formally supposed to have his tariffs approved by the municipal council but in fact this control is just a formality.

As a modern port, the business community (private port authorities in cooperation with the Chamber of Commerce) took the initiative in late 1986 to install an electronic message system and make it accessible to as many companies and other systems as possible. The intention was to replace all paper correspondence and telephone communication by electronic messages. This is called the Seagha system which is the Dutch acronym for System for Electronically Customised Data Exchange in the Port of Antwerp. This private initiative is now fully developed and works in combination with many other systems in the port but also outside it. An important link, for example, is with the port authority's electronic systems APICS (Antwerp Port Information and Control System). Thus it has been compulsory since 1 July to provide notification of dangerous goods electronically. With the aid of APICS it has also become possible for ships to notify of their arrival in advance (sixty percent of ships already do this) so that the port authority's Vessel Traffic Service is more efficiently and rapidly informed of a ship's cargo and route. The port authority can also ascertain a ship's financial commitments far more easily. APICS is also used by the dock pilots among others.

The port authority does not formally concern itself with the consultation between employers and employees. Despite occasionally serious disputes, solutions have always been worked out together and joint labour agreements have been made. The few cases of industrial action in the past involved the port authority employees who operate the bridges and locks: they effectively rendered the port inaccessible and/or completely closed it. Discussions on a whole range of port matters are conducted via the Consultation Committee, on which the port authority, the port business community and the unions are represented. In the past, the municipal council has always acted on recommendations from this committee.

5.2.5. ANALYSIS OF PORT MANAGEMENT IN ANTWERP

From the mid-nineteenth century until 1988, the port of Antwerp was accorded national importance. This meant that the national government took care of the costs for major infrastructural works without this being reflected in the port dues and without institutionalised national public consultation being required. As a result of national changes in the eighties, the responsibilities for port planning were accorded to the Flanders Region and the port of Antwerp underwent a change of function. The port was no longer judged on its national-economic contribution but instead regional development was given priority. The Flanders Region continued the previous national policy but appeared initially to want more payback than the former national government. Since the late eighties, there have been numerous reorganisation proposals for the Antwerp port administration until finally, in 1996, it was decided to establish an Antwerp Port Authority plc, comprising only local political and sectoral representatives. Through the privatisation of the Antwerp port authority, the autonomous (sub)regional function of the port took on a clearer shape. The privatisation was also intended to strengthen the commercial element.

These days, Antwerp boasts about the efficiency of its port management and that there is competition between the various service providers in the port. The formal public task of the Port Captain is also explicitly formulated as a component of the service provision and is regarded as a toll good. It can be stated that the most important regulatory principle in the role division between port authority and market parties is *market orientation*. That is to say, the market takes the initiatives in principle, the port authority creates optimal conditions for them, and the costs made by the port authority are passed on as much as possible to the port users.

The nautical management in Antwerp, in contrast to many other ports, is regarded as a toll good to which the profit principle applies. The land is the port authority's most important property which must be utilised as strategically as possible, in view of the competition with other surrounding ports. It may be expected that a great deal of consultation and cooperation with the business community takes place. Finally, the port service provision is regarded explicitly as a private good which should, as far as possible, be offered in competition.

Reviewing the description of the port of Antwerp in the previous sections, the following picture of the institutional format of port activities emerges. The Port Captain, as a component of the Antwerp Port Authority plc, has become a service which in the future will be judged increasingly by shipping companies on its client-oriented approach and service level in relation to the costs. In view of the public nature of nautical management, where policing powers must be exercised irrespective of person, it may be expected that this will create internal difficulties. After all, it is not clear how the imperium is separated from the dominium (the statutes provide scope for the development of both public and private activities) and by formulating the activities of the port captain as a toll good, it is highly likely that the dividing lines between responsibilities will become blurred.

In the past it was customary for the Antwerp port authority to undertake major port planning projects in a contractual relationship with the national and, later, regional government. There were no firm conditions attached to the financial contributions since it was mainly infrastructural works that were involved and their necessity was never in doubt. Nowadays it seems increasingly as if the regional government is reconsidering its function in port affairs and it remains to be seen whether the port of Antwerp can substantiate its own vitality in the long term. The principle of domestic competition between seaports still prevails and there is still no regional seaport policy so that financial participation by the Region is decided on a project by project basis. How does such an informal approach relate to the need for limited coordination with regard to, e.g., infrastructure to and from the port, capacity (sites and terminals) and scant financial resources? In other words, how does the choice for scale reduction in Belgian port administration relate to the often huge (financial) efforts which are needed to effectively manage the product 'port'?

The port planning for the right bank area is in the hands of the Port authority which develops it in consultation with private actors, in so far as anything remains to develop. After all, it might eventually be possible to buy back the sold-off sites but the market itself will determine that. On the other hand, decisions about the granting of sites on the Left Bank area are made by various public players who are keen to weigh the pros and cons of the use of land. The influence of companies in this is more diffuse than in the granting of sites for the right bank area. This is also the reason why the Development Company for the Waasland port is a government monopoly. How will the privatised Antwerp Port Authority, that is increasingly accountable for its own vitality, relate in the future to the Development Company which advocates a careful weighing of interests? Is it not highly likely that privatisation will have served to set relationships on edge so that decision-making processes will become more difficult?

The port service provision is a hotchpotch of control relationships and this is partly caused by the historical distinction between the two management areas of the Scheldt and the port. This is why various public and private monopolies still exist whose purpose is not entirely clear and the coordination of which occurs on an opaque basis. Only in the field of container handling, where concessions have been divided among various private actors, is the competition consciously preserved by the port authority. This has so far been productive but in this sector, too, signs of mergers crop up occasionally and only time will tell if this consistent policy can be maintained politically. Moreover, the privatised position of the port authority may backfire on this point. It has now become possible for the Antwerp Port Authority to undertake subsidiary interests in private activities. If it participates in a container handling company, however, the Port Authority will get burned because concessions for the new terminals can no longer be granted from its present independent position. After all, this would involve a conflict of interest which is unlikely to be accepted by the port business community.

The Port Authority itself has become a private monopolist whose day to day activities will no longer be under direct political control. In terms of control, the Port Authority is still categorised as a government monopoly because the ownership and final control lies with the Antwerp municipal council. The purpose of privatisation was to provide greater freedom under the conditions mentioned above. Only time will tell to what extent strict political enforcement of these conditions will be possible.

In principle, it can go one of two ways, neither of which will lead to productive relations. The first possibility is that the political arena will continue to keep a close eye on port management. This is how the situation was, only now the political executive board can no longer interfere in all the decisions. But the general decisions which nevertheless entail consequences for secondary, adjacent issues can take just as long as in the old situation. Now it is no longer the procedure which determines the decision-making line, but the political willingness to decide.

Another situation is also conceivable, i.e. one of complete trust between the executive board and the port management so that in time the information asymmetry will become so great that the port authority has a completely free hand. It would then in fact no longer be regulated by anyone so that in the long term it would be highly likely to become even more introverted than it already is. There is a great likelihood of this, particularly because the effects of port projects only become evident in the long term.

5.3.   Port Management in Hamburg

The port of Hamburg lies in the north of Germany about 100 kilometres upstream on the river Elbe. It is a real rail port; between 40 and 50 percent of the goods transhipment is brought in or out by rail. The port traditionally performs an important trade function; until the sixteenth century it formed part of the Hanse network and in more recent times Hamburg has been strongly oriented towards trade with Asia. Since the nineteen-sixties the primary trade and industrial function of the port has been enriched by the establishment of high-grade technology, and logistic and financial services. When the Iron Curtain disappeared, a golden future was predicted for the port. The boom has not yet occurred since the economic regeneration of East Germany has not been as rapid as was anticipated. In 1995, fourteen percent of the total transit trade handling was to destinations in Poland, the Baltic States, the Russian Federation and Scandinavia. Another important reason for the decrease in growth in handling volume was and still is the long-standing uncertainty regarding the actual implementation of the city's port expansion and/or maintenance plans which have run up against considerable opposition.

The port covers a total area of 8,700 ha. of which 3,085 is water. In 1998, nearly 76 million tons were handled, half of which was shipped in and out in containers (3.547 million TEU). The percentage of general cargo that is transported in containers (the 'containerisation level') is around eighty percent. The bulk goods consist mainly of ore, coal and oil products. Per year, about 11,500 ocean-going vessels put into Hamburg. 2.5 million people live in the Hamburg metropolitan region and the port generates 140,000 direct and indirect jobs.

Although the port of Hamburg is the largest in Germany and the second largest container port in Europe, it does not formally fulfil a national interest. The federal Constitution contains no separate provision that the Federation should have responsibility within the port boundaries. It only has responsibility for the inland water and road links, and the link between the sea and the ports. The port forms an important link in the route to and from the former Eastern bloc but the port administrators are primarily held accountable for /judged on /by the regional-economic and environmental impact of the port. The emphasis on expediency and the tough confrontation which this causes with other interests is apparent e.g. in the plans to deepen the Elbe although this is against the wishes of neighbouring states.

5.3.1. PORT MANAGEMENT AS A COMBINATION OF MUNICIPAL DEPARTMENTS

In contrast to the other ports we have examined in Europe, the port of Hamburg is not managed by an independent, quasi-autonomous port authority with its own bookkeeping. As far as it is possible to speak of 'port management' this comes under various departments of the city-state of Hamburg (finance, traffic, economic affairs, etc) and port affairs fall under general administrative activities. Initiatives relating to port development are weighed against other important political matters by the government and parliament of the city-state. The coordination of tasks concerning port management also takes place at this level. In other words, the 'port of Hamburg' is not an autonomous legal and economic entity.

This institutional situation has not always existed in this form. Behrendt (1994a) distinguishes three periods in the port management of Hamburg: the period until 1935, from 1935 to 1970 and from 1970 until now. In the nineteenth century it was customary for the city to take care of quay management (construction and leasing) in order to give smaller companies the opportunity to develop activities and compete against the larger companies. In 1885 the city, together with the then Norddeutschen Bank, set up the Hamburger Freihafen und Lagerhaus-AG (HFLG) which was responsible for the construction and maintenance of the superstructure in the free port, while the city remained responsible for the construction of infrastructure. The HFLG was the only lessee of land which it then subleased to other private parties.

In 1935, the port management changed because the quay management merged with the HFLG to form the Hamburger Hafen- und Lagerhaus - Aktiengesellschaft (HHLA). From 1935 until 1970, the port of Hamburg was administered by the type of port authority which is still found in other European cities. A treaty was drawn up between the city and the HHLA in which agreements pertaining to handling were made. The HHLA was free to regulate the handling activities itself or to lease them to other companies. The HHLA started to function as an important port actor; it was even consulted regarding the leasing of sites in the port area which were not covered by the treaty. It was clear, however, that this city company did not run a single entrepreneurial risk nor did it vie for goods packages in competition with other private companies. On the authority of the city, the HHLA also performed nautical management activities.

The advance of the container in the sixties also meant Hamburg had to make tremendous investments. The city's economic affairs department had reached agreement with private handling companies that they were in principle favourably disposed towards financing the superstructure. In practice, this did not occur. In the mid-sixties, the private handling companies began to insist to the Hamburg politicians that a reform of the economic port structure was the best solution. They had come to the conclusion that if they did not participate in container traffic they would eventually be squeezed out of the international market. The container would in any case oust general cargo from the market. In this context two desires were expressed. The first was for a *liberal port structure* with an autonomous port authority and the second was to be able to take care of the acquisition of shipping companies themselves.

In 1967, a government report was published which announced the liberation of the HHLA from its 'hoheitlichen Aufgaben' by handing these over for the future to a politically controlled *Hafendirektion*. The HHLA would be restructured into a private company that from now on would have to compete with others. In other words, the economic structure which had formerly applied had now been renovated and from that point new game rules would structure the expectations between the city-state and the private parties. Central points of that port structure which is still in operation today can be broken down as follows. In the first place, there is a consistent division of tasks between the investments for infrastructure (opening up of sites, waters and harbour basins, the construction of quay walls[1], construction of roads up to the leased site and nautical matters) the costs of which are paid by the city-state and the construction of superstructure (buildings, streets, pipelines, moving equipment, etc.) for which the private sector is responsible. The direct consequence of this division was that once the new structure was in place numerous private companies immediately started investing, and this is still the case today.

A second central concept in the policy document is that a specially-established public body (the *Hafendirektion* which came under the *Wirtschaftsbehörde*) should be charged with public matters which are aimed at improving the competitive position of the port, such as nautical management and the analysis and planning of port development. Thus in Hamburg a deliberate choice has been made not to place various port-related public

---

[1] Quay walls appear to have a variant form: normally the quay walls are considered as part of the infrastructure for which the city has full (financial) responsibility but apparently even Germans are not always consistent because the private companies have to pay a rent of 3% of the average new construction costs.

responsibilities under one functional port management organisation. This lays heavy claims on the coordination between the various public departments, in particular for port planning. The *Finanzbehörde,* which has the same formal position as the Wirtschaftsbehörde, coordinates the joint budget and partly determines the amount of the annual investment in the port. A department of this Ministry is also responsible for the leasing of sites.

A final result of the restructuring of the port management was that the treaty which had been in force until that time between the city-state and the HHLA on the use and the leasing of the port sites was terminated and new leasing contracts were entered into between the city and the HHLA and other private entrepreneurs. From that time, it became possible for the HHLA to participate in other companies.

Until 1989, the wharfage was laid down in consultation with representatives from the business community by the Wirtschaftsbehörde, to which the Hafendirektion belongs. After this time, this construction was discontinued for competition reasons so that unique and customised tariffs could be agreed for each company.

5.3.2.   NAUTICAL MANAGEMENT AND PORT PLANNING

The principle of division of ownership between city-state and Federation determines the responsibility for the nautical management and port planning activities. Outside the formal boundaries of the port both activities are taken care of by the Federation while in the port the city-state looks after them. This means that the Federation is responsible for the transport connections to and from the port (road, rail and water) and that the city-state itself develops the port area within environmental limiting conditions and directs shipping. The physical boundary of the city-state also represents the boundaries to port expansion.

Although formally a strict division of responsibilities exists, circumstances are conceivable which would induce the two authorities to work jointly. This is the case, for example, with the deepening of the Elbe. Formally, this is the responsibility of the Federation but the city-state will very probably participate financially because the project has already been delayed enough due to environmental and other objections and to the Federation's shortage of funds as a result of reunification with East Germany. In general it can be said that the content and the priority of port planning is determined by the procedures for the political decision making concerning the federal and city-state *Budget Acts.* In that sense, port plans are not only assessed at regional but also at national level. The 'checks and balances' of the German port management thus lie mainly in the number of official controls. In Germany, apparently, official prudence is not felt to hamper effectiveness in this modern age.

Nautical management is one of the responsibilities of the Wirtschaftsbehörde and comprises all nautical and shipping-related issues within the port such as the VTS management, granting building permission, the monitoring of the water depths for arriving and departing ships and the nautical supervision of the port pilots. In addition, Hamburg's Ministry for Home Affairs performs an important role in the control of dangerous substances, the river and port police and the fire services.

The parliament of the city-state of Hamburg has adopted a separate Act for port planning in which a number of planning principles are laid down: 'The underlying principle of the Port Development Act is that port development including construction and maintenance of port infrastructure is considered a public task of the City-state of Hamburg' (Behrendt, 1994b:3). The total port area of 75 km^2 (a tenth of the total area of the city-state) is formally delineated under the Act. Moreover, in the Act details are given of which economic activities may take place in which part of the port and it specifies that no habitation may take place in the area. The only activities and enterprises which are allowed to be situated in the port area are those directly related to the port. These are companies which need a waterfront and profit financially from it.

Nearly all the sites in the port area are city property and if sites become available the city has the first right of purchase. This principle was included in order to provide an institutional barrier to port sites being bought on speculation and to facilitate long term planning. The duration of the contracts which are entered into vary per company but may have a maximum duration of thirty years.

The parliament of Hamburg has adopted a number of political principles for port planning. In the first place these concern the political formulation of the port's function: 'The Port of Hamburg is a universal port. It handles all sorts of goods without restrictions' (Behrendt, 1994b:4). Moreover, in 1970 the economic structure of the port was redefined, as has been mentioned above, which meant that the public HHLA corporation would have to compete with other private providers and that a strict division between infra- and superstructure would be made.

The influence of environmental groups is channelled via a separate department for environmental affairs. This department implements a planning requirement that ten percent of the total port area must consist of green spaces. The Groenen have a great deal of power and influence in Hamburg. They were able, by means of numerous petition procedures, to delay the *Altenwerder* port expansion project (construction of a container handling and distribution centre in the present green area) for years. The initial plan that was launched in 1982, and which attained its definitive shape in 1989, was finally approved in 1996 by the highest court of appeal in Hamburg after a good deal of deliberation.

Port planning in Hamburg is influenced not only by the Groenen, but also by what the neighbouring states will allow. The further deepening of the Elbe is essential for Hamburg in order to be able to accommodate the larger container ships when fully loaded. There were objections to this infrastructural project, however, from the two neighbouring states of Schleswig-Holstein and Lower Saxony both of which have their own ports located on the banks of the Elbe (Brünsbuttel and Cuxhaven respectively). The two states produced ecological arguments against the further deepening and moreover expressed concern for the safety of local residents along the Elbe. In this conflict it cannot be denied that the smaller ports along the Elbe had their own agenda: Brunsbüttel particularly wanted to increase its share of bulk goods handling. The environmental minister of Schleswig-Holstein felt that it should also be possible to shape some coordination with regard to land-use policy following Hamburg's problems with the Altenwerder project. Hamburg did not want to know, however, and insisted that such coordination should come from the business community.

Hamburg is also under pressure on the Elbe issue from the shipowners because they have no desire to be hindered by infrastructural problems. Maersk sent a letter to the Hamburger Abendblatt newspaper in which it asked the city-state why the port had still not begun construction of the new container terminal on Altenwerder. Moreover, Maersk has already cut back on the number of containers being handled in Hamburg by several thousand.

5.3.3. PORT SERVICES

Most shipping-linked companies (bunker facilities, maintenance docks, etc.) are private. The pilotage of ships on the Elbe and in the port has been carried out since 1980 by more or less independent pilots associations. The pilots on the Elbe are appointed by the Federal Ministry of Transport and the pilots in the port of Hamburg come under the jurisdiction of the city-state. Until 1980 the pilots were part of the city-state and had the same status (working hours and salary) as other public servants. There came a point when this system no longer worked; the pilots were dissatisfied with their income and loathed working overtime during peak hours so that there was often a shortage of pilots.

The present two pilots associations are regulated by means of several methods. The German Ministry of Transport is authorised to regulate compulsory pilotage and pilotage tariffs by ordinance. The administration of the national pilots association and Hamburg council must be heard in advance. The fact that this power is more than a formality is apparent from the recently-implemented six percent cut in pilotage costs. The number of pilots will also be reduced now that many ships no longer need more than one pilot on board.

In addition to the regulation concerning tariffs and quantities, training requirements are being imposed on the profession of pilot (six years experience and a captain's licence for ocean-going ships). The consequence of this regulation is that the pilots have almost no freedom of action. A ship cannot choose its pilot who has an advisory role during the trip and the price is fixed down to the last cent by the government. The nautical aspects of concrete action are monitored by the harbour master by means of the VTS.

The towing services are private and the recent arrival of a Dutch tugboat company has created real competition. This meant that although five towing companies had been active all along, they had to reduce their prices by 35 percent when Kotug arrived in Hamburg as an extra provider with more modern and cheaper equipment.

The cargo handling is in principle a private affair which is provided under competition in the port. There is one big exception to this, however. The HHLA has emerged from the past as a major (container) handling company whose shares are entirely owned by the city-state. Until recently, the HHLA (and its subsidiaries) had captured around sixty percent of the Hamburg market and two other companies (Gerd Buss and Eurokai) dealt with most of the remaining cargo handling. In August 1996, however, it was announced that the Bundeskartellamt in Berlin had no problem with an HHLA takeover of the container activities of the Buss group. This gave the HHLA a market share of seventy percent. Only time will tell whether the city-state will allow its own company to compete with Eurokai in the long term now that negotiations are about to start on the granting of new sites on the Altenwerder location which should be ready early in the next century.

5.3.4. ANALYSIS OF PORT MANAGEMENT IN HAMBURG

The port of Hamburg performs a regional-economic function. The city-state of Hamburg has full control of the port area but is dependent on the Federal government for investments in connections to and from the port. Its Federal character ensures that these national investments are weighed against the economic interests of other member states. In the port itself, the regulatory principle of market orientation expressly applies: a choice consciously made in 1970. The port administration is part of the general administration; there is no specific, functionally-organised port authority. The role division between the city-state and the private actors is mainly based on the functional division of ownership between infra- and superstructure. Private parties are expected to function in competition and to take care of the construction of superstructure themselves. It is expressly laid down that nautical management and port planning should be a public matter whereas the port service provision should be private.

Within the framework of the theoretical analysis, the position of the 'port authority' and concrete activities may be questioned. Although the requirement of 'checks and balances' is amply met, as regards port management the question is how the effective coordination between various public departments is safeguarded. Might not the separation of nautical management and port planning draw too heavily on the integrating and coordinating capacity of government? Is there not a strong likelihood that the coordination of plan formation will be needlessly delayed because the various public departments do not all assign equal priority to the port? Or is it not conceivable that the division of direct political responsibility for various port-related activities (spatial planning, environment, etc.) will greatly hamper decision making, causing delays and deadlocks, which will not promote confident behaviour towards the business community?

Although the pilots are a private organisation, they perform a strictly regulated public function. This government conditioning makes it clear that such an institutional solution makes heavy demands on public planning capacity (quantity and training) in the long term. How can this be effectively sustained? The institutional position of the HHLA raises similar questions. Container handling in the port of Hamburg is seen as a private activity. In the context of the explicitly-formulated liberal port regulation, its present status as a state enterprise is baffling. Apparently in Hamburg, too, arguments regarding the container sector are based on goal-rational considerations, i.e. that it is no longer a question of competition within ports but rather between ports and for this reason concentration is advisable. The granting of the Altenwerder sites will be the acid test of German port planning.

5.4.   'European Seaport Policy'

Europe has about 2,000 ports which vary greatly as to character and institutional structure. European port administration can be broken down into three geographically determined administrative traditions. The ports in Northern Europe, with the exception of Great Britain, are characterised by a decentral municipal port administration and are part of the *Hanseatic* tradition. The various countries of Southern Europe, such as France, Spain and Italy, have a centrally-organised port administration and belong to what is known as the *Latin* administrative tradition. Finally, an *Anglo-Saxon* port administration tradition can be distinguished which fits the autonomous and private law-based ports of Great Britain and Ireland. Although this outline should not be interpreted too categorically, the various administrative traditions reflect the essentially different socio-economic functions and regulation of ports and port administration. This is clear from the discussions on the options for a European seaport policy which have been going on for thirty years.

Since the sixties, the European Commission (Directorate-General VII for Transport) in cooperation with representatives from the major European ports has been seeking an appropriate working relationship and format for a potential European seaport policy. First of all, the *Seaports Working Party*[2] needed to map the various institutional structures and it had to reach consensus on the definitions and questions which would be used. In 1977, the first tome was published containing descriptions of the major ports of the former Common Market. At that time there was no question of an internal market and no need was felt to formulate a European seaport policy. In 1986, a follow-up report was published to update the available information and also to include new member countries; Greece joined the EEC in 1981, and Portugal and Spain became members in 1986. In the 1985 situation, the European Commission was still claiming that there was no reason to harmonise the port tariff systems and that there was no evidence available to suggest that state contributions have a disturbing effect on the competition between ports (Suykens, 1995). For the time being, it was to be business as usual for the port administrations.

The original purpose of European unification was to achieve one big European market in which transaction costs for economic activity between the member states would be as low as possible. In the mid-eighties, under pressure from the globalisation of world production, the motivation for European unification was given a new impetus by European major industrialists. On the political front, the *Code for Economic Unification* was devised which will function as the scenario for the *European Union*. Since 'Brussels', in addition to its preparation and approval of what has turned into highly-detailed agricultural legislation, has chiefly been appointed to standardise the rules for economic activity, major efforts are being concentrated on the implementation of competition legislation. On that pretext, since the late eighties it has renewed its involvement with the port administration of the member states.

Sectoral Significance of Ports

Ports are seen by Brussels as a component of the *transport infrastructure* which in turn is considered to be an essential condition for achieving and implementing economic and social cohesion in the EU. 'In order for citizens and businesses to derive full benefit from the internal market, new communications routes and distribution channels must be developed across borders between Member States' (Blonk, 1994:481). Great efforts were

---

[2] The working party consisted of representatives from the major ports of the then member states (Germany, Belgium, the Netherlands, Luxembourg, France, Italy, Great Britain, Ireland and Denmark). The secretariat is under the control of DG VII.

required before the European Commission finally achieved a joint transport policy in 1993. Traditionally, road transport can count on strong support from the European Commission and the railways are still run by national companies. With the increasing congestion on the motorways and pressure from environmental groups concerning the economical use of energy sources, it was no longer possible to avoid the options offered by other forms of transport. The criticism of old, featherbedding institutional structures also had to be answered in the European forum in the context of the free movement of goods and services.

The aim of DG VII is to establish physical and virtual *interconnection and interoperability* between countries and, at a later stage, between various modes of transport. It concerned one of the most important lines of policy agreed in 1993 in the Maastricht Treaty. Since then, the format of the *Trans-European Networks* (TENs) has been discussed. The European Commission can provide guidelines for these which have implications for the aims and priorities of projects, and furthermore it identifies the joint projects. For the financing and final implementation of the TEN projects, the European Commission remains dependent on the member states' willingness to help.

In view of the fact that maritime transport accounts for as much as ninety percent of the total intercontinental transport and the shortsea sector ships 35 percent of intracontinental cargo, the European ports play a major role in realising the TENs (multi- and intermodality) (and see Erdmenger, 1996). In the framework of the TEN policy, the European Commission has also prepared a number of *directives* intended to expedite the implementation of port projects. The purpose of these projects is to improve the position of ports in the transport chain, increase the efficiency of their operations, e.g. through the use of telecommunication (Maritime Information System), and to promote shortsea transport. In this framework, ports are given a number of *goal-driven functions*, such as facilitating the growth of trade, dredging silted up transport corridors[3] and minimising negative external costs by the use of (inland) waterways, and improving the access to and strengthening social cohesion with islands and other remote areas (Blonk, 1994; Kinnock, 1996). With regard to this last element, mention should be made of the *Cohesion Fund* projects with which the European Union is trying to improve the transport infrastructure of Greece, Spain, Portugal and Ireland. The ports in particular are used to provide a regional economic boost.

---

[3] From discussions with transport economists and representatives of shippers and carriers it became clear to me that strategic concepts such as *intermodality* and *transport corridors* are difficult to operationalise. There is as yet very little qualitative know-how about the technical and management aspects of the various types of terminals and a translation into more generic concepts is not yet available. Translating them into more concrete projects and variables on a European scale will remain the major challenge in the coming years.

Economic Significance of Ports

It is in Brussels' interest, based on the 'single market' concept, that *fair* competition should exist between the European ports. On this pretext it has prohibited the special low rates that German rail used to offer to the German rail ports. Ports in Europe are in fact defined as *toll goods* to which a profit principle applies that must be as low as possible due to the competition between the parties. This also means that member states are expected to pass all investment costs on to the port users and not influence the price mechanism through nontransparent support measures which could create unnecessary capacity surpluses which would in turn influence the goods flow. In the framework of the economic unification of Europe, DG VII was interested from the outset in the financing of infra- and superstructure, but also in the economic power positions of the various port services. In the early nineties, the exploratory reports of the seventies and eighties were supplemented by policy-initiating studies into the economic position of tugboat, pilotage and linesmen's services and into the port accounting systems. The Price Waterhouse report of 1992 on the accounting systems of various ports (Hamburg, Rotterdam, Southampton, Zeebrugge, Marseilles and Valencia) achieved notoriety and has still not been made public. Each port was provided with the chapter relating to its own port but not those relating to any of the others. The more conciliatory report that followed in 1993 was compiled by Marconsult of Italy and Ocean Shipping Consultants of the United Kingdom, and made recommendations concerning the format of a European seaport policy although its real purpose was to clarify once again the relationship between European port authorities and DG VII.

For many port administrators, a study such as that by Price Waterhouse made the advent of the European Union seem increasingly threatening. In the near future they would have to meet expectations which they had never had to worry about before. The studies by Price Waterhouse and Marconsult/Ocean Shipping were intended to form merely the initial impetus of what was to be a formal European seaport policy. In 1993, the European Seaport Organisation (ESPO) was set up in which port administrators from 13 member states were represented. In May 1995, the European Commission and ESPO issued a joint declaration which comprises four principles for the conditions of this cooperation:

'1.   Ports and the commission agree that there should be no designation of Ports of Community Interest - only ports of common interest;

2.   There is no need for an all embracing European port policy;

3.   The General Terms and Articles of the European Treaties (competition, state aids and freedom to provide services) are also applicable to the ports and should be respected by them;

4.   The Commission and ESPO agree that there is no need for coordination at the European level of port infrastructure investments'(Suykens, 1995:11).

The agreed principles appear to have restored harmony and clarity regarding mutual expectations. Formally, there is no question of a European seaport policy. European influence on transport flows still takes shape chiefly through sectoral (agriculture and transport) and facet (trade, environment, social) policy. It remains to be seen, however, how long Brussels will continue to go along with the agreements with ESPO and keep to its conditioning and limited steering role in the field of transport.

5.5.   Summary

This chapter looked at the ports of Antwerp and Hamburg as Rotterdam's major competitors in Western Europe. Striking differences came to light pertaining to the institutional format of port management and the impact of regulatory principles on various activities. Antwerp has recently opted for a company with legal status which as port authority has both public and private powers. The city has thus chosen scale reduction in line with the national, administrative restructuring which created the regions. The main question remains of how the division between imperium and dominium will be safeguarded in the long term. Since 1970, on the other hand, Hamburg has held fast to its chosen regulatory principles which had even been explicitly laid down and in contrast to the Belgians appears unworried by the dynamic changes and the complex problems facing it.

The final section dealt with the 'European seaport policy'. In Europe, ports are seen as important components of transport policy and they are expected to operate as efficiently as possible. In fact, there is no European seaport policy but major efforts are being concentrated on determining the role of governments in ports (passing on investment costs to users, market position of port services, etc.). It was partly due to the setting up of the European Seaport Organisation that in 1995 agreements in principle were reached on the position of the European Commission with regard to seaports.

CHAPTER 6

Ports in the United States

'In today's economic climate, it is doubtful that there will be any change in the port management philosophy of maximizing economic activity in the region served by the port' (United States Department of Transportation, 1996).

6.1. Introduction

An important focal point in the study of port management in various North American ports was the widespread conviction that the United States would be at the forefront of privatisation trends. After all, President Reagan, together with Mrs Thatcher, had achieved worldwide notoriety in the early eighties with his 'New Federalism' and the 'cut backs' in numerous federal activities. It must be said, however, that port management in the United States is traditionally a local or regional matter on which the federal government can exert influence 'merely' in a conditioning sense through e.g. the construction and maintenance of waterways and means of navigation or by the creation of fiscal facilities. In the ports studied, there was frequently no trace of extensive privatisation initiated at the federal level. The regional or local involvement in port management has also ensured that some ports, e.g. New York, have had a high level of private input for years, while others still very much retain the public character of port management (for instance Seattle).

The United States has a wide variety of port management types, ranging from 'bi-state authorities' (New York is the best-known example), 'state port authorities', 'navigation or port districts', 'unified port districts', 'city port commissions or departments' (Los Angeles, e.g.), 'specialized port authorities' (particularly in the Great Lakes area) and 'public corporations' (Seattle, e.g.). The distinction between the various forms of port management is not always sharply defined. The port management in New York, for example, may be termed a bi-state authority but, in exactly the same way as a port district, it has a clear-cut area within which activities may develop. More or less the same could be said of the port of Seattle: although this is managed by a public corporation it also has a clear-cut management area from which the port authority can obtain tax revenue. Moreover, both ports are steered by commissioners. New York, Seattle and Los Angeles were chosen because they are (traditionally) the largest and best-known multi-purpose ports in the US and represent different regulatory principles in the role division between the public and private sectors.

Ports Studied in North America

A short account of the economic regulation in the United States is followed by a description of the position of the various port authorities. The section on nautical management is a uniform one because the responsibility for nautical management lies with the federal government. The sections on port planning and port services once more address the three ports individually. Finally, port management is analysed and conclusions drawn.

6.2.   Economic Regulation and Significance of Ports

The United States is a highly industrialised country with wide cultural, geographical and industrial variation. Despite its large geographical area, with a population of 250 million it can be called an urbanised country. It is rich in a wide variety of minerals and a broad range of modern technological applications (defence, software, electronics, automobile and aircraft industries, etc.). With the exception of crude oil and a number of other minerals, as far as raw materials are concerned the US is entirely dependent on foreign imports.

The economic regulation of the United States is designed on the principle of the market and private enterprise. With the exception of a number of sectors, the federal government does not participate in market activity and only determines the conditions under which it should occur. Thus, for example, government control exists in the field of private utility company tariffs and banking is subject to further control provisions. There are also strict controls regarding the formation of trusts, taxes are levied and there are labour protection laws and consumer protection provisions. From 1982, a policy of ' supply-side economics' was pursued under the Republican presidents Reagan and Bush. This economic policy is based on the theory that the government can only support the economy by making the conditions as advantageous as possible, e.g. by tax reductions and deregulation. The election of the Democratic president Clinton, in 1992, brought an end to this policy but the government's position in society was still questioned, as the restructuring of the health care system shows, for example.

In the United States, the responsibility for port management lies at the level of the individual states. 'Seaports as such, however, are primarily a State rather than a Federal matter (...) there is no Federal policy on ports, but there are a number of Federal programmes within ports' (Goss, 1979:310). The federal *Department of Transportation* (Maritime Administration Office of Ports and Domestic Shipping, Coast Guard) and the *Army* (US Army Corps of Engineers) are directly involved in the ports for nautical and defensive reasons. Ports originally played a regional or local role, but the further development of intermodal transport in the eighties created competition between the west and east coast ports, which meant that the west coast ports began increasingly to perform a national economic function (Goss, 1990a). Moreover, right from the start the port of Seattle has performed an important import/export function for the region.

In contrast to Western Europe, ports in the United States are not controlled by the anti-trust laws. This means, in fact, that ports are judged more by socio-economic standards (employment) rather than operating in a primarily commercial way (cf. Goss, 1979). In 1994, in a lengthy report on the financial methods employed in American ports, the Maritime Administration Office of Port and Intermodal Development and the American Association of Port Authorities described how the majority of ports play a regional role in which the emphasis lies on maximising economic activities rather than focusing on direct revenue for the port authority. And despite the fact that in the late seventies port authorities were asked by cities and states to stand on their own two feet (in the framework of New Federalism), the 1994 report concluded that no trend towards increased financial self-sufficiency had taken place. Recently, the Department of Transportation similarly concluded that ports are more concerned with their own direct environment and less worried about their function as a link in a greater whole. Compare the quotation which opens this chapter.

In its annual reports to Congress for 1994 and 1995, the Office of Ports and Domestic Shipping wrote that there has been a gradual decline in profitable ports in the United States, and that many ports which are not self-sufficient cover their costs with revenue from other sources (cross-subsidising), with tax levies or by receiving subsidies. The fact that the political scope exists to judge American port management by its impact on employment means that in fact, and depending on the economic situation, many different external success standards (employment, service provision, promoting trade, etc.) can be found simultaneously in the national system. This is probably also the cardinal reason why American port management displays a multiplicity of forms and that specific mixes of government and commercial sector can operate freely. In the past, for pragmatic reasons, a different form of port management was found in each state.

The main problems for various American port authorities at present include: the financing of port development and the revenue from it, environmental regulation, the dredging of the port and the storage of (polluted) dredging sludge, the accessibility of the country by means of intermodal transport, the next generation of container ships, and the container alliances. The growing tendency to balance the books by means of tax revenue and/or cross-subsidies, means that port management in those ports will increasingly become a component of political decision making. The Office of Ports and Domestic Shipping points out that it will thus become essential in the future for some ports in the same region to make price agreements on the leasing of terminals because otherwise ports will be increasingly played off against each other by internationally-operating shipowners and shippers (United States Department of Transportation, 1996).

6.3.   The Position of the Port Authority

In the United States I visited the ports of New York, Seattle and Los Angeles (May/June 1995). Until 1962, New York was formally the largest port in the world and, with the advent of the container age, provided the location for the first container terminal, constructed by Sea Land. At present, New York is still the largest port on the American east coast; its main competitors are Halifax, Norfolk, Miami and Philadelphia. The port of New York is managed and operated as a separate division by *the Port Authority of New York/New Jersey* and traditionally has a liberal role division between public and private sector. Over the last ten years, as a result of falling growth figures and labour and environmental problems, some changes have occurred.

The port of Seattle, on the northwest coast, is a major competitor of Vancouver, in Canada, and is managed by the *Port of Seattle*. The port authority is also responsible for the management and operation of the airport and has a remarkable mix of public and private activities and powers. It is administered by an *elected* Board of Commissioners. The port of Los Angeles was visited due to its locally-funded structure and the increasing importance of its trade with Asia. Moreover, Los Angeles provides a good example of a port whose entrance and hinterland have to be directly shared with another port authority, i.e. the port of Long Beach, which has a different role division between public and private.

Table 6.1. Multi-year Overview of Port Performance in New York, Seattle and Los Angeles (in million metric tons and TEUs). *revenue tons

Port/year		1975	1981	1985	1990	1995	1998
New York	ton	110	48.3	52.1	48.9	44.9	56
	TEU	1.0	1.8	2.4		2.3	2.5
Seattle	ton		8.0	8.1	12.0	18.8	13.54
	TEU	0.48	0.7	0.8	1.2	1.5	1.54
Los Angeles*	ton	25	38.4	45.1	67.9	74.6	
	TEU		0.5	1.1	2.1	2.6	3.2

6.3.1. THE PORT AUTHORITY OF NEW YORK/NEW JERSEY

The port of New York is managed by the *Port Department* of 'The Port Authority of New York/New Jersey'. In 1921, the Port Authority was established by law by the states of New York and New Jersey, and approved by federal Congress, to prevent or find solutions to infrastructural problems between the two states from a politically independent position. A separate Act was necessary to set up The Port Authority of New York/New Jersey because it involved cooperation between two states and a guarantee was needed that states would not cooperate with the purpose of excluding others. And the principle still applies even now that **both** states must gain some benefit from each project (backscratching).

The principal motives for establishing this *interinstitutional cooperation* were the continuing political conflicts and deadlocks which could not provide a definite answer to the increasing congestion and poor layout of the existing infrastructure. An integrated development and planning concept was lacking but was sorely needed in order for the region to develop its commercial and industrial potential. Awareness of this mutual dependence existed but it was not known how to overcome self-interest and achieve joint long-term planning. 'Between 1900 and 1915, the temper of the times in the bistate New York metropolis meant commercial rivalry and political conflict - between city and town, between machine politician and reformer, between Manhattan's business leaders and the New Jersey outlanders, who associated New York's dominance with wrongful imperialism. The story of the Port Authority begins in this hotbed of rivalry and contention' (Doig, 1993:32).

The creation of the Port Authority is linked with values and ideas which were being promulgated at that time by advocates of the *Progressive Era*. Central to their view was 'the need for efficiency in economic relationships and for government action to help attain a vigorous and efficient economy. One implication of this view was that rational, regionwide planning was essential to provide transportation facilities needed in a vital, expanding economy' (Doig, 1993:32). The distinguishing features of the Port Authority are its business ethic, the open formulation of its targets and its *Board of Commissioners* comprising a number of members.

The governors of the two states each *appoint* six Commissioners who together form the Board of Commissioners and who are in charge of the Port Authority's myriad activities. There is a historical division of tasks between the two governors. The governor of New Jersey appoints the president of the Board and the governor of New York appoints the Executive Director. The appointing of the Commissioners must be approved by the Senates of both states. The extraordinary thing about this bi-state authority is that Commissioners can never be called to account by just one of the governors. After all, matters always involve the joint interest of the two states. This also means that strategic decision-making becomes more complicated when the two governors belong to different political parties. The decision making in the Board of Commissioners takes place by means of a majority of votes, in which at least three people from each state must give their approval.

The Commissioners are not paid for this prestigious part-time job. They are mainly prominent and wealthy business people. They are appointed for a period of six years and take up an independent position vis-à-vis politics because they are not elected and maintain no direct relations with any lobbyists. The governors retain the right to veto decisions made by the Board. This rarely occurs, however, and functions as an escape hatch, e.g. in the event of disagreement on price increases. The governor is elected for a period of four years, so it can happen that a governor is confronted with Commissioners who were appointed by his predecessor. The Commissioners meet twice a month in a public meeting and have delegated various powers to the Executive Director. They monitor general policy and approve the sundry departmental budgets. This is also the most important instrument they have to directly impose their will.

The Executive Director is in charge of the Management and Budget Department which keeps an eye on the diverse financial consequences of the departmental programmes. The Budget Department is the one which principally makes recommendations to the Board of Commissioners. The framework for the recommendations and the options for control form the five-year *business plan* in which, after consultation, (financial) objectives are formulated for each department. The directors of the other departments have no direct formal contact with the Board. The director of the Port Department consults with the Executive Director about the dredging activities in the port, for instance, and then reports back to the Board of Commissioners. The Port Authority staff do not have *civil service protection* but they do have a contract similar to a collective labour agreement.

The Port Authority's primary aim was to make the fragmentation in the rail links manageable. It could not make any impression on the railways, though, which continued to give priority to promoting their own interests. The construction of the George Washington Bridge, which beat both its deadline and its budget, established the position of the Port Authority for good. From that time, the principal legitimation for its position would always be linked with the realising of transport connections (bridges, tunnels, public transport system) between New York and New Jersey.

The present Port Authority manages a far broader range of activities than just the port. In addition to the interstate transport connections, it also operates two commercial heliports plus the J.F. Kennedy International Airport, Newark International Airport and LaGuardia Airport. Moreover, it manages various pieces of real estate of which the World Trade Center is the most famous. The construction of all these projects was financed using the growing flow of revenue from the tunnels and bridges.

The port of New York is scattered over various locations within the legally delineated district of the Port Authority, so it is logical that in 1984 this infrastructural activity joined the larger Port Authority conglomerate. Before that time the Port of Newark, for example, was managed by the city of Newark. The construction of the Elizabeth Marine Terminal was an exclusive Port Authority activity: the terminal was opened in 1958. At present, about five hundred people work at the Port Department and at the Port Authority a total of around 8,500. Although the Port Department is the smallest Port Authority activity (the port handling comprises about six percent of the Port Authority's total turnover), the port has recently acquired great regional-economic importance once more, not only for the consumption of the 17 million people who live in and around New York, but also for the 14,350 direct and 119,000 indirect jobs which are linked with the port. The port, with its 46.2 million tons of cargo, generates two percent of the total economy of the city of New York. The total port area scattered over New York and New Jersey comprises about 1,462 ha. of dry area, and about 4,500 ocean-going vessels per year put in at the port.

Financially speaking, the Port Authority is entirely autonomous and can meet its strictly financial needs by means of the revenue from its infrastructural projects and bond loans. The bondholders do not have to pay income tax on the bonds which, (depending on the time of repayment), are issued for a term of anywhere between 10-35 years. The Port Authority is not financially covered by the two states so that all revenue can be allocated to expenditure and investments as its own discretion. The states exercise no control over the allocation of the money. On the basis of the profit principle, interest is made on various bridges and tunnels which can then be used to cover losses on other projects. At present, the operation of the port is not profitable due to the tremendous costs of dredging. In 1993, the Port Department again made a loss of 44 million dollars. The revenue of the Port Department mainly comes from the leasing of the sites, seven cranes and distribution centres. The leasing profits form more than 80% of the gross revenue while the rest is made up of 'cost recovery', parking fees and other revenue. The total revenue amounts to about a hundred million dollars.

Important conditions for success are seen to be the absence of political interference and being able to outline a long-term policy. Direct political interference is kept to a minimum in this enterprise. The independent position must stand surety for a rational long-term analysis and result in a programme development which appeals to the imagination. The governors can only appoint the Commissioners and the Executive Director and ultimately veto Board decisions. The downside of this independent position is the minimum amount of political interference which gives rise to the permanent debates on this form of public management. 'Although distinctive in its wide geographic scope and in its financial strength, the experience of this agency illustrates recurring themes in the debates on the creation and evolution of public authorities in the United States: the reluctance of elected officials to loosen the reins of democratic accountability; jealousy and suspicion among cities and across state lines; undermining cooperative efforts; private corporations working with the public authority only when their own profits or other specific interests seem likely to be enhanced; and the tendency of public authority leaders, invested with some entrepreneurial freedom, to build alliances with specific private groups and to adopt programs that then generate opposition from citizens outside the dominant coalition' (Doig, 1993:31).

The Port Authority of New York/New Jersey was originally created by the two states to bring an end to the political conflicts about joint infrastructural regional development. In that aim lies the justification for the politically independent Commissioners and the options for drawing on their own financial resources. The permanent debates on the position of the Port Authority make it clear that the recurrent challenge is 'to encourage entrepreneurial planning and action while maintaining the sustained vigilance by elected officials and the citizenry essential to a democracy' (Doig, 1993:41).

6.3.2. THE PORT OF SEATTLE

Seattle is a city of over half a million inhabitants in the northwest United States which, since the fifties, has been famous principally for its Boeing factories (in 1994 providing 110,500 jobs) and, more recently, for the computer software giant Microsoft (over 15,000 jobs). Seattle lies in the King County region (1.6 million inhabitants), has the fifth largest container port in the United States and maintains sister ties with the ports of Kobe and Rotterdam. In 1996, 1.474 million containers were handled in the port, which showed a 4% drop compared with 1995. A total of 16.1 million metric tons were handled in 1996. Of the 18 million tons handled in 1995, no less than 10 million were containerised.

Seattle exports mainly hides, industrial equipment, paper, frozen fish, beef, chicken, poultry, grain and grain products, heavy machinery, automobile parts, electrical and electronic components and resins for the manufacture of plastics. Between 25 to 30 percent of these exports is shipped in by rail from the interior of the United States (Midwest and Northeast). The most important imports via the port are equipment, office machines, automobile parts, electrical and electronic components, telecommunications equipment and hifi, (video) games, heavy machinery, shoes, cars and snow vehicles. According to estimates made by the Port of Seattle, 75% of its total imports are destined for the interior of the United States. Its most important trade partner (import and export jointly) is Japan, followed by South Korea, China, Taiwan, Hong Kong, Australia, Thailand, Canada, New Zealand and Singapore. It is clear that the trade via Seattle concentrates on the countries in the Pacific Rim with which it maintains close relations via marketing offices.

Due to the growing opportunities in intermodal transport and the development of more southerly Asiatic countries, Seattle is encountering increasing competition from the ports of Tacoma (which lies only thirty kilometres south of Seattle), Los Angeles, Long Beach and Vancouver. In the past, Tacoma succeeded in enticing the major container shipowner Sea-Land away from Seattle. It looked for a while as if American President Lines (APL) would also leave Seattle for Tacoma or Los Angeles but, after lengthy negotiations involving the whole Seattle port community, this was prevented. At present, Los Angeles and Long Beach handle about 45 percent of the total Asiatic cargo and Seattle and Tacoma about 25 percent. The container division between Seattle and Tacoma has fluctuated over the last ten years, but roughly speaking Seattle handles about 15% of the total international share of the market and Tacoma the rest.

Port of Seattle as Port Authority

The landside port management in Seattle is the responsibility of the Port of Seattle which also manages and operates the Seattle-Tacoma International Airport, the Fishermen's Terminal and the Shilsole Bay Marina. In contrast to many other American states, the state of Washington in which Seattle lies has chosen to have the ports managed by *'special purpose districts'*. There is a historical reason for this. Until 1911, the port of Seattle had private ownership and use of the waterfront. This private ownership was controlled by various railway companies who were not prepared to invest if they could not have exclusive rights to the revenue. At the beginning of this century, a private stalemate had been reached regarding the further development of this area. Moreover, the economic profits only benefited the rich while it was the poorer inhabitants of the area who were hardest hit in the depression. Towards the end of the nineteenth century this led to the rise of a Progressive Movement and the creation of a political people's party which claimed the public ownership of the waterfront. It was not until 1911, however, that a city enterprise could be established within legal frameworks (*Revised Code of Washington*). 'In September 1911, the Port of Seattle became the first autonomous city corporation in the nation to engage in port terminal operation and commerce development' (Washington Research Council, 1990:5).

The Port of Seattle as a 'special purpose district' is linked with the residents of King County. These residents elect the five commissioners who manage the Port of Seattle. In the same way as mayors and governors, the port commissioners run a campaign for their own election with the aid of funds from the business community.

The residents of King County must be consulted by the commissioners on matters relating to expanding the administrative scope, using tax monies for specific purposes, increasing tax percentages and floating loans, but they may also be consulted about politically-sensitive choices. 'Washington's port commissioners typically serve six-year terms, with one commissioner elected every two years. (...). Washington's commissioners are not salaried but receive compensation for each day they attend port-related meeting or perform other services for the port district. The limit is $50 dollars per day up to $5,800 per year, plus expenses (...). Commissioners may also be covered under a port's group health and insurance policies. The legislation allowing the coverage indicates that it "shall not be considered" compensation' (Washington Research Council, 1990:6-7). The principal task of the commissioners is to maintain contacts with layers of government, the (international) business community, interest groups, the Chamber of Commerce, etc. They are constantly engaged in generating community support for taking decisions or in underlining the port's importance to the economy. Decision making within the board of commissioners occurs by majority ruling.

In the early years of the Port of Seattle's existence, the movements of the commissioners were carefully followed by the critical business community of Seattle with its laissez-faire attitude. Although they were initially supported by the newspapers, after ten years they could no longer sustain their objections to public decision making which had been motivated by fear. The import/export values grew, in response to the First World War, from 84 million dollars in 1911 to 627 million dollars in 1918 (Port of Seattle, 1995:44). In the years that followed, the outside environment became increasingly cooperative. During the depression, two commissioners were discharged due to a crisis of confidence. Special action had been taken by the local residents to achieve this. In the years after the Second World War, when the Port of Seattle was undergoing a period of tremendous economic growth due to the expansion of the airport, the residents had great confidence in the commissioners. In their planning, on the other hand, they did not take much account of the local residents. This method of working was highlighted when attention began to focus on the environment. The Port of Seattle had to hold a referendum on the further development of the port and a number of hearings were held regarding its economic growth. In carefully balancing the economy and ecology, the public character of this function was completely restored once again. The function of port commissioner is primarily one of refined lobbying, where one works on the principle of: 'You scratch my back and I'll scratch yours'. The principle reflects the combination of accountability and trading.

Although the port commissioners have exclusive administrative power, the responsibility for the day-to-day running of the Port of Seattle is delegated to the port authority by means of guidelines and procedures. This has been the case since 1933. Until that time, the commissioners had had direct control of policy making and implementation. The growth of the port saw the introduction of this concept and of a general manager who was selected by the board of commissioners and could also be discharged by them. The latter only happened once when the commissioners felt they were losing their grip on the general manager. Formally, the port commissioners have only direct contact with the general manager, but it also happens that they go looking for information in the organisation behind the general manager's back. The general manager makes frequent use of ad hoc 'task forces' which advise him on a whole range of matters. Once a month, the port commissioners meet with the port director in a closed meeting and twice a month there is an open meeting in which the public may participate. The chairmanship of the board of commissioners rotates on an annual basis.

The original aim of the Port of Seattle was to make the waterfront accessible and develop it on behalf of every potential user. The legislation of the state of Washington allocated clear powers to this function. 'The original law, RCW 53.04, gave port districts authority to acquire, construct, maintain, operate, develop and regulate within the district all types of facilities and services needed to transfer and transport people and goods. Those purposes were later expanded to include industrial and economic development' (Washington Research Council, 1990:8). Due to its great trade potential, however, it was a long time before Seattle oriented itself towards industrial-economic activities. At present, the important regional-economic function of the port and the amount of employment which it creates with its activities are being given particular emphasis. The Port of Seattle considers it a great achievement that the region is no longer wholly dependent on employment at Boeing. When the first redundancies occurred at Boeing in the early seventies, this had a huge economic impact on the region. That is no longer the case, since there is now a greater diversity of companies in the region, partly thanks to the port.

The ports in the state of Washington are unique not only because of their clear-cut structure, but also because of the fact that as 'special purpose districts' they are allowed to collect a part of their revenue from *tax levies*. 'Port taxing authorities in Washington are also unusual among state taxing districts - the state, cities, counties and special districts - in that ports are not subject to some of the constitutional and statutory limits on property taxation' (Washington Research Council, 1990:9). In addition to wharfage and berth dues, the tax revenue comes from the leasing of property, interest yields, subsidies and loans. The profits from wharfage and berth dues, leasing of property and interest yields, also called 'operating revenues', form the largest source of income. When the port authority makes a profit this does not have to be surrendered to the city or other public body, but is used to promote new port enterprise. The oppportunity to levy a property tax can 'provide Washington ports and their bondholders greater insulation against the risks of their investment decisions' (Washington Research Council, 1990:9). In 1995, the Port of Seattle received over 35 million dollars in tax revenue. Its turnover from activities that year amounted to 71 million dollars. However, even the tax levy could not prevent the port making a loss. In the future, there is a desire to be less dependent on the tax benefits. To achieve this, a reorganisation programme has been initiated in which a number of corporation components, among other things, were sold off (Port of Seattle, 1996).

There are a number of formal restrictions to the Port of Seattle's levying of taxes which are not only determined by the size of the figures but also by the duty to listen to the citizens of King County. 'Without voter approval, ports may levy property taxes for the following:

- General port purposes, to supplement operating revenue and to establish funds for future capital improvements. This can be done by levying an annual tax on the district's taxable property of up to $0.45 per $1.000 of assessed valuation.

- Industrial development purposes, up to $0.45 per $1.000 of assessed valuation for a limited period of up to 12 years. (...)
- Principal and interest payments on its general obligation bonded indebtedness.

With voter approval, ports may levy an additional tax specifically for dredging, canal construction and land-leveling or filling purposes' (Washington Research Council, 1990:10). Although the tax revenues in the past were mainly used to enable the undertaking of large-scale projects, they will play an increasingly important role in the future in meeting environmental requirements and contending with competition from Vancouver and the Californian ports (BST Associates, 1991).

The Port of Seattle annual accounts are audited by an independent private accountant. Banks in the United States have their own form of audit in which the status of the management, the organisation, the market, etc. is examined and this is disaggregated into a credit rating (the Port of Seattle has always been awarded the highest: AAA). Once a year, a more formal financial audit is also conducted by the Washington State Auditor who mainly checks the balanced economic growth of the various ports on behalf of the state interest. The state of Washington has only recently shown financial interest since the ports have always operated as autonomous 'special purpose districts'. Under the Revised Code of Washington, the state can formally create the legal conditions under which ports are allowed to borrow. Five years ago, a task force was also set up to examine whether it would not be possible for ports to cooperate with each other and harmonise their policy to prevent overcapacity in the state of Washington. To date, although cooperation exists between the ports of Seattle and Tacoma, the negotiations on the establishment of companies in the port are carried on separately. In a broader context, consultation between the various ports takes place via the *Washington Public Ports Association*. In the early eighties, a bright future was forecast for this form of regionalisation aimed at preventing potential overcapacity (Olson, 1980).

As an independent public corporation, the Port of Seattle is situated between state and market and forms a very interesting case in the context of the demand for control relations. Such a mixed position, however, makes it extremely difficult for the Port of Seattle to clarify its expectations towards the citizens of King County and the business community. In the seventies, for example, this led to serious conflicts between various commissioners which had an impact on the whole organisation (Port of Seattle, 1995). In 1990, therefore, eighty years on, the port's rationale was explicitly formulated anew (Port of Seattle, 1990) and the formal mandate of the citizens was once more obtained. Doubts remain, however, which cause more unrest in one period than another. For instance, how can an organisation say that it provides employment when 75% of the freight is shipped to the hinterland by rail? And how many commissioners should there actually be on the board, who do they represent and who should they represent? Why should both the port and the airport be run by the same organisation? We will return to these questions in the final chapter.

6.3.3. THE WORLDPORT OF LOS ANGELES

The port of Los Angeles lies on the bay of San Pedro, 25 miles south of the city of Los Angeles. On this same bay lies also the container port of Long Beach which is managed by the city of Long Beach. The two ports are competitors up to a point, but in view of the growth in freight volume there is enough work for both. They perform both a regional and national economic function: 259,000 direct and indirect jobs in Southern California. Although both are city ports, they are managed on different principles. These differences are addressed in more detail in the section on port services. The emphasis of this subsection lies on the port of Los Angeles. It has a longer history and forms an interesting case against the backdrop of the recently proposed reorganisation.

The port covers 3035 ha. (land and water) and in 1996 handled 68.8 million (revenue) tons of cargo of which the major part was transported in the 2.49 million containers (measured in TEU). There were also 945,180 passenger arrivals and departures via the cruise terminal. Los Angeles is clearly feeling the effect of the ever-larger container alliances: the number of ships putting in at the port over the last few years has been steadily dwindling. In 1991, 3,414 ships put in at the port and in 1998 this had fallen to 2,569. The principal import products are oil, iron, steel, automobile parts and bananas. The export comprises coal, oil products, scrap, steel and waste paper.

In contrast to many other ports, the port of Los Angeles is not directly adjacent to the city of Los Angeles but is situated 25 miles further south in San Pedro which is the 15th district of the city of Los Angeles. In 1876, with the help of the private rail company Southern Pacific, a raillink was made through the city of Los Angeles connecting the city and the port of San Pedro. Until 1885, Southern Pacific was able to capitalise on its monopoly position for the goods cargo from the mines of Nevada by retaining ownership of the line to the port of Santa Monica. This meant that the prices for transport via Los Angeles were kept artificially high, to the great annoyance of the city which could not undertake any direct measures against this. When the federal government decided to construct a breakwater along the Los Angeles coast with the intention of further developing San Pedro Bay, Southern Pacific's ownership and control of the waterfront became an increasing subject of discussion. Just as in Seattle, a reform movement arose which advocated that the use of the waterfront should be called a public good which should subsequently come under public control. The transition from private to public control took over ten years once the judge had intervened to annexe the land.

In 1908, the *Harbor Department* of the city of Los Angeles was established and from 1911 it was enabled to invest in public facilities. The political agreements which were made at that time with various bodies to bring them under city ownership, had substantial consequences which still function as institutional limiting conditions for the port management of Los Angeles. 'The early struggles with the Southern Pacific have had a lasting influence on the Port of Los Angeles. To buy off a number of influential actors, the Reformers cut political deals that in some cases still affect port policy. One was the promise to the fishing industry of San Pedro to purchase its support for annexation. This deal set the pattern for a relationship between fishermen and port that the latter is forced to maintain even under changed circumstances. Another is the set of restrictions written into the Tidelands Trust Act. To buy their acquiescence to city control, the Trust cities, (Los Angeles, Long Beach, Oakland and San Diego) guaranteed to inland shippers that urban waterfronts would remain public and port revenues restricted to the promotion of commerce navigation, fisheries and recreation' (Denning, 1983:2).

Before examining the organisation and position of the Harbor Department in more detail, some points about the political and administrative organisation of the city should be mentioned. The political structure of Los Angeles is decentralised and fragmented which dates back to the forced cooperation between the city and the neighbouring communities in the region which were dependent on the city of Los Angeles for their water. In order to survive, they had to agree to the political control of the city but have always remained individual communities and this is still apparent in the districts' representation on the city council. The port comes under the control of the political representative from the 15th district of San Pedro/Wilmington.

Formally, the city is managed by an elected mayor with limited powers, a strong city council (full-time members) with a whole range of regulatory powers and autonomous departments which are managed by appointed commissions. This configuration is a result of the highly conservative nature of the reform movement which held political sway at the beginning of the twentieth century. It comprised mainly business and professional interests and contained virtually no agricultural or union representation. Administrative forms were consciously sought which would exclude political interference and partiality (corruption) as far as possible in favour of public administration and professionalism. Success was achieved by developing the relationship in question between mayor, city council and departments. In order to introduce some form of 'checks and balances', various control instruments were created for the citizens, such as the right of initiative, the holding of a referendum and the revoking of decisions. Because in the past a system based on professionalism had been chosen, in Los Angeles there are no well-organised or influential political parties, unions, minority coalitions or representatives of business interests.

The impact of the administrative fragmentation of the city and the emphasis on professionalism was twofold. Due to its institutional autonomy and its professional character, for a long time the Harbor Department was able to function independent of political interference. It was seen as the authority in the field of port issues and there was a great deal of political confidence in it. And by representing the individual character of the San Pedro/Wilmington district on the city council, port projects were usually assessed from the perspective of this district interest. The representative from the 15th district is able to exert a relatively large amount of influence on the Board decisions via the recommendations of the *Commerce, Energy & Natural Resources* commission which deals with port affairs and has three members. This representative also plays an important role in the appointing of the Harbor Commission members who represent the district of San Pedro/Wilmington. 'The influence of the council member from the 15th is not invariable, however. It tends to be greater when the issue affects the constituents of San Pedro, as in the case of waterfront revitalization and the health of the fishing community, and less when directed at matters of industrial importance, such as major lease negotiations' (Denning, 1983:4).

The Harbor Department is managed by five part-time commissioners who are appointed by the mayor for a maximum of five years. The appointments must be approved by the city council. The council also takes decisions on a number of other subjects, including services with respect to legislation and regulation, tariffs, leases, franchises and licences, granting or paying off loans, contracts and the choice of a foreman for the industrial and administrative survey should the mayor so request. The commissioners are responsible for strategic port policy and the appointing of a port authority. Due to the part-time nature of their jobs, they also have a 'commission staff' who provide them with the correct information about port activities and supervise the Harbor Department. The board of commissioners meets twice a month.

In the past, the appointing of commissioners formed the mayor's most important instrument for directing policy. For example, it was an open secret that when the commissioner was appointed, the mayor asked him to write an undated letter of resignation so that he would have a means to control him in the event of any disagreement. This instrument also disappeared a number of years ago, however, when the mayor by mistake approved a proposal giving the city council the right to overrule decisions made by the commissioners.

Nowadays, there is a fear in the Harbor Department that in the future there will be more political interference. An amendment in the City Charter requires a decision via a referendum of the people, although this is a very infrequent occurrence.

Since the port is one of the most cost-effective city components, it not only has to contribute to the city treasury without a quid pro quo but it also operates on the profit principle with regard to the provision of city services (fire department) for the port. The contribution to the city is the recent outcome of an amendment in the state of California regulations which is aimed at trying to balance state income and expenditure as a result of a reduction in property tax. Cities such as Los Angeles were able to compensate this by appropriating monies from the port via the *Harbor Revenue Funds*. In 1993 this was the sum of 44.2 million dollars and in 1994 the sum was 25.3 million dollars. In 1995 nothing was taken from the funds. An increasing amount also has to be paid for city services. Both in 1995 and 1996 an additional twenty million dollars was paid to the city for services provided (fire department), which did not go down at all well with the private sector (esp. the shipowners). At present, they are attempting to prevent such practices via the state of California which, through the California State Lands Commission, supervises correct compliance with the Tideland Trust Act.

The Harbor Department operates on the financial principle of 'self sufficiency'; there is a financial requirement of a 10% *rate of return*. So far, it has always managed to deliver this. Due to its unique location for trade with Asia, since 1932 its monopoly position had always enabled it to make a surplus profit until the expansion of the port of Long Beach in the seventies and eighties. Even during the recession in the early eighties, the port was still able to achieve limited economic growth (Denning, 1983). The Harbor Department's principal sources of income (totalling 196 million dollars in 1995) come from the provision of various services to ships (chiefly the leasing of berths, cranes, pilotage of ships; total 75.6%), leasing (of land, buildings, distribution centres, recreational sites; total 22.1%) and royalties (concessions, oil royalties; total 2.3%). The staff of the Harbor Department are civil servants with the exception of the top managers who are appointed and may also be discharged by the commissioners.

In the summer of 1995, the Harbor Commissioners announced a reorganisation which would lead to a change of job description for a third of the existing personnel. The pilots and technical personnel (designers and maintenance staff), in particular, would from that time function from a distance (externalising). The plans were apparently developed during a meeting of the Harbor Commissioners with minimal input from the (management) staff. The presentation of the report at the public meeting was thus greeted with howls of derision when its 'fundamental' approach was described. Although the report made the port authority aware of the fact that a more customer-friendly strategy was required and that less time and energy should be spent on interaction with the political environment, the pilots have still not been privatised (April 1997).

For the time being, the Harbor Department remains a city corporation with a Board of Commissioners appointed by the mayor that can be called to account by the city council more frequently than in the past. The most important question remains: how will the changed system of 'checks and balances' affect the management of the port in the long term? How will the long-term planning be guaranteed? How will the relationship between the city council and the Board develop in the long term now that the post of commissioner is no longer an honorary title?

6.4.   Nautical Management

The nautical management of the United States ports falls under two federal departmental divisions, i.e. the Coast Guard, which devotes itself to the guidance of the shipping traffic and port state control activities, and the Army Corps of Engineers, which is responsible for the civil engineering maintenance of the waterways. The reason behind the federal involvement lies in a historical agreement that the various states would surrender their customs revenue in exchange for federal maintenance of the waterways. The agreement was that no money would have to be paid by the individual ports for the services of these departmental divisions, thus making them public services which would be financed from the general funds. This is no longer the case. At present, a financial distinction is made between the dredging of the channel and the dredging of the berths. Only the costs of the first activity are paid by the federal government, while those of the second have to be passed on to the users via the port authority.

Coast Guard
The management of the water boundaries of the United States has been a federal affair right from the start. In the late eighteenth century a 'fleet of cutters' was already active which, at the outbreak of the war of independence against the English, fought as a component of the navy. During the nineteenth and twentieth century the fleet of cutters, which on 28 January 1915 became the Coast Guard, came alternately within the remit of the navy and the *Treasury Department*. Finally, on 1 April 1967, the Coast Guard became a part of the federal *Department of Transport*. The budget has to be approved on an annual basis by Congress.

The organisation of the Coast Guard is divided into two areas of command: one for the Atlantic coast (head office in Portsmouth, Virginia) and one for the

Pacific coast (head office in Alameda, California). It comprises ten separate districts along each of the two coasts, 44 groups and 185 multi-mission units which perform a whole range of nautical functions for shipping and the American ports. The major ports all have their own Coast Guard Captain of the Port who is responsible for the coordination of shipping traffic and for conducting special operations (environmental pollution and rescue operations). This person is also represented on a number of public commissions pertaining to port development, safety and environmental policy. The Coast Guard is formally responsible for enforcing maritime law, maritime safety, the quality of the maritime environment and national security. On the basis of these four main tasks it is concerned with the construction and maintenance of navigational aids. For a large country like the United States, this comes to about 50,600 federal and 48,000 private resources. In addition, the tasks include guaranteeing the safety of shipping traffic in American waters (sea and ports) and protecting the coastline. There is close coordination between the Coast Guard and the navy. Traditionally, the Coast Guard is involved in combating smuggling and pollution. It takes a preventive approach to pollution and cleans up actual pollution at sea and in the ports. Chemical waste collection in ports is regulated by the federal and state Environmental Protection agencies. The fire department in the ports (mainly city) must comply with the Coast Guard's fire and safety rules. All port state control activities are also placed under the Coast Guard and it issues licences to ships. Finally, it is the Coast Guard's responsibility to conduct rescue operations and to render frozen shipping lanes in the polar regions accessible for trade.

Army Corps of Engineers
Since 1824, the US Army Corps of Engineers has been responsible for the construction and maintenance of the waterways (channels and rivers). This comprises not only the care for the dredging activities but also for the construction of dams, tidal management measures and irrigation activities. Before starting to dredge, the Corps first conducts a detailed cost/benefit analysis into the necessity for and format of the work. This is subsequently submitted to Congress for approval after which the work can begin. Congressional approval can mean, on the one hand, that the decision making is sometimes a long drawn-out process and on the other, that some ports have to wait longer before it is their turn for dredging and other navigation works. For example, it took fifteen years before the Corps could issue the port of New York with a '40 feet mean low water' certificate. As regards the second aspect, in the past it sometimes made a difference if the port concerned was well-represented in Congress or not (Goss, 1979).

The main problem for ports like New York these days is not so much the dredging of the ports as such, but being able to come up with a suitable location to dump the (polluted) dredging sludge without causing too much damage to the natural environment. This is what makes it prohibitive for ports like New York, which have to be permanently dredged, to maintain the correct draught at the berths or to further deepen them in order to be able to accommodate the new generation of container ships. The Port Authority already pays twenty million dollars per year to maintain the appropriate draught at all the berths at the scattered terminal locations. Due to the existing environmental regulations pertaining to dumping, decision making is delayed because the Corps cannot directly approve the dredging of the berths. The opacity of the decision making has resulted in the association of shipowners labelling the port of New York 'unpredictable' and it has warned the port authority that the port will become unsafe and unusable in the long term (Via Port of New York - New Jersey, 1995).

This is why the Port Department in New York is of the opinion that this is gradually turning into a national problem. It is not able to produce the extra funds which are required as a consequence of environmental quality requirements. Moreover, its market position will suffer if a clear dredging policy is not pursued at federal level. The Port Department has the support of the American Association of Port Authorities which has already made a number of policy recommendations. In October 1996, the American Congress decided that 800 million dollars would be earmarked for the dredging activities (chiefly the storage of polluted dredging sludge). The city of New York will have to contribute 130 million dollars, and the two states (New York and New Jersey) will have to release funds for this operation which will take several years. This decision seems to have ended the protracted uncertainty regarding the accessibility of the port in the next century.

In Seattle and Los Angeles these problems do not apply, or to a lesser extent, because Seattle is a naturally deep harbour and Los Angeles has enough resources to finance the maintenance work in the port basins. Seattle only has to carry out dredging activities and to remove minimal sedimentation from the Duwamish Waterway. Moreover, in Los Angeles the sludge is not polluted and is used for land reclamation. This does not alter the fact that ships putting in at the port of Seattle still have to pay a Harbor Maintenance Tax which is used to finance dredging work elsewhere in the United States. So although there is no national port policy, the various ports do underpin each other.

6.5.   Port Planning

In the past, port planning in the United States was primarily a matter for the ports themselves but with the rise of intermodal transport, the federal government is expected to play a role. By means of *The Intermodal Surface*

Transportation Efficiency Act of 1991, the federal Department of Transport attempts to promote the integration of various modes of transport. The federal contribution retains a conditioning character and provides an initial impetus to innovation which in this case will also be financed with funds from the ports. The differences in the economic significance of the various American ports can be clearly seen if we compare the various forms of port planning. This also reveals a totally different relationship between government and the business community.

6.5.1. PORT PLANNING IN NEW YORK

The port of New York is primarily a consumption port for the 17 million people living in the New York region. The major import products are alcoholic beverages, organic chemicals, automobiles and bananas/fruit and fruit juices. The most important export products, which come from the Midwest region via the excellent railway links, are (waste) paper, plastic materials, automobile parts, waste products and paperboard. Due to the limited economic growth in the last ten years and the moderate expected growth for the coming five years, New York is mainly concentrating on improving its service level by e.g. developing intermodal transport, purchasing new container cranes, deepening the shipping lanes, improving roads, utility facilities and safety measures and renovating distribution centres. After years of absolute and relative recession, things appear to be looking up for the port of New York. As a result of the improved trade relations with Asia, in the first half of 1996 a growth percentage of no less than 18% was achieved.

Until a few years ago, the only contact the Port Department had with the business community was to sign a new leasing contract or renew an existing one. Due to the economic growth which continued until the late seventies, and the fact that the Port Department did not have to worry about attracting cargo, there was no direct reason to change this situation. In the failing economic climate of the eighties, the amount of cargo declined. The port also appeared to have lost some of its attraction due to the labour conflict that had been sparked off by containerisation. Since employers and employees could not resolve matters on their own, the Port Department successfully mediated (see section on port services). As a result of this situation it also gradually became apparent to the Port Department that if they wanted to keep the port attractive to cargo they would have to increase joint enterprise with the regional and international business community (shippers and carriers) and other parties concerned. The Port Department acts as initiator and facilitator in this and is entirely dependent on the enthusiasm of the participants and public recognition of the socio-economic function of the port.

In the last five years, great efforts have been made to develop the *intermodal transport* concept and the Port Authority has constructed a rail terminal so that cargo can be transported directly to both Canada and the Midwest. The awareness also exists that this makes the port accessible to cargo from Asia which can now reach New York directly by rail via Los Angeles and Long Beach. Attempts are being made through the port's own offices in Asia to convince shippers that transport by ship via the Panama Canal is cheaper. The Port Department initiated the setting up of the *Shippers Council* which functions as a consultation platform and gives the Port Department more insight into the wishes of shippers. But it cannot actually do much more than initiate consultation, provide information and other methods of marketing. It is the shipowners who, in giant alliances, opt for a particular port and New York is not the only port on the northeast coast.

Earlier, the dredging problem was discussed in detail. Such projects clearly demonstrate the way in which the Port Department is tied to the federal government for its maintenance activities. Options for the further physical expansion of the port are not being addressed at present. This is not only due to the dredging issue but also to the broader decision-making framework of the Port Authority as a whole. On the one hand, it is a good marketing instrument to sell the port as part of a larger transport infrastructural whole (chain thinking). On the other hand, the projects of other departments (airports and World Trade Centre) undeniably produce a higher yield, while increasing the tollage to enable more investment is certain to be vetoed by the governor so that at the present time new plans for the port are finding a very limited audience. The picture is one of an established port which is having to pull out all the stops to stay alive.

6.5.2. PORT PLANNING IN SEATTLE

The Port of Seattle has acquired the legitimation for its existence chiefly by winning a whole string of protracted legal battles regarding the compulsory purchase of port sites which had been the property of the railway companies. There was no other means to get the port planning underway because prior to the creation of the Port of Seattle the land had been sold by the state and city, and the railway corporations were not willing to consent to a transfer. The compulsory purchase option still exists. As was mentioned earlier, it was not essential to Seattle's economic growth that there should be industrial locations in the immediate vicinity. For the Port of Seattle it was fortunate that the First World War broke out shortly after it was established. As a result of the German submarine movements and the closure of the Suez Canal, the French and English were forced to find another route to deliver their arms supplies to the Russians. Seattle turned out to be the only port on the Pacific with the facilities they required. Even the Canadian port of Vancouver, e.g., was dependent on a heavy crane from Seattle.

In addition to this, a number of other factors were responsible for the rapid economic growth of the port under its unique public administration. 'Seattle was the closest major American port to Asia, and given the critical necessity for speed in wartime, was quite important to international shippers. Seattle was also the cheapest port on the coast. (...). A very important element in the success of Seattle's harbor was the aggressive sales campaign that the Port commissioners waged throughout the country to convince shipping agents that Seattle was the superior port on the West Coast' (Port of Seattle, 1995:45-46). In the 1920's, for example, the shipyard in Seattle negotiated important contracts with the federal government.

For the development of the port in the early years, substantial loans were taken out which, from 1918, gained the approval of the whole port community. The money was used to purchase land and construct piers and gantries. The port continued to grow rapidly, partly as a result of the increasing trade with Japan. In the twenties, the competition with other ports along the west coast increased, leading to the creation of price agreements to which in fact only lip service was paid. Since Seattle did not have any industrial activities, the port was wholly oriented towards export which made it vulnerable. Until the Second World War, hardly any port planning took place at all. 'At the end of World War I the Port owned the most modern berthing, docking and storage facilities on Seattle's waterfront; and yet it controlled only slightly more than three percent of all the wharfage space in Elliott Bay. Following a small building program in the 1920's and almost none in the 30's, the Port embarked in the late 1940's on a construction program costing millions of dollars that would result in the Port owning the vast majority of Seattle's docks and wharfs by 1959' (Port of Seattle, 1995:72). By purchasing various parcels of land (in the sixties, too) the Port of Seattle now has 28 commercial marine terminals, a grain terminal, a distribution complex and a customs inspection office, including a free trade zone of 1400 are which is used for both port and airport activities. 'Soon after the purchases were finalized, dredging and filling operations were begun, and once submerged tidelands were reclaimed and transformed into storage facilities and terminals with extensive back-up space, with every imaginable modern handling and loading facility' (Port of Seattle, 1995:90).

The development of the container in the early sixties ushered in a challenging and innovative period for the port of Seattle. Here, too, the tremendous costs and uncertainties involved in this new method of transport were beyond the financial capacity of many private actors. The investment in container cranes in particular, which were now no longer fitted on the ships themselves, called for tremendous financial efforts. In 1960, with the help of a loan of 10 million dollars approved by the electorate, the construction of the terminals commenced. The construction of the container terminals was part of a more comprehensive improvement plan for the port that was to cost a total of 32 million dollars. Fortunately, containerisation brought great prosperity to Seattle. Between 1963 and 1972, foreign trade via Seattle grew by 111 percent, which put it way ahead of ports such as San Francisco and Portland.

In the late sixties, the tremendous economic growth of the preceding years (44,000 jobs per year between 1965 and 1969) came to an end which prompted the authorities to embark on new investment. The vitality of the port in particular proved to be a good marketing resource. The port commissioners pursued a policy of port expansion and industrial development. 150 million dollars were pumped into the airport, for example, and important land purchases were made for the port. The site of Boeing factory Number One (25 acres) was acquired for three million dollars in order to build a new terminal (terminal 115). The grain elevator on Hanford Street was also transformed into a container terminal. 'As one major project followed another and Port profits and income continued their spectacular climb, general community approval for the Port increased' (Port of Seattle, 1995).

From the mid-seventies, port planning in the Port of Seattle was highly dependent on public opinion. In the state of Washington this was motivated not only by the increasing attention devoted to environmental issues but also because so many terminals and facilities had been built that overcapacity had been created (Olson, 1980). In response to research into the expected growth in freight and the existing handling capacity in the state of Washington, a Cooperative Port Development Committee was set up to look into the options for joint planning. This commission was to be sponsored with monies from, among others, the *Washington Public Ports Association* which had already won its spurs as a voluntary association through joint consultation and acting for the joint port interests. The Cooperative Port Development committee (CDC) was to function primarily as a forum for the exchange of information between ports (master plans and so forth). 'In some instances it took the form of the larger ports assisting the smaller ports in plan preparation, and improving the content of proposed projects. In other instances simple communication about the merits or demerits of a proposed facility development project as it made its way through the review and evaluation process disseminated information to member ports of CDC. Port representatives to CDC also learned earlier on and in more detail than otherwise would be the case the degree and kind of facility development occurring among other ports within Washington State. Finally, information was shared on barriers ports had confronted in the permitting process and the means for resolving them' (Olson, 1980:32-33).

The citizens can express their opinions on policy by means of a referendum and this must be respected by the commissioners. In the early seventies, the port authority was not in the habit of lending an ear to the local community, so a separate programme was set up 'to "people-ize" the port and attempt to make it more responsive to environmental and community problems' (Port of Seattle, 1995:94). These days the commissioners mainly provide a forum for dealing directly with complaints. They even organise informative discussion evenings in King County.

In the mid-eighties, the port of Seattle was shocked by the report that their principal container shipowner was moving to Tacoma. 'Seattleites with little knowledge of maritime affairs were rocked when pioneering container freighter Sea-Land moved its business to its neighbor to the south. It was, after all, "our" company; who in Seattle hadn't driven behind one of its familiar trucks? The loss of Sea-Land seemed as unlikely as losing the Space Needle to Portland' (Port of Seattle, 1995:105). According to formal reports, Tacoma had the advantage of new sites and new cranes. The fact is that in the Port of Seattle people were absolutely scared to death when in 1991 American President Lines indicated that would probably move out of Seattle. This container carrier was not only innovative but also contributed around 20% of the total cargo handling which gave it 'anchor tenant' status. Seattle pulled out all the stops to try and keep American President Lines (APL). 'The Port of Seattle worked with APL for nearly three years before arriving at a final proposal. But it wasn't working alone. Meeting APL's needs and selling it on a new Seattle facility became one of the largest cooperative efforts ever seen in the city. It included not only the efforts of the Port, but of Seattle Mayor Norm Rice, members of the Seattle City Council, business leaders and labor representatives. The needs of the project also involved the City Parks Department, the City Engineering Department, the federal government's Environmental Protection Agency and national and state resource management units concerned with underwater lands' (Port of Seattle, 1995:106). The APL project made the Port of Seattle realise that it could only achieve productive port planning through cooperation with others, in order to prevent duplication and facilitate processes.

6.5.3. PORT PLANNING IN LOS ANGELES

The development of the port of Los Angeles is the responsibility of the Harbor Department. It has its own engineer's office and until a few years ago, when the total maintenance and construction costs were still around twenty million dollars a year, it carried out all the work at the port itself (from drawing board to actual implementation). Work on the current major Port Plan 2020 project is mainly being contracted out, however. This project is providing the construction of large sites in the middle of the port (Pier 300/400) for which, among other things, a coal terminal on pier 300 is planned. This coal terminal was devised in cooperation with 41 private American companies and 15% of it is financed by the port authority.

Over a period of 17 years, the Federal Army Corps of Engineers dredged the port draught from 35 to 45 foot. The project cost a total of 148 million dollars, of which the Harbor Department ultimately had to pay 62 million to accelerate the process. For this acceleration a 'credit agreement' from federal level was necessary to provide private dredgers with a guarantee that the money would be paid. Environmental problems are not really an issue due to the strong emphasis on container transport. The dredging sludge is not polluted. In the case of major projects it is usual to buy off environmental interests prior to actual implementation.

In Los Angeles at present four major projects are in implementation. An important project that is being undertaken equally with the port of Long Beach is the construction of the *Alameda Corridor* for which purpose the two ports set up the Alameda Corridor Transportation Authority. This functional organisation is responsible for the implementation and coordination of the project. The project was set up with the purpose of 'establishing a comprehensive transportation corridor and related facilities consisting of street and railroad rights-of-way and an improved highway and railroad network along Alameda Street between the Santa Monica Freeway and the Ports of Los Angeles and Long Beach in San Pedro Bay linking the two ports to the central Los Angeles area. The Port of Los Angeles and the Port of Long Beach share income and equity distributions equally' (Worldport LA, 1995). The success of the project depends primarily on the cooperation of cities and communities along the route. The two ports bought rights of way from these communities for the sum of 370 million dollars.

Another project on which the port of Los Angeles participates jointly with the port of Long Beach is the Intermodal Container Transfer Facility Joint Powers Authority (ICTF). For 2.5 million dollars apiece the two ports participate in this organisation whose purpose is to finance and build the intermodal facility for the truck-to-train transfer of container cargo. Since 1986 the actual work has been carried out under long-term contract by a private provider who also developed the facility. The ICTF is managed by a board of commissioners, two of whom are appointed by the port of Los Angeles. The ICTF is able to take out loans independently which falls within the risk of the private provider. Furthermore, the two ports support this authority by providing staff and other non-material resources.

Chiefly as a result of this participation and the decision making concerning it, both the Harbor Commissioners and the port of Los Angeles management gradually began to take more interest in the institutional differences between the two ports. The most fundamental difference in the decision making on port projects between Long Beach and Los Angeles is the fact that in Los Angeles all decision are appraised by the city council and take considerably longer than in Long Beach. This is highly frustrating for the Harbor Commissioners, on the one hand, who in fact no longer form a counter power and the port management in Los Angeles on the other, which has to build and maintain an extra relationship of trust.

6.6.   Port Services

In an economy controlled by the market mechanism it may be expected that the port services will be entirely supplied by private providers. The situation is in fact a little less straightforward than this.

6.6.1. PORT SERVICES IN NEW YORK

The various port services (pilots, tugboats, distribution centres, stevedores, etc.) are all offered by private providers. The pilotage of ships is taken care of by the Sandy Hook Pilots and is largely in the hands of a few families who have been engaged in the profession for generations. There are various tugboat companies who compete with each other and do not need a licence. The linesmen and dockworkers ('longshoremen') need a licence from the Waterfront Commission of New York and New Jersey. This is a sort of harbour police organisation.

The Port Authority itself does not provide any services and has no longshoremen in its employ but it does occasionally facilitate for the companies. For example, it owns seven container cranes which it leases to two different stevedores. It also mediates in consultations between employers and employees. In the old days, every time there were labour negotiations the longshoremen would go on strike. As a result of a generous settlement for the longshoremen, there has been no mass labour unrest at the port for 15 years or so. An agreement was made that all longshoremen who had worked at the port before 1969 via the International Longshoremen Association would be given an income guarantee for a minimum of 1900 hours work per year, even if there is not enough work on. Moreover, they were also offered early retirement at 55 years. This settlement is the reason why the handling costs at the port are high. Of the present 3400 longshoremen, no less than 1100 people have an income guarantee on the basis of this settlement (40,000 dollars a year).

In autumn 1996, a collective labour agreement was signed in New York which is unique in the history of the port. It was signed for the unprecedented period of 5 years, which has brought peace to the port and will provide the opportunity for increasing productivity and a general reduction in costs.

6.6.2. PORT SERVICES IN SEATTLE

Due to the legal character of the Port of Seattle it is possible for it to take on both public and private activities, and this it does. This is principally the case in the sphere of handling and distribution. The service provision in the context of nautical management is offered by both public and private parties.

The pilots all belong to an independent pilot's association over which the Port of Seattle has no control. They come under the regulation of the state of Washington. From 70 miles outside the port of Seattle a pilot is compulsory and the Coast Guard Vessel Traffic System is in operation. The towing and assisting of ships is carried out by four separate private tugboat companies which also arrange for the linesmen who tie up the ships. The processing of ship's waste is carried out by private companies which possess the relevant licence from the state of Washington or the Seattle fire department. The Port of Seattle collects the waste from fishing and pleasure craft.

The Port of Seattle is empowered under the Revised Code of Washington to set up its own police force and it has done so. Order and safety on the landside of the port (the scattered terminals) and at the airport are looked after by a separate police force. The fire department of the city of Seattle will also come into action for the Port of Seattle in the event of emergencies at the port. There is no financial contribution for this service and this led to friction because the fire department had to purchase new equipment specially for port emergencies.

The handling activities are offered by both the Port of Seattle and private providers. The Port of Seattle owns four smaller and two larger container terminals, a fruit terminal, a public terminal, various warehouses and a fish terminal. Until recently, the handling of these activities was taken care of by Port of Seattle staff. For some activities, such as container handling and warehousing, the Port of Seattle competes with private providers. On inquiry, it was apparent that the management personnel of the Port of Seattle themselves were unclear as to which activities did and which did not have to be undertaken by the public corporation. Much of it had simply grown over the years and a clear strategic design was absent. In the past, the container terminals had been entirely constructed and operated by the Port of Seattle. The same was true of the distribution centres. In the recent reorganisation, the Port of Seattle has privatised the employees at the container terminals and the warehouses.

Due to the fact that the Port of Seattle itself employs longshoremen, regular consultation with the King County Labour Council is necessary. The 24 labour unions which are involved in the port in one way or another are represented on this council. The port authority consults with the council once every three months. No collective labour agreements are signed, but it is essential for the

Port of Seattle to keep on the right side of the Labour Council which is very powerful in this part of the United States. It cannot directly influence the council's agenda, however. What it can do is take union representatives on trips abroad to bring them into direct contact with various shipping companies and observe the production methods used elsewhere. Because the unions have a lot of influence, potential lessees are screened by the Port of Seattle regarding their attitude to employees. The Port of Seattle wants above all to prevent strikes and avoid an unreliable image: it thus mediates between employers and employees in other port sectors.

As far as the industrial section is concerned, this has always been run on the principle that the Port of Seattle develops the sites for building and installs any other desired facilities after which sites are let on long-term lease (25 years) to private parties.

6.6.3. PORT SERVICES IN LOS ANGELES

For a long time, the main priority of port management in Los Angeles was to generate the best possible service. That also meant that the Harbor Department was a jack of all trades, carrying out numerous activities itself which could equally well have been done by the market. Given the economic growth this had never caused a problem until last year when the research consultancy Booz-Allen and Hamilton published an efficiency comparison between the ports of Long Beach and Los Angeles in which Los Angeles was depicted as a great (bureaucratic) Moloch. The comparison with the port management in Long Beach was a logical choice because Long Beach is organised along entirely different lines and over the years has become just as large as Los Angeles.

The institutional format of Long Beach is very similar to that of Los Angeles (a city port authority) with one important difference in political-administrative relations and a number of differences in the definition of the port authority's role. The difference in the political-administrative relationship concerns the control which the city council has. In Los Angeles this is far greater and means the decision making takes longer. The difference in role definition involves the quality criteria which the port authority lays down for service provision. In Long Beach the quality criterion is not the service level but commercial considerations. The only thing preventing the privatisation of the port of Long Beach is the Tideland Trust Act which stipulates that the port must serve trade, the pleasure cruising sector and the fishing sector. The city council does not have much control over the port authority of Long Beach because the port director has a considerable degree of policy freedom. This also results in there being very little on paper which significantly reduces options for public audit and involvement. This was apparent from the sharp contrast in public attendance at the Board meetings, among other things. In Los Angeles, an average of between fifty and a hundred people attend the meetings whereas in Long Beach the public interest can be counted on the fingers of one hand.

The Los Angeles Harbor Department employs a staff of about 700 which includes the technical section, the port pilots and the port police. Unlike the pilots at Long Beach, the 32 pilots in Los Angeles come under the Harbor Department. At Long Beach this service is performed by a private monopolist under licence from the city. The port police (59 people) in Los Angeles are regarded as an essential component of the Harbor Department and thus will not be included in the reorganisation plans. The fire service in the port is a separate city unit for which the Harbor Department has recently started to pay a charge to the city. For each of the fiscal years 1994/95 and 1995/6 the Harbor Department will pay twenty million dollars for services provided in the past. This has caused indignation among the shipowners who have protested to the California State Lands Commission to test this practice against the principles in the Tideland Trust Act. The outcome is not yet known but this indicates that the city has managed to get more of a grip on the Harbor Department.

The towing services for both the port of Los Angeles and the port of Long Beach are taken care of by three competing tugboat services. The linesmen's are members of the labour union. Although the port has a very orderly layout and the weather is generally good there is a Vessel Traffic System in operation. This system is operated by a private provider in cooperation with the Coast Guard.

The stevedoring and handling work is a private concern. The Harbor Department itself has nothing to do with negotiations between employers and the unions. Apart from one terminal (NYK), the Harbor Department owns all the container cranes which it leases to container shipping companies. It also carries out its own maintenance work on the cranes, again in contrast to Long Beach where such activities are contracted out to private providers.

6.7.   Port Management in the United States: Analysis and Questions

In the United States, an important role is set aside for ports as links in transport chains, as stimulators of trade and as engines which drive employment. In 1992, over fifteen million people were (directly and indirectly) linked with the port sector in the United States. They contributed 780 billion dollars to the gross national income and paid 154 billion dollars in tax (Maritime Administration Office of Port and Intermodal Development in cooperation with the American Association of Port Authorities, 1994:5). The format of port management is a matter for the individual states and there is no national port policy. The federal government nevertheless plays an important role in laying down the (transport) conditions, e.g. by prioritising dredging and levying taxes but also by the development and co-implementation of the intermodal transport scheme. It is also nationally regulated that ports do not fall under the trust laws and this enables the cross-subsidising of components and public investment in superstructure.

The individual states each have their own institutional solutions to port management. In New York, port management comes under an organisation which is aimed at providing transport facilities in a broad sense. At present it provides financial underpinning for the port but it was originally set up to promote activities between the states of New York and New Jersey. Due to the mechanism of the commissioners mentioned earlier, the emphasis is on long-term policy and expertise. The state of Washington, on the other hand, has elevated a uniform mode of 'special port districts' to a legal norm. These are in the unique position of being able to levy taxes and their commissioners are elected to underscore the public interest of the port. And California has brought port management under the control of the cities but has stipulated conditions for use (Tideland Trust Act). This strategy was originally aimed at curbing corruption and promoting expertise but this ended recently when the city council gained the right to overrule decisions of the board of commissioners.

Although the United States wants to provide as much scope as possible for market forces, it is difficult for American port management to find a clear-cut regulatory criterion. This is not only due to the different state forms that exist but also to the fact that the American port authority is a functional organisation. On the pretext of maximising economic activities to create employment, in principle any activity which appears at first sight to be beneficial to the regional economy may be undertaken. As long as the market in which the port as a whole finds itself is undergoing substantial economic growth and/or those parties who take care of the revenue for the port authority do not cause any real difficulties no fundamental review of the position takes place and mutual expectations in specific projects are clarified. If the point of departure of the national market regulatory principle is as much competition as possible, however, then the institutional format of American port management can be criticised.

In the United States, a clear institutional distinction is made between nautical management, construction and maintenance of the shipping lanes and channels, both of which come under federal responsibility, and the port planning and service provision which may be a state, city or private matter. As far as nautical planning is concerned, therefore, institutional clarity exists and a cooperative arrangement between the Coast Guard and a private provider (offering extra service) was found only at the nautically orderly port of Los

Angeles/Long Beach. In the empirical description of port planning and port services, however, other configurations of state and market were encountered.

6.7.1. NEW YORK

The port of New York is operated by the Port Authority/Port Department which must bear its own financial risks and which only leases the sites and a few cranes and distribution centres to private users. The port is the smallest component of the Port Authority's numerous operations. The Port Authority is expected to stand on its own two feet and initiate transport activities which will deliver an economic advantage to both states. In the last decade, however, many activities have been undertaken in the real estate market of which the World Trade Center is the crowning glory. Due to the fact that these activities are financed through public loans, the success of the Port Authority stands or falls by their profitability.

Of the ports which were studied, New York appears to be the only one which has implemented its chosen regulatory principle in a reasonably consistent way. For the port planning it is largely dependent on initiatives from private corporations. Terminal projects are agreed with each other by means of contracts. The leasing of sites, a few cranes and distribution centres, gives the Port Department the opportunity to tie private providers to the port. It owns very little superstructural property so that little can be expected in this area. The bulk of the terminals is in use by major international shipping companies (Sea-Land, Maersk, Maher, etc.) and a tightening of requirements can only occur through renegotiating the contract or on its expiry. The towing services are taken care of by two competing tugboat companies and there are a number of pilotage companies.

The tragedy of the port of New York/New Jersey is that is has been hard hit by environmental requirements which can be passed on to a decreasing number of ships. In order to keep the port attractive to increasingly large container ships, the appropriate draught needs to be maintained which entails tremendous costs. This is primarily a federal responsibility in which the Port Authority cannot intervene. The Port Authority maintains no formal relations with environmental groups, for example. Although the functional division does justice to 'checks and balances', coordination, however, appears to lose out. The public functional division seems to lead to a situation in which, when new problems arise, responsibilities are not immediately assumed but first have to be determined in an extensive political debate. A great deal of time is lost in this way. The question is, in what institutional way does the Port Authority stay in touch with other public interests?

The Port Authority's own options for making the port attractive are very limited. Until recently, due to the (long-term) leases it was not well-informed about what was happening in the companies. It was only through information from customs and the US Army Corps of Engineers Navigation Data Center that it was possible to find out which goods were being handled and to detect trends. This is why, in 1991, a Shippers Council was set up so that the port authority could stay in better touch with the wishes of the shippers and shipowners. Moreover, the Port Department mediated for years in the negotiations on labour contracts which in 1996 effectively resulted in a long-term contract unprecedented in New York's history. The consultation instrument thus appears in the long term to be highly significant. In the short term, responsibility for the development initiatives and commercial risks will continue to lie with the shipowners and companies. The port authority primarily performs an important conditioning role, for example by making terminals suitable for intermodal transport, although this also makes the port of New York vulnerable to competition from ports in Canada and Los Angeles/Long Beach.

The dredging problem has put the function and position of the Port Department as division component under pressure. The Port Authority's main concern is to make the most efficient possible use of (financial) resources which the port management has been unable to do for the last few years in succession. The autonomous position of the Port Authority makes it possible in principle for it to question the position of the port management in the overall transport activities. The fact that the Port Authority decided in the past to assume the responsibility for port management does not necessarily mean that it will continue to do so in the future. Other components of the Port Authority, such as the management and operation of the international terminal at JFK Airport, have also been contracted out for a number of years to other and more specialised organisations. Why should the same thing not happen with port management? In other words, the autonomous position of the Port Authority means that it determines for itself which activities it will undertake and how. For the time being the extra costs for the port, which actually squeeze other divisions, are accepted because the port performs an important regional function and there is a chance that costs will be covered. But what will bind the Port Authority to the port management if it becomes apparent in the long term that it is not possible to make the port cost-effective? How much influence will the politicians then be able to exert on the independent management of the Port Authority? How much public money is involved in such relations and how can this be controlled?

6.7.2. SEATTLE

Port management in Seattle is organised along highly unusual lines. For historical reasons, the Seattle harbour is managed and operated by the Port of

Seattle. This public corporation is managed by directly-elected commissioners whose task it is to procure the goodwill of the port towards the community and the commercial sector. The Port of Seattle is expected to act as facilitator for attracting trade and economic activities to promote employment. In other words, in a broad sense it is all about achieving economic growth which is measured by the number of jobs created and the import and export values. Economic growth is also an important success condition for the Port of Seattle itself in order to be able to steal a march on direct competitors such as Tacoma and Vancouver. Whether Seattle has been able to succeed in this up to now is debatable. After all, Tacoma (with the same institutional structure) proved able to work its way up from being a relatively unknown port to being a formidable competitor. Against the (historical) backdrop of the Port of Seattle's position, it remains to be seen whether the port authority will be able to achieve the public objectives. The fact is, these objectives do not have concrete handles which would enable the citizens to have some control over the public corporation. This raises questions about the position of the elected port commissioners. How do the public objectives correlate with the port authority's own interests? And when public interest is really at issue, which safeguards are then in operation to protect it? Is the port commissioners' extensive political consultation sufficient to safeguard the coordination with other interested parties? Does the referendum function as an effective additional political instrument and how does this relate to the political authority of the port commissioners?

In addition to the terminals scattered over the various sites, the Port of Seattle also owns the superstructure and various distribution centres which, until recently, it managed itself and leased to private corporations. This meant it was in direct competition with private corporations. Last year, the port commissioners decided to allow the actual handling and storage activities to pass into private hands. It remains to be seen, however, whether such outsourcing will be enough to improve the efficiency of the port. After all, the tax revenues still function as financial featherbedding so that the Port of Seattle remains a hybrid organisation. How does the financial control by the state of Washington relate to the financially autonomous position of the Port of Seattle and what will be the effect of this in the long run?

A characteristic feature of the Port of Seattle's institutional format is its constant search for, and the considerable attention it devotes to, the shaping of its own identity. Each anniversary sees the publication of yet another historiography in which the reasons for establishing the Port of Seattle are recited in minute detail. The story often stops there because senior management does not know which activities now characterise the port authority. This increases the opportunities for a political job description. One of the respondents commented, for example, that the Port of Seattle sometimes tends to be more of a 'general purpose' organisation than to be primarily concerned with the port and the airport. What safeguards the functionality of the Port of Seattle? Why does the port not form part of the city of Seattle since, when it comes down to it, it is the city which has to fight for clients?

6.7.3.   LOS ANGELES

Until now, the port management in Los Angeles has aimed primarily at professionalism and/or expertise, service provision and consensus building. This was why private activities were carried out by its own staff. Now that the further development of the port is happening mainly in cooperation with Long Beach, the operating methods are being compared with the institutional format of Long Beach which in some ways appears more efficient. Another reason for scrutinising its own position is the decrease in the number of ships putting in at the port and the declining revenue from wharfage and berth dues. The port commissioners have now decided that the port management needs to be stripped of all its frills. Might the traditional opting for expertise and indirect political control have led to technical dilettantism? What changes will occur in the relationship between the port commissioners and the city council in the long term now that the city council has primacy? Why has the port commissioners' internal staff not been able to prevent this unintended state of affairs? How can the port finances in the present institutional format be prevented from becoming part of the political game? How does the political craving for more efficient port management relate to the interests which are safeguarded under the Tidelands Trust Act?

In spite of the comparative studies and the political objectives there has as yet been no fundamental change in position. The need for this is also not directly felt because the port is still scrupulously generating the ten percent net profit required of it. The most important question which may be asked of the Los Angeles port authority is to what extent the port would be able to keep itself commercially afloat in the event of reduced growth? As long as the trade and transport flows with Asia keep on growing, the competition with the port of Long Beach seems of minor importance, but what will happen if this growth (suddenly) comes to a standstill? Is there not then a strong chance that the city will have a major social problem on its hands?

6.8.   Summary

Although the United States is always associated with the economic miracle of the market, that picture needs some adjustment as far as port management is concerned. Nautical management as a public good comes under the federal Coast Guard. The port authority is the owner of the sites, the terminals and usually the cranes, too. And sometimes the port authority is even in direct competition with private providers (Seattle). As the federal Department of Transport also indicates, most ports perform a regional-economic role and they are more interested in capturing as many economic activities as possible which will then generate as many jobs as possible. The options available to the American port authorities in our study are marginal, however. Their sites are already occupied and they have to wait until the lessee either terminates his tenancy or is not in a position to pay. If the port authority sees growth opportunities it may decide to purchase or construct new sites. This is an expensive business, however, necessitating a long decision-making and implementation process. And finally, a port authority simply lacks data about the business community in its port, as was the case in New York until very recently.

The most important question raised by American port management is to what extent is it realistic to expect the port authority to provide economic activities and employment unless entirely different criteria are used (cf. Long Beach).

CHAPTER 7

Canada: the Port of Vancouver

'Although the possibility of developing a unified national ports system exists, this has never happened in practice. In fact, in many instances, this uncoordinated set of ports compete with each other so that an inappropriate development of port facilities results' (Ircha, 1993:51).

7.1.   Introduction

The port of Vancouver is the most important port in Canada. Despite the fact that Canada's economy is wholly intertwined with that of the United States, the port of Vancouver is managed on entirely different lines from ports in the United States. In Canada the law stipulates that the responsibility for the ports is a *national* matter. Despite this clear choice, the central theme running through the political history of Canada is the recurring discussion on the desired divisions of authority between the national and local levels. The next section provides a broad sketch of the economy and function of the Canadian ports. The third section concentrates on the position of the port authority of Vancouver. Sections four, five and six shed light on the institutional format of nautical management, port planning and port services. Finally, section seven contains the analysis.

7.2.   Economy and Functions of the Ports

Canada has a highly-developed industrial *market economy* which is strongly export-oriented and closely interrelated with the economy of the United States. This finds expression among other things in the fact that nearly all non-agricultural activities are financed by American capital. The signing of the *North American Free Trade Agreement* removed the last trade barriers between the two countries and they are moving rapidly towards economic unification. The economic dependence on the United States is the principal reason that the Canadian government is only able to pursue a limited independent economic policy. This is reinforced by the high degree of autonomy of the provinces which possess powers in the sphere of wage and price movements and tax levies. A general economic objective which is held in high regard in the country is the attempt to create equal opportunities for the different regions.

189

Canada, with its less than thirty million inhabitants, has an enormous potential in energy sources, raw materials and minerals which are still largely unexploited. The economic activity in the various provinces is dependent on the natural circumstances which are to be found in the area. The province of British Columbia (3.5 million inhabitants in an area measuring 892,677 km^2) in which the port of Vancouver lies, is rich in minerals: gold, silver, lead, copper, coal, gas and oil. The industry, in combination with the wealth of forests and the rich fishing, is the most important means of existence and comprises blast-furnaces, fish canneries, fish oil, fishmeal and fertiliser factories, and numerous sawmills, paper and pulp mills. The fishing (salmon, halibut, herring and shrimp) in this area accounts for about half of the total Canadian fish production. The high humidity of the coastal region has encouraged the growth of dense conifer forests which produce very good quality timber (incl. cedar). Other industries include shipbuilding, aluminium manufacture and foodstuffs.

Due to the extremely long and indented coastline of the country (23,700 km) there are many natural harbours and inlets to be found. In the last century Canadian ports were seen as providing a natural opportunity to conduct trade. In that sense little has changed. The fact that the ports formally fall within the federal level's remit has a historical basis in the *military threat* posed by the US in the last century. At that time, in contrast to the US, Canada had a strong federal government and relatively weak provinces. The Canadian ports had to be protected as a national interest. At present, the relationship between the federal and regional governments in the two countries is the exact opposite of how they were last century (Ircha, 1993). Over the years, seven Canadian ports have acquired a national economic function and the majority perform a regional or local function in various capacities. Until recently, there were three different port systems in Canada, all of which nevertheless adhered to the commercial principle.

Vancouver is one of the seven ports of national interest. It is by far the largest port in Canada and, as far as its exports are concerned, counts as one of the three largest ports of North America (together with Houston and Louisiana). Vancouver is of national importance to Canada because no less than 60% of its exports come from the *'prairie provinces'* which are Canada's principle source of grain for export. Vancouver transports chiefly to the countries in the Pacific Rim (Japan, Taiwan, South Korea, China, Indonesia). The port of Vancouver encounters considerable competition from the American ports of Seattle, Tacoma and Los Angeles. In relation to Japan, however, Vancouver's location is more favourable than that of its American competitors. The port of Vancouver maintains 'Sister Port' relations with a number of ports in the Pacific Rim countries (Yokohama/Japan, Inchon/Korea, Kaoshiung/Taiwan and Dalian and Shanghai/China). The purpose of these is to exchange knowledge, establish a relationship of trust and deliver an expansion in trade. Vancouver, for example, points out the fact that they were the first to have a Mandarin-speaking representative in China.

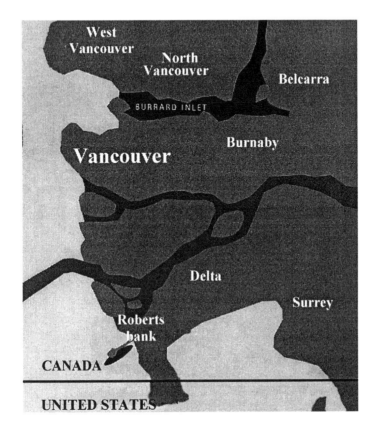

Nautical Management Area of Vancouver Port Corporation

The port has a number of terminals which contribute to Vancouver's strong position. 83% of the total freight handled comprises bulk (coal, grain and sulphur). In addition, 11% of the turnover consists of pulp and other waste materials and the remaining 6% of the cargo is transferred through the port by means of containers. In the coming years, rapid growth and a sharp increase in competition with the American ports is anticipated in this last sector in particular. This is principally a result of the fact that Canada's interior can also be accessed via the American railway links (intermodal transport). The port of Vancouver itself is also directly linked to the railway network which is run by two (quasi-)governmental companies: Canadian Pacific Railways and Canadian National Railways. Vancouver is the chief port for cruises to

Alaska; in 1998, no less than 873,102 people took advantage of this. In the Vancouver region, 17,300 direct and indirect jobs are linked with the port. In 1998, 71.9 million tons were handled, a small part of which was in the 840,098 containers.

Table 7.1. Multi-year Overview of Port Performance in Vancouver (in million tons) (Source: Vancouver Port Corporation, 1996:6-7 and website).

Handling/year	1994	1995	1996	1997	1998
Total tonnage	67.6	71.5	72.0	73.5	71.9
Bulk	56.3	59.7	61.1	62.3	60.6
General cargo	7.0	7.6	5.8	5.3	4.3
Containers (TEUs)	494,004	496,000	616,692	724,154	840,098
Passengers	519,409	596,724	701,547	816,537	873,102

7.3.   Port Management in Canada

Canada has 540 harbours which at present can be broken down into three separate institutional structures but which all come under the federal Ministry of Transport (Standing Committee on Transport, 1995).

The seven ports of national-economic interest are managed by a federal government body called the *Canada Ports Corporation* which is located in Ottawa. It is represented at the seven local levels by *Local Port Corporations*. This first group of ports is covered by the Canada Ports Act of 1983.

A second group of nine harbours is covered by the *Harbour Commissions Act* of 1964 and they perform a regional and local economic role. In comparison with the Local Port Corporations they have a high degree of autonomy and are managed by Harbour Commissions.

The majority of Canadian harbours (524) have local importance and are managed directly by the *Ministry of Transport* (Harbours and Ports Directorate) on the basis of the *Public Harbours and Port Facilities Act*. They handle about 20% of the total Canadian tonnage. Canada has 52 private industrial port and wharf facilities which account for no less than 17% of the total tonnage. And finally there are over 2,100 fishing ports and harbours for smaller ships which are owned and managed by *Fisheries and Oceans* (Ircha, 1993; Standing Committee on Transport, 1995). Relatively precise and topical information is available regarding the institutional format of Canadian port management. The reason for this lies in the recurring discussions on the rationale underlying the nationally designed port administration and the disputes about the exact division of authority between the federal and local levels. In March 1997, a Bill to reorganise the national port management was presented in the House of Commons for the fourth time.

Objectives and Effects of Nationally Organised Port Management

In 1867, it was regulated under the *British North America Act* that shipping and the ports would fall within the remit of the national government. Later, this federal responsibility was embedded in the Canadian constitution. No list was compiled designating the ports concerned and no clear description of the boundaries of the various ports was made; these omissions were to lead in later years to regular *demarcation conflicts* between the federal and provincial levels. Moreover, a shortage of staff and facilities at federal level meant that the activities of the various ports could not actually be controlled. In times of economic prosperity a plethora of construction projects were carried out at the ports. This meant that in periods of economic decline they had to contend with tough competition as a result of the surplus capacity thus created. Many local public services thus went bankrupt in those years whereupon the federal government took over their debts. Such practices, in combination with the patronage system, meant that a time came when no-one felt responsible for port management any more (Ircha, 1993).

In the nineteen-thirties, proposals were made to regain more control over the port management. The basic philosophy behind these proposals was that national ports should serve more than local interests and, due to their national importance, should be managed by means of a nationally-coordinated policy (formulated by the *National Harbours Board*). In addition, a case was also made out for the embedding of local autonomy. The following were the first to qualify as national ports: Montreal, Quebec, Vancouver, Halifax and Saint John. Later, Trois Rivières and Chicoutimi were added.

The National Harbours Board (NHB) was the first institutional recognition of the differences in economic position between the Canadian ports. However, it neglected to embed the proposed local autonomy in the system whereby a highly centralised system arose. 'In the course of time the NHB ports became noted for their rigid centralized financial control system, isolation from any form of municipal and provincial outputs, and the absence of any communication with Commission ports' (Ircha, 1993:55). The creation of separate port commissions guaranteed to some extent the input of municipality and province. They were financed with municipal and provincial taxes and were intended to promote the growth of trade through their port. A consequence of this, however, was that some provinces had discovered in this an instrument to by-pass national policy.

Little came of the idea for a nationally-coordinated port policy. Because the need continued to be felt (with a view to capacity planning and cost savings[1]), in the late sixties new reorganisation proposals were submitted. The actual implementation of the central ideas had to wait, for political reasons (other important issues and delays), for another fifteen years. For the time being, the introduction of local advisory committees from the National Harbours Board had to suffice.

The creation of these local advisory committees proved to be a good move. In 1973, the chairmen of the advisory boards for the various national harbours came up with a proposal to classify the national harbours into what is called in Canada a *Crown Corporation*. This type of public corporation functions virtually autonomously vis-à-vis the public administration but is dependent on national government and parliament for the approval of its annual report and major infrastructural projects. The Minister of Transport appoints the commissioners. The advantage of this institutional model was the opportunity it provided to find a definite solution to the local autonomy for the day-to-day running of things. Once again, it was not until 1983 that this model was adopted. This was also the institutional format which I encountered in Vancouver in 1995.

Port Management in Vancouver: Vancouver Port Corporation
In 1983 a solution was found for the ports which served a national interest and which were able to support themselves financially. At national level they are coordinated by a crown corporation called the *Canada Ports Corporation* (CPC). At local level the CPC is allied with seven different *Local Port Corporations* (LPC) each of which has the characteristics of a federal agency. Both the national and the local level have the same organisational structure: both are supervised by a *Board of Directors* appointed by the Minister of Transport. The chairman is appointed for a period of three years and this may be extended by a further three years. The local port authorities enjoy a high degree of autonomy but are subject to various regulatory provisions.

For example, the national ports are expected to be self-supporting. The federal port authority may undertake action on its own initiative or on the instructions of the Minister of Transport and in the framework of the national interest. The local port authorities hand over a certain proportion of their own revenue to the federal level to help finance this. They also have to consult the federal level regarding the appointment of a new Chief Executive Officer. The national port authorities have to obtain the Canada Ports Corporation's approval of their commercial name, their own ordinances or when entering

---

[1] Due to the national responsibility for the harbours, the financial risks were also passed on to the Canadian Ministry of Transport. In 1980, the National Harbours Board's debt to the federal government had risen to 475 million Canadian dollars, while this should have been one of Canada's most financially resilient harbours.

into long-term leasing contracts. Finally, the LPCs have to submit their corporate plans and annual budgets to the CPC for the CPC's perusal and comments. The local port authority also pays for the CPC police unit.

'LPCs are Crown corporations in their own right and therefore governed by the Financial Administration Act. It requires that operations are managed economically, efficiently and effectively; that capital budgets and corporate plans be submitted; and that internal and external audits, and special examinations be undertaken' (Standing Committee on Transport, 1995:3).

Vancouver is one of the seven ports to be given LPC status and is also called *Vancouver Port Corporation* (VPC). It has not departed from the original intent when it was established in 1983 although the recent policy document 'A National Marine Strategy' may not give that impression. 'The original intent was that VPC would operate with commercial discipline, be financially self sufficient and proactive in promoting the Port and enhancing its competitive position' (Vancouver Port Corporation, 1995b:4). VPC, with its staff of 185 (including port police) is fully autonomous in its daily operations. It manages its own receipts and expenses. The determining of policy, port tariffs and the leasing prices of sites and terminals falls within the remit of the Board of Directors (part-timers). The same applies to the planning and promotion activities. The Board of Directors meets ten times a year; it has 7 members, two of whom are appointed by the federal level and one of these is the chairman. The members are chosen for their *independence* which means they do not directly represent a port actor (carrier, shipper, union, etc.). The Board delegates various powers to the general director, such as entering into contracts within the budget and selling assets up to a certain amount.

The CPC in Ottawa has considerable control over the Vancouver Port Corporation with regard to important strategic matters. For the purchase or sale of sites, projects costing over five million Canadian dollars, and signing contracts with a duration exceeding twenty years, the approval of the federal port authority and the Minister of Transport is required. The annual financial report is audited and approved each year by an accountant designated by the federal level. In accordance with the provisions in the Financial Administration Act, a financial audit is carried out every month and every five years an extensive internal audit takes place into the efficiency of VPC. In short, the emphasis lies on financial controls.

The revenue of the VPC comes from the leasing of sites and terminals, issuing licences, port dues, berth dues and wharfage, levies on cruise ships and profit on invested assets. The breakdown for 1995 was: 32 million dollars in leasing revenue, 28 million in tariffs, 4 million in investments and 1 million from other sources. Per year, VPC pays about 30% of its net income to the federal government which is the only shareholder. Between 1986 and 1994, VPC contributed 143 million Canadian dollars towards the joint federal fund;

54 million dollars in the form of dividends and 89 million in the form of special payments. It is these latter payments in particular which the VPC wants to do away with because the federal hand is being put ad hoc into the VPC till. During this same period, however, the federal level cancelled 103 million dollars in VPC debts. For new initiatives, such as the current construction of the container terminal at Roberts Bank, the federal government acts as guarantor for VPC (Vancouver Port Corporation, 1995b:6).

The 1995 Report: 'A National Marine Strategy'
In May 1995, in response to numerous discussions with the nautical sector and various ports in the country, the Standing Committee on Transport presented a report entitled '*A National Marine Strategy*'. There were a number of reasons why this report was compiled. In the first place, after twelve years there was still no national port policy and there was also no national coordination between the different types of harbours in Canada (port commissions, public ports and private industrial ports). Due to regional political interests, this even led to *competition* between Canadian ports. The most glaring example is the construction of an underused container terminal at Fraser River which is run by a Harbour Commission and which competes with similar facilities at Vancouver. Fraser River, however, is not a natural harbour for the larger container ships so that dredging costs were also incurred.

A second reason for publishing the report was the intense dissatisfaction felt by the LPCs regarding the functioning of the CPC. It should act as an advisory agency and facilitate in decision making on port development, but in fact the opposite is the case. Almost overnight a federal bureaucracy of 120 people was created providing services that no-one wants. This dissatisfaction has prompted the port of Vancouver to reduce the 2.2 million dollars which it paid to the CPC in 1994 to 1.5 million dollars. For the port of Halifax, for example, this contribution (including CPC police) represented 19.6% of its total expenditure in 1994 (Standing Committee on Transport, 1995:5). In the third place, the report was produced in response to distinct dissatisfaction with the pilots' organisation. And finally, there was a desire to start up a discussion on the introduction of a profit principle for nautical service provision by the Coast Guard.

The general conclusion, on the basis of which the proposals were formulated and on which the report is built, is: 'What has evolved over the years is an uncoordinated collection of federal entities which are all trying to accomplish similar objectives under different sets of rules' (Standing Committee on Transport, 1995:6). For the national ports, in particular, contrasted with the greater autonomy of the regionally-oriented Harbour Commissions, this was sometimes frustrating. After all, the Harbour Commissions could negotiate directly with the Ministry of Transport whereas decision making for the LPCs always had to take place via the intermediary CPC. Moreover, this also meant that CPC was in full control of the annual budgets and corporate plans. In the third place, Harbour Commissions are able to expand or reduce their port area directly, whereas the LPCs need federal approval for such matters. On the other hand, however, it is easier for LPCs than for Harbour Commissions to lease federal land and LPCs can display greater price flexibility because the Harbour Commissions need the approval of the public authorities concerned.

The recommendations of the Standing Committee on Transport were finally expressed in the Bill presented to the *House of Commons* (Bill C-44) that in November 1996 was just entering its second session. If the Bill is adopted, it will mean fundamental changes in the port management of the Canadian national ports. In the first place, the federal Minister of Transport will no longer select all the members of the Board of Directors. The surrounding municipalities and the province will each get one delegate who must not already hold any other public function. In Vancouver this means that the prairie provinces may also appoint a delegate. The Bill clearly states, however, that port users may not have a representative on the Board which goes some way towards safeguarding the principle of independence. Furthermore, the chairman is no longer selected by the Minister of Transport but is chosen by the members themselves. The maximum number of members on the Board has been increased to 11 (formerly 7).

The Bill expressly states that the national ports should be commercial and that, with the exception of a few articles, the Financial Administration Act no longer applies. On the other hand, the port authorities will henceforth be governed by the *Canada Business Corporations Act*. The port authority's accountant will carry out a special audit into the efficiency of the company at least once every five years and *publicise the results*. The minister may decide to have the audit carried out by a second accountant. The only condition attached to taking out a loan is that federal property may not be used as collateral. There is also an arrangement that the new port authority may make its own reserves. A lot of procedural obligations are thus cancelled and the ministry regulates far more ex post and on the basis of financial criteria.

In the new situation an annual payment will still be made to the national level, the size of which will be determined by the federal Minister of Transport. The payment will no longer be allocated to the Canada Ports Corporation because this body will be disbanded. Its property will accrue to the Ministry of Transport.

In the Bill, the task of the port authorities is clearly described: 'the power of a port authority to operate a port is limited to the power to engage in the port activities of shipping, navigation, transportation of passengers and goods and handling and storage of goods, as well as activities necessary to support

port operations' (art. 24(2) Bill C-44, p. 15). In this context, the national port authorities may also participate in the construction and maintenance of railways within their own district and they may even participate in a broad sense in transport activities which relate to the port (art. 25(1b+c)).

The democratic accountability for the past and future port policy pursued is embedded in the provisions which deal with the organising of an annual meeting for the parties concerned and port users. Linked with this is the provision that spatial planning programmes must be drawn up for the port and these must also be made public. For Vancouver, these provisions under the Bill are already used in practice; close relations were already maintained with the hinterland and surrounding municipalities and a spatial plan for the port had already been set down in the Land Use Management plan (Vancouver Port Corporation, 1994).

The national port authorities remain responsible for nautical management and safety in the port. They can draw up their own rules for regulating shipping traffic in the port. However, the federal level determines the content of the rules with respect to *public order, the environment and the control of dangerous substances.* The Minister lays down the requirements for the enforcers and issues certificates or licences.

If the new Act comes into force, the name Vancouver Port Corporation will disappear and it will become *Vancouver Port Authority.* This public port authority will be able to operate far more autonomously than is the case under the present institutional format. The involvement of the CPC will disappear and there will be far less stringent requirements with regard to taking out loans and making investments. It remains to be seen whether the formal involvement of various stakeholders and the obligations with regard to the financial reporting and control will induce this regional legal monopolist to furnish accountability.

7.4.   Nautical Management

10,000 ships per year put in at the port of Vancouver. The exclusive responsibility for nautical management in Canada lies with the Canadian Coast Guard which is a division of the Ministry of Transport. It provides services in the framework of the navigation system (navigational aids, dredging shipping lanes and traffic services; over 50% of the budget), icebreaking, shipping traffic regulation and 'search and rescue' activities. The Coast Guard's annual budget is around 580 million dollars of which up to now about 5% has been passed on to the actual users by means of the profit principle. In the future, the intention is to increase this percentage but clarification is first needed as to exactly what the Coast Guard spends its money on. Until now it has functioned under the conditions of a purely public service.

The VPC has its own Harbour Master who utilises the Coast Guard's Vessel Traffic System. The Harbour Master and the Coast Guard generally work together closely. The division of authority between the two organisations is between internal and external port business but in the event of serious accidents the port authority calls in the Coast Guard. The federal police of Ports Canada are also active in the port 24 hours a day, upholding safety and order. As mentioned earlier, Vancouver makes a financial contribution to this organisation.

The port of Vancouver is a very deep, natural harbour with no heavy industry which means there is little risk of environmental problems. From the viewpoint of service provision, Vancouver employs two people for environmental affairs. For a number of years, environmental projects have been tackled jointly with the federal environmental ministry and the surrounding municipalities. For example, among other things the VPC has made sites available for a nature park.

7.5.   Port Planning

On behalf of the federal government, Vancouver Port Corporation manages the area that formally belongs to the port. The Board of Directors has a separate commission for port planning and development which makes recommendations to the board. Given the fact that the VPC still needs its approval for the purchase and sale of sites, this area is formally the property of the federal government. The primary task of the VPC, however, is to commercially develop the area and keep it attractive which is no easy task in view of the increasing competition from the American ports of Seattle and Tacoma. Until now, it was customary for the VPC to have to submit its long-term plans for approval to the Ministry of Transport and Finance via CPC. For major expansion plans in particular, such as the construction of the container terminal at Roberts Bank to the south of Vancouver, this meant additional rounds of decision making which tended to cause delays. On the other hand, the fact that the VPC is an extension of the federal government means it can get cheaper loans and that the federal government will stand surety for any miscalculations. This is also why the VPC wishes to remain a federal agency in the future (Vancouver Port Corporation, 1995b).

Until five years ago the VPC was able to develop the port in relative isolation from other interests. In fact, as a functional organisation it was wholly autonomous. It was only at the Board of Directors level that any input from the local level took place. With the region around Vancouver becoming increasingly densely populated and with social problems escalating (unemployment, criminality, etc.) for the last five years or so this has no longer been possible and the VPC feels that it is no longer desirable, either. There is now a growing awareness by the VPC that for the development of the port it is dependent on the actors who are involved in the port for whatever reason. 'The key to the successful operation of the Port of Vancouver is interdependence. Stakeholders - labour, terminal operators, shipping lines, railways, truckers, etc. - rely on each other to perform their respective roles effectively and at competitive cost. In this sense we are like a chain - as strong as our weakest link' (Vancouver Port Corporation, 1995b:7). In 1995, the port authority 'voluntarily' donated 7.5 million dollars to the surrounding municipalities because the VPC does not have to pay any local taxes.

In its long-term development plan PORT 2010, which includes advisory input from the various stakeholders, VPC has created a framework for the spatial development and design of the port (Vancouver Port Corporation, 1994). The plan functions more or less as an agreement between the VPC and the region on what will be developed or expanded and where. The laying down of this plan was of great importance to the surrounding municipalities because although they have no formal control over the land use they now at least know where they are. In preparing the plan it became clear that there was a section of the port (Maplewood waterfront area in North Vancouver) which had to be kept in its natural state for environmental reasons. The management of this area was thus transferred to the national environmental ministry (Environment Canada).

In order to give structural form to the contact between the port and the region, for a number of years now various people in the VPC organisation have also been designated to act as permanent '*municipal liaisons*' for the surrounding municipalities. In this way direct communication on new developments in the region takes place. The liaisons make recommendations to the Board of Directors. The Board itself also has a separate committee for Public Affairs because there are eight separate municipalities which adjoin the water of the port. The creation of the liaisons has ensured that people now know who is responsible for what within the various municipalities which means matters can be settled more quickly. *Public meetings and public forums* are also held to provide information to the port community.

Formally, the VPC is still able to develop port planning autonomously but for reasons of mutual dependence (linking up with other transport infrastructures and the goodwill of the various municipalities) it does not actually do this and allows scope for other views. For a number of years, for example, the amassing of ideas for new initiatives has taken place during regular consultation with the stakeholders concerned. 'In addition to regular contact on specific projects, VPC meets four times yearly with the mayors and city managers of the municipalities to foster good communications and understanding of the Port's activities' (Vancouver Port Corporation, 1995b:7). Specific projects relating to the competitive position of the greater Vancouver region are taken up in the *Greater Vancouver Gateway Council*. This advisory council, which was set up in 1994, makes recommendations to the various levels of government on policy proposals in the field of transport and regarding the attractiveness of the region. 'Its sole purpose is to ensure the on-going development of Vancouver as a key North American gateway to the Pacific Rim. The group identifies and works to resolve any problems. The impact of taxation on Canada's railways has been one area of concern, and the Gateway Council has worked closely with the Province of British Columbia towards solving an on-going problem in the assessment of property taxes on railway main line trackage' (Vancouver Port Corporation, 1995a). In addition to VPC, other groups participating in the Gateway Council are the British Columbia Chamber of Shipping, employers and union representatives, the major airline companies, the manager of Vancouver airport, the railways, road transport, the dockworkers and the Fraser River Harbour Commission. As a result of the participation of the port of Fraser River in particular, the coordination on port development actually occurs at a regional level rather than a national level. The VPC also regularly puts a lot of time into the shippers from the prairie provinces. It organises meetings in these provinces to exchange ideas about specific problems concerning the port. Basically, these meetings are partly an arena for marketing the port and partly for airing ideas for improvements. The individual municipalities and the province do not contribute financially to the development of these projects, however.

The PORT 2010 plan opted to elevate the container, timber products and cruise ships sectors to economic spearheads in the nineties. In concrete terms, this involved an investment of 230 million dollars over five years for the construction of a container terminal on Roberts Bank which was completed in late 1996. The VPC contributed 180 million dollars towards this project and the remaining fifty million was financed by the private terminal operator (Empire Stevedoring) and the two railway corporations. The second project is the renovation of the Ballantyne Pier which will serve both as a distribution centre (for pulp) and a cruise terminal. VPC has invested the sum of 49 million dollars in this project. The third project concerns the expansion of the Lynnterm which specialises in the handling of pulp and lumber (total cost to VPC 13 million dollars). In 1995, VPC took out a loan of 139 million dollars to cover the funding of these projects. This was the first time since 1970 that the VPC had had to take out a loan to make investments. Until then it had been financially *self-sufficient*.

7.6.   Port Services

The port of Vancouver is a commercial port. This means that as far as possible all activities are managed and operated by private companies. This is not the case with the pilotage of ships, however, to many people's annoyance. The towing of ships is in fact also carried out by a private monopolist but this has not (yet) led to dissatisfaction among the shipping companies. With regard to goods handling, the port of Vancouver has a large private share but the VPC definitely intervenes.

The Pilotage of Ships
In Canada, the pilotage of ships is broken down into four separate regions and categorised under four different organisations[2]. The organisation for Vancouver is *British Columbia Coast Pilotage* which is an executive component of the Pacific Pilotage Authorities (PPA). The PPA is empowered under the Pilotage Act of 1972 to lay down rules concerning: 'the establishment of compulsory areas, the prescription of ships or classes of ships that are subject to compulsory pilotage and the circumstances under which this may be waived, and the classes of pilots' licences and pilotage certificates that may be issued, as well as the qualifications and examinations required to acquire same' (Standing Committee on Transport, 1995:22). Apart from the tariffs which must be approved by the Minister of Transport, the PPA basically regulates its own market. The PPA is a Crown Corporation which is expected to be self-sufficient, but there is an arrangement that any losses will be met by the federal government. The main problem with the present institutional structure is that whereas the pilots have been able to use their monopoly power to raise their tariffs, other port parties have been forced to keep the costs down and this has caused a good deal of aggravation. Moreover, the idea of granting exemption to captains who regularly put in at the same port has never really got off the ground because this would mean that the port was acting against its own interests. For the time being, the pilotage of ships as such is not up for discussion by the Standing Committee on Transport. Even though there is an increasing amount of increasingly modern communications equipment, this remains supplementary to the present pilotage profession. The 1995 report contained a number of proposals to make the pilots more cost-conscious, ranging from privatisation to having them join the port authority, to appointing an arbitrator or introducing certification.

---

[2] There are four pilotage regions for the whole of Canada: the Atlantic, the Laurentian, the Great Lakes and the Pacific Pilotage Authorities.

Operating the Terminals
Over sixty percent of the terminals are under private ownership. They only lease the site from the VPC and pay for licences. By means of the leasing contracts, the VPC has a number of options for laying down conditions for the use of the site or the terminal. The remaining 40% of the terminals are owned by the VPC which leases them on the basis of commercial criteria to private terminal operators. The normal procedure is that when a client presents himself, the VPC develops the chosen location in accordance with the wishes of the client. The actual work of preparing the site is contracted out to private companies under the supervision of the VPC. This generally involves developing the site for construction or building a quay wall. In a number of cases, however, the VPC also finances the terminal equipment (cranes and so forth). This is particularly the case with the container terminals, the cruise terminal and the terminal for timber products. The reason for undertaking the (almost) total construction is linked with the tremendous *initial costs* of such a terminal for which private actors cannot raise the money. Another reason lies in the fact that these are public terminals at which any shipping company can moor although this in turn is inherent in the cargo flows which put in at Vancouver. The actual terminal handling is contracted out by VPC to a private provider. This private provider is not selected by putting out a tender but by application in response to an advertisement. VPC's primary concern is not cost benefits but the quality of the service provision and reliability. The establishment of *service contracts* is the key element in this and not the direct leasing of the sites and the cranes. In the service contract, the VPC may profit from the land price but it can also employ it in a competitive way. The operation of the container terminal can thus be supported with the profits from other VPC activities (cross-subsidy).

Unions
The dockworkers are represented in the *International Longshoremen's and Warehousemen's Union* which is a full partner in the Greater Vancouver Gateway Council. The VPC employees are also represented in this union via their own section (Local 517). All but 40 of the VPC staff are union members. There are good relations between the VPC and the union; union members are regularly taken abroad by the VPC to enable them to enter into direct contact with foreign shippers and carriers. 'Some people with long memories believe that the labour situation at the Port is lacking in stability and efficiency. In truth, the reality is far better than the perception. [There are, for example, 41 collective agreements in the Port's jurisdiction. Over the past 9 years, only 22 days have been lost to labour disputes with the I.L.W.U., and the foreman.]' (VPC, 1995b). Since 1987, 28 joint training programmes have been created

between the employers and employees. These include, for example, simulator-training for container cranes.

An example of a situation in which the VPC was not clealy aware of its position and in which its limited managing capacity came to light, was the longshoremen question. For a brief period, the port of Vancouver operated a 'container clause' (Ircha, 1993:60). This stated that it was compulsory for containers which were destined for or had come from an area within a 90-mile circumference of the port to be stripped and stuffed by the Vancouver longshoremen. This provision led to a substantial loss of freight to Seattle and Tacoma and at present the VPC operates a positive incentive scheme to try and attract containers again.

7.7.   Port Management in Canada: Analysis and Questions

Canada has an economy based on the market mechanism. The Canadian ports have to be directly commercial (read: self-supporting) whether they perform a national or a regional and/or local economic function. For reasons of defence it was determined in the past that the responsibility for the organisation and functioning of the ports should be a federal matter. This division of responsibilities is formally embedded in the Canadian constitution. But up to now a nationally-coordinated port policy has not materialised, whatever its organisation. And despite all the frustration with the federal approach, highly profitable ports such as Vancouver also continue to cling to the federal coordination principle. The sole reason they have for this is the fact that in times of economic decline the federal level will foot the bill for the financial risks. Apart from this, the ports would like nothing better than for the federal level to keep out of their affairs (except for direct financial controls by the Ministry of Finance).

On the basis of the regulatory principle that Vancouver is a commercial port of national interest it may be expected that, with the exception of nautical management, all other port activities will be undertaken as far as possible by the market. Even port planning could be undertaken by the market since the initiative for making a terminal operational lies with a private client. After all, the VPC is only going to develop a site once a client has presented himself.

If we study the port empiricism in Vancouver it is apparent that the nautical management operates entirely in accordance with the expectations of the Canadian Coast Guard but that the port planning for the spearhead sectors of containers, cruise and timber products are designed along different institutional lines. The VPC facilitates the construction of these terminals by building the superstructure itself and negotiating service contracts with a stevedore. Moreover, the VPC has been undertaking the broader port planning activities on its own for years. For the last five years or so there has been more direct contact with shippers and the surrounding municipalities. It also appears that some port services are not offered in competition but enjoy highly exclusive rights. Pilotage activities provide the clearest example of this.

A number of discrepancies exist between norm and empiricism; on closer examination, the port of Vancouver is not as commercial as it appeared at first sight. There is very little payment for the Coast Guard service provision both to the ships and also to the VPC. The VPC used its own risk-bearing capital to invest in the construction of a container terminal and the restructuring of the cruise and forest terminals. For the leasing of a container terminal on Roberts Bank the site and the terminal were not directly leased but a service contract was negotiated with a stevedore. The VPC thus retains control of the speed with which it wants to recover costs. Its primary concern is service provision; commercial arguments are not always at the forefront. The VPC can compromise because it has income from other sources and is aware that ultimately the federal level is backing it. The construction of its own terminals and the leasing of these to private stevedores is in fact the only solid instrument that the VPC possesses with which to actively deal with port planning. As regards making the port as a whole attractive, the VPC has become far too dependent on other actors in the port community. This is why substantial investments are being made in such things as marketing and public affairs.

The fact that Canadian port management is not as commercial as it might seem has its origins in the regulatory principle of federal responsibility. The formal regulatory principle of national responsibility is essentially at odds with the economic regulatory principle of the market. Canada's economy is increasingly intertwined with that of America where this type of federal responsibility does not exist. The port of Vancouver is also encountering increasing competition from the American ports of Seattle and Tacoma which have no federal safety net to catch them if they get into financial difficulties. Moreover, the principle of federal responsibility appears to have been superseded now that the coordination of port planning in fact already takes place at the *regional* level, as the case of the Greater Vancouver Railway Council illustrated.

7.8.   Summary

Port management in Canada is considered to be a federal responsibility; the task is embedded in the constitution. The fact that it was not easy to achieve the desired administrative coordination and to do justice to the specific nature and economic function of individual ports was made clear by the committees of inquiry which have been regularly set up over the years. Fairly recently

(1995) proposals were made to formulate a redefining of the federal and local responsibilities.

The port of Vancouver in southwest Canada mainly handles bulk cargo and cruise passengers. In recent years, tremendous efforts have been made to increase its container share for Asia. Due to the national-economic importance attaching to the port, it is managed by a federal agency that must operate in a self-sufficient way. Although Vancouver in particular has argued in favour of increased financial control and policy freedom, the port authority continues to cling to the federal relationship to have insurance against economically uncertain times.

It is striking that the port authority has recently devoted considerable time and effort to foster goodwill towards citizens and companies. Another striking point is the service contracts which have been negotiated with terminal leaseholders; it is not profit which is the primary factor but the reliability of the port.

CHAPTER 8

Japan: the Port of Kobe

'In the coming era of continued stable economic growth, there will be no big change in the primary role of the ports. Ports will continue to be regarded as indispensable social capital by which we can realize a very stable and prosperous national life and economy' (Japanese Ministry of Transport, 1995a).

8.1. Introduction

On 17 January 1995, the Japanese city of Kobe was partially destroyed by the major Hanshin-Awaji earthquake (7.2 on the Richter scale). More than 5,000 people were killed, 240,000 were made homeless, virtually all the telephone lines were down, the electricity, gas and water supplies were cut off and only a fraction of the 116 kilometre-long quay survived the earthquake undamaged. It may be wondered why a researcher would want to visit a port in these exceptional circumstances a year after the event. The purpose of this study is to gain a clear picture of the ways in which ports are managed. In times of disaster and reconstruction in particular, authority and dependency relationships are more clearly visible than under normal circumstances. After all, the necessity for rapid reconstruction is clearly felt because valuable market shares could otherwise be lost. A typical feature of the reconstruction of Kobe is the way the mayor was publicly blamed for the fact that the aid was so badly organised and that it took so long before aid operations could begin. There are two reasons for this: on the one hand an earthquake in this part of the country was totally unexpected and on the other hand there is the idiosyncratic way in which the political arena, administrative authorities and the business community are interlinked so that sudden unexpected events cause a *power vacuum* (cf. Chen, 1995). It is this last element which makes it very interesting to include a Japanese port in our comparative analysis.

The port city of Kobe lies in the Osaka Bay area and until the earthquake in 1995 it was the most important (container) port in Japan. As a sister port it has maintained close ties for a number of decades with the ports of Rotterdam and Seattle. The institutional structure is dealt with in the same way as in the previous chapters; first of all, Japan's specific economic regulation is described and the general and specific importance of the port for the country is elaborated. The third section focuses on the organisation and powers of the port authority and the relationship with other port actors. The fourth section

207

explores various nautical management activities. This is followed in section 5 and 6 by a look at port planning and port services. Finally, the descriptions are analysed and questions posed.

8.2.   Economic Regulation and Function of Japanese Ports

For a long time foreign researchers thought that Japan was a market economy, as we in the West know it, but on closer inspection this appeared not to be the case. The political economy of Japan (and also countries such as Korea and Taiwan) can neither be compared with free capitalism nor with centrally controlled systems. The Japanese regime is more of a system of leadership by groups: the strength of the Japanese system, which is also its weakness, lies in the close cooperation between politicians, bureaucrats and industrialists. The chief reason underlying the creation of this arrangement, which dates back to the nineteenth century, lay in the fear that Japan would become a new colony of the West. The Meji dynasty thus decided to take on the development of the major economic sectors (mining, shipbuilding, communication technology, the arms industry and the textile industry) and, in the late nineteenth century, to sell them to the private sector at a bargain price. 'The large-scale transfer of major state-owned enterprises in 1880, the first attempt to privatize, paved the way for the formation of the large industrial and financial combines called *zaibatsu*' (Chen, 1995:152). From this time stem such famous names as Mitsubishi, Mitsui and Sumitomo.

At a later stage, thanks to the goodwill it had thus acquired, the government could bank on financial aid to subdue political opposition and to foster colonial expansion. When Japan was defeated after the Second World War, the former top figures in industry, who had been directly involved in the war effort, were replaced by a more democratic system and other industrialists. 'Although the family-dominated and vertically-linked zaibatsus were dismantled under the charge of war crimes, they were quickly replaced by large horizontally-linked *kereitsus*. These new groups were not strict monopolies because they competed against each other, but they together formed an oligopoly' (Chen, 1995:153).

After the war there was no fundamental change in the close relationship between the government and the business community. Defeat in the war was put down to a lack of the necessary technology and the economy which had fallen behind. With united forces and combined efforts an economic growth policy was initiated for the same reason as in the previous century with the government supervising progress. This provided important coordinating and initiating roles for the Ministry of International Trade and Industry (MITI) and for the Ministry of Finance which took care of subsidies and tax payments. All the policy measures were aimed at promoting domestic industry and hampering foreign industry. Examples include: trade tariffs, import and foreign currency restrictions, a taxation system that favoured domestic products, low-interest loans, tax exemptions, subsidies, a compulsory authorisation for the import of new technology using foreign currency and exemptions from import levies for essential machines and equipment.

On the pretext of the economic construction of the country and the protection of industry which was still in its infancy, Japan was able for many years to stave off foreign competition from the domestic market. Agricultural products in particular were neither effectively nor efficiently produced and over time became very expensive. Increasing internationalisation, however, also requires an increasingly open attitude from Japan vis-à-vis foreign currency. The international community is becoming increasingly intolerant of official and unofficial competition restrictions.

In the way it thinks and acts, Japan is at times still very much an island. Until 1868 it was able to effectively close itself off from the outside world and the country has always tended to vacillate between an open and an isolationist position. With the development of telematics and the international financial integration of Japan's capital, the possibilities for isolating itself from the outside world are rapidly decreasing. Moreover, Japan is undergoing increasing competition from Korea, Taiwan and Singapore which are entering the same sort of markets that made Japan so successful after the Second World War (cars, consumer electronics, shipbuilding, photography, etc.). After a boom which lasted for many years, Japan's economy has recently started to show signs of decline. Former foreign investments are being sold off, and even the yen has taken some hard knocks. In the years of consolidation and economic growth, the sustained fraternisation in numerous co-arrangements between government and the business community appeared a good institutional concept, but it remains to be seen whether the Japanese economy will not have to be completely restructured in the face of the continuing advance of international competition. It may be expected, however, that the retreat of government will be a very slow process while the links between industry and the authorities remain so close (Chen, 1995:164).

Significance and Role of the Ports
Geographically speaking, Japan consists of two large islands surrounded by a large number of small islands. There has always been shipping traffic to enable the transport of goods between these islands. The way in which the administration of the Japanese ports is organised depends on the *economic function* earmarked for the port concerned. In total there are 1,100 ports (excluding 2,950 fishing ports) which fall into a number of categories under the 'Port and Harbour Law'. There are 21 '*Specially Designated Major Ports*' (including Kobe) which play a special role in the promotion of foreign trade.

The second group consists of 112 '*Major Ports*' which play an important role in the national economy and finally there are 968 ports which come under the heading of local harbours (some 35 of these harbours simply provide shelter for smaller ships and there are 68 others with no special function at all).

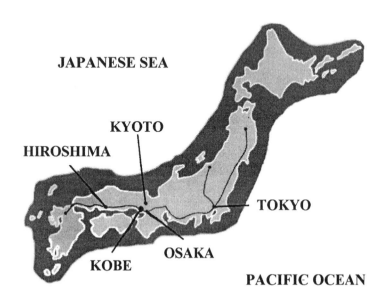

JAPANESE SEA

KYOTO

HIROSHIMA

TOKYO

OSAKA

KOBE

PACIFIC OCEAN

The Position of Kobe in Relation to other Japanese Ports

The ports are categorised according to their economic function. Other public bodies are involved in port administration depending on the function. A complete overview is provided in Table 9.1.

Table 8.1. Forms of Port Administration in Japan
(Source: Japanese Ministry of Transport, 1995a:1)

Port type	Prefecture	Municipality	Port Authority	Cooperative System	Other	Total no.
Specially designated ports	10	8	0	3		21
Major ports	90	19	1	2		112
Local ports	524	376	0	0	68	968
Total	624	403	1	5	68	1101

Of the twenty-one major ports, six are situated on the bays of Tokyo, Ise and Osaka: Tokyo and Yokohama, Osaka and Kobe, Nagoya and Yokkaichi. These ports, which are located in densely populated, industrial areas, not only perform a terminal function but also an industrial function. They also play a role in the storage of energy stocks or waste, in the redevelopment of the city and in recreation (marinas; Japanese Ministry of Transport, 1995a)[1]. By constructing two man-made islands off its coast, for example, Kobe has promoted the development of both the port and the city. In Japan, ports form one of the essential economic preconditions, along with electricity, gas and water, to enable industry and trade. After all, Japan has no raw materials or other natural resources of its own, so much of this must be imported and subsequently exported. Partly as a result of international development, ports will start to function increasingly as industrial production centres. 'Ports will continue to be regarded as indispensable social capital by which we can realize a very stable and prosperous national life and economy' (Japanese Ministry of Transport, 1995a:i).

Table 8.2. Multi-year Overview of the Major Japanese Ports (Source: H. Stevens based on Institute of Shipping Economics and Logistics, 1991 and 1996. Figures in million freight tons).

Port	1985	1990	1994
Kitakyushu	92.0	95.1	97.1
Kobe	160.4	171.4	171.0
Nagoya	112.8	128.9	137.2
Osaka	86.1	97.3	91.1
Tokyo	59.5	79.3	77.9
Yokohama	115.9	123.8	128.2

The principal feature of Kobe's economy is the presence of port-related industry such as the iron and steel industry, shipbuilding, the shipping branch and the distribution sector (amounting to 39% of Kobe's Gross Industrial Product). Foodstuffs, beverages and electrical equipment are also produced for the (inter)national market. The importance of iron and steel production for the construction of vehicles has decreased in recent years while machine construction and the manufacture of electrical equipment has increased. The investment of 900 million guilders in Kobe High Tech Park is an attempt to attract high-quality labour. Another feature of Kobe's economy is the abundance of small and medium-sized enterprises. Some of the industries represented in Kobe have captured the major market share among which is the clothing industry. Kobe is the fifth-largest city in Japan and has a population of over 1.5 million. 17% of the total employment is located in the port (the service sector forms 70% of the total; Economic Bureau Kobe City Government, 1993).

In the last decade the import volume via the port of Kobe has increased (measured in tons) while the export volume has remained more or less the

[1] The surface area of Japan is 380,000 km^2. Only 30% of the country is suitable for residential and work purposes for the over 120 million inhabitants.

same. A port city like Kobe is also gradually coming to realise that labour in Japan is more expensive than elsewhere so that industries are being transferred abroad. The inward and outward volumes (national trade) have followed the same pattern as that of the import/export volumes. The latter could indicate greater competition from other Japanese ports. The exact ratio as regards the value of the goods is not known, but it is expected that these will show a similar trend. The declining economy of the last five years has certainly contributed to this (disadvantage of the hard yen and more competition from other Asian countries); there are relatively fewer foreign business transactions and Japan remains dependent on foreign raw materials and semi-manufactured products.

In 1994, (the year before the earthquake) Kobe handled nearly 3 million TEU. The total transhipment of goods from abroad and from other Japanese ports amounted in 1994 to just over 171 million freight tons. More than 55 million tons of this came from or was destined for foreign markets. In late 1995, Kobe was already back to handling 70% of its original freight volume. In the first half of 1996, Kobe's freight volume was back in second place behind that of Tokyo (Lloyd's List, 7 November 1996). The following Table shows a list of Kobe's major trade partners. In the text, figures for 1994 have been consistently used since the use of figures for 1995 and 1996 would not do justice to Kobe's potential as a port.

Table 8.3. Port of Kobe's International Trade Partners in 1994 (Source: Port and Harbour Bureau Kobe City Government, 1995:7). *measured in freight tons.

International trade partners 1994	Export (in %)	Import (in %)
Asia	59.4	49.7
North America	24.4	34.2
Europe	7.3	5.6
Oceania	4.0	4.7
Africa	1.8	1.2
Middle East	1.7	1.9
South America	1.4	2.7
Total	25.860,000 ton	29.368,089 ton

A remarkable fact about the port of Kobe is that approximately one third of its total transhipment derives from foreign trade (transported by 10,836 ships in 1994); the remaining two-thirds is domestic trade (76,872 coasters). Kobe thus performs not only an important international but primarily a national function. In addition to this, the port has always played an important role for the city of Kobe in regional economic development. It has become even more highly integrated with the city as a result of the construction of two man-made islands (Port and Rokko Islands, 436 and 580 ha. respectively). Container terminals have been constructed along the edges of these islands while inside this fringe, urban activities have been set up for such things as housing, offices, shopping centres, etc. There are plans to construct another island just off the coast of Kobe to house a regional airport which was to have been completed in 1998. Due to the earthquake this project has been delayed, however.

The construction of these man-made islands in the sea uses a sustainable method. By levelling the mountains behind the city ('bringing the mountain to the sea') and dumping the stones off the coast, Port and Rokko Islands were created (Port Island cost the city 9.5 billion guilders). The mountain is quarried, the stones are transported via an underground conveyor belt to the shore where a terminal has been set up to unload the stones onto dumb barges. The conveyor belt runs 24 hours a day and is a unique system in the world. The dumb barges are then brought into position and the stones are discharged by simply releasing the bottom of the barge. At the same time, hoppers spray sand around the man-made island. The conveyor belt system is a permanent feature for the present because the airport is also in the pipeline. Moreover, houses and other facilities can be built on the mountains which have been levelled in this way. For Japan, this is a two-edged sword in a positive sense; houses on the island and houses in the mountains. After all, 120 million people live in Japan on an area 16 times the size of the Netherlands, but because the country is so mountainous the area available for residential purposes is in fact only twice the size of the Netherlands. This means that expansion into the sea not only serves an industrial-commercial purpose but also creates living space.

Competition with other Ports

Prior to the earthquake, the port of Kobe was the largest container port in Japan. At present, it is undergoing strong competition from the ports of Osaka (located on the same bay), Pusan (South Korea), Shanghai (China), Kaoshiung (Taiwan) and Hong Kong. Transhipment cargo from China for the US is particularly under fire from the various ports. In this period of reconstruction Osaka has emerged as a domestic competitor because in the Kinki district a lot of cargo lies closer in both in distance and time to Osaka than Kobe and it provides the only opportunity to permanently snatch cargo from Kobe. Once shipowners are located in a particular port it is difficult for Japanese ports to entice them away again. In fact this is only possible with new clients. 'Once settled, the major lines seem to become "glued" in place by their labour agreements, making it very awkward to move about unless their circumstances undergo a major change. Consequently the hardest line for Kobe to attract, for instance, is one already settled in Osaka' (Containerisation International, 1986:65). Shipowners like low prices but they also value relatively fixed schedules and it costs extra effort to change a route around after two years. This is why it was considered a must to get the port of Kobe back on its feet again as quickly as possible because of the ever-present danger of privateers (in a literal sense). Kobe appears to have succeeded in its aim as far as cargo volume is concerned; in the first half of 1996, the cargo volume of Osaka decreased by over 20% (Lloyd's List, 7 November 1996).

8.3.   Port Organisation and the Role of the Port Authority

A number of administrative functions at the port of Kobe are controlled by various (quasi-)government bodies. The *Kobe City Port and Harbour Bureau* (PHB) operates at the local level and is responsible for the planning and construction of new port components, the management of public berths and the marketing of the port. The *Kobe Port Terminal Corporation* (KPTC) is also active at the local level and is responsible for the management of those container and cruise terminals which have an international character. At the national level, the Ministry of Transport (MOT) is involved in the port of Kobe in a number of ways. As harbour master through the *Maritime Safety Agency*, as coordinator of port planning in various ports through the *Ports and Harbours Bureau* (civil engineering) and the *Maritime Transport Bureau* (for transport technical purposes). The Ministry of Transport has various district offices which function as interfaces between the local and national level, to provide direct comment on Kobe's plans, to aid in the construction of harbour works or to oversee the pilotage and stevedoring of ships. The sections on nautical management and port planning will address the functions of the various bodies and the coordination between them in more detail. Furthermore, the Ministry of Transport has a Maritime Technology and Safety Bureau which devotes itself to the regulating of ship's equipment and the requirements which the crew must meet. Since this bureau concerns itself mainly with shipping itself, it will not be further discussed in this volume.

Role Division Between Public and Private
The general principle for the role division between public and private actors is that the various port authorities are not supposed to become involved in those port activities which the business community can perform just as well. 'Due to its unique "closed" history prior to 1868, Japan never really developed a port industry in the usual way. Always it was the shipper, whether feudal lord or trading corporation, which provided the necessary labour to load or discharge the ship, using his own or a public pier. For this reason, and also by law, port authorities have always tended to be an extension of the local or municipal authority, providing and administering the necessary common-user facilities but never becoming directly involved with more commercial aspects of port operations, specifically never employing stevedores' (Containerisation International, 1986:63). Because ports are part of the social capital of the country, the Japanese administration considers it more important that the ports

function well than that they should be fully cost-effective. 'Port development in Japan, however, is determined on the basis of its contribution to the social and economic development of both nation and region rather than on the basis of a direct return from port operation' (Japanese Ministry of Transport, 1995a:13).

8.3.1. POSITION OF THE PORT AND HARBOUR BUREAU KOBE CITY GOVERNMENT

It is only since 1951 that the port of Kobe has had a municipal port authority. Before then the Ministry of Transport was responsible for the administration and development of the port. The port is run by the Port and Harbour Bureau (PHB) whose director is appointed by the elected mayor (in consultation with the three deputy mayors). The approval of the municipal council is not required. In contrast to other municipal agencies, the PHB has its own budget, because as a municipal corporation it can specify its receipts and expenditure relatively easily. The municipal council, which is directly elected by the population of Kobe, approves the PHB's budget. The function of the PHB in fact boils down to the planning and construction of new port components and the management of the public berths. Public berths are leased per assignment. The present Chief Executive Officer has been employed at PHB for thirty years and so knows the company inside out. Although it is not essential that the CEO should come from inside the company, he is usually from a government agency. The CEO has no delegated powers. Once a week he meets with the mayor and the other departmental heads. Any policy freedom is not formally granted. A lot depends on the particular matter and the person himself. The CEO receives recommendations and proposals from his staff.

Within the PHB there are two categories of function: the management functions and the technical specialists; the managers who can be more broadly deployed are rotated every four years. This rotation is intended to increase the integration between the various departments. For the technical specialists the situation is different; they are less broadly deployable. Due to progressive quality improvements, however, it is becoming increasingly difficult to know a bit about everything, so that the system of rotation will also come under pressure. The PHB has 148 management jobs and 176 engineers. Moreover, it has placed a number of staff with other port organisations, such as the Kobe Port Terminal Corporation (43) which is responsible for the international container terminals and the Port Service Association (22).

The PHB's port plans are assessed by a *Local Port Council* under the provisions of the Port and Harbour Law. It acts as a compulsory advisory board on major harbour works and comprises a number of groups:

- the academic world and various citizens' groups;
- representatives of the municipal council;
- representatives of the private section of the port and the unions;
- representatives of other public bodies;
- municipal staff in the fields of environment, spatial planning, etc.

The local advisory council does not deal with financial matters.

The policy freedom of the PHB is more or less regulated under Article 8 of the Harbour and Development Law in which a particular *port area* is defined as such. There was much discussion with the Planning Bureau regarding the boundaries of this area. The boundaries were finally laid down in an ordinance and the functions specified per sub-area. The Planning Bureau is advised by the city planning council (cf. port council) and the PHB sometimes enters into direct consultation with the planning council. There are separate environmental guidelines involving cooperation with other departments.

Given the complex knowledge and huge investments needed for the construction of container terminals in the late sixties, the national government at that time determined to share the responsibility in cooperation with the local port authority and the business community. Until 1981, there was one separate management organisation functioning for the whole Osaka Bay area. After this time, the ports of Osaka and Kobe split up and established their own management organisations. For Kobe this became the Kobe Port Terminal Corporation in which national government, the municipality and the business community (the shipping companies concerned) participate in a 40%, 40%, 20% split respectively.

8.3.2. KOBE PORT TERMINAL CORPORATION

The Kobe Port Terminal Corporation (KPTC) is responsible for the construction and management of container terminals pertaining to foreign trade and the terminals for the various ferry services and cruises. It is a body that was set up with funding from the Ministry of Transport, the city of Kobe and private companies, and it operates formally independent of the PHB. The KPTC makes use of PHB staff, however, but is expected to generate its own income. The terminal corporation manages a number of larger container yards on Port Island and Rokko Island. This port body originated at the dawning of the container age when the national government felt that the individual ports were not able to meet the requirements of the container cargo trade. In those days it actively intervened in both the construction of the container terminals and in the acquisition of clients through the former Hanshin (Osaka & Kobe) Port Development Authority; in August 1981 this became the KPTC which only serves the terminals for Kobe.

In the early days, it was primarily allied to national government because this provided the bulk of its financing, but later this body has been manoeuvred increasingly towards the municipal bureaucracy through the exchange of staff and the provision of interest-free loans by the city among other things. KPTC still also relies on interest-free loans from the Ministry of Transport, from loans carrying interest from banks or from leasing to the city. Repayment is possible because KPTC itself receives revenue from the leasing of terminals. In principle, the leasing period is ten years and may be extended.

Due to the orientation towards foreign trade, a number of requirements must be met by the lessees of the KPTC terminals. In the first place, these terminals may only be used by shipowners who concentrate on foreign trade and who use the port of Kobe as port of entry, exit or transit. A second condition for the use of these terminals is that the shipowners must have a licence to participate in the general transport organisation of the port of Kobe. The company must thus be oriented towards port transport. For cruise ships only the first condition applies (Kobe Port Terminal Corporation, 1994).

Now that the terminals have long been realised, the management relationship between PHB and KPTC is advantageous to PHB. KPTC has had to surrender important tax advantages and PHB has overall planning authority in the port. Moreover, the port of Kobe has a total of over 239 berths of which 150 are public, 36 belong to the KPTC and 53 are private (for industry)[2]. KPTC only supervises the layout of specific berths for international private use. 'As the providers of each port's public-user facilities, the municipal authorities do, therefore, control the investment balance within a port. If too many private berths are built (...), leaving some without lessees, the option of redesignating them as public berths could seem attractive as a means of avoiding embarrassment. The local authorities would, however, be well aware of the attendant danger of subsequently making the public-user option so attractive that private-user investments were discouraged even further. It is always important to remember that the major port authorities are run by men whose background is in local government administration, mostly in departments unconnected with the port. For them, administrative solutions to commercial problems will invariably seem more manageable than anything designed to harness some of the forces of free-market competition' (Containerisation International, 1986:63).

In view of the fact that the private part of the port is run either by the KPTC or by the stevedore who uses the public terminal, the only creative tasks left to the PHB are plan preparation and maintaining good port promotion. To

---

[2] The private berths for industry (Mitsubishi, Kawasaki, etc.) were constructed partly with the aid of subsidies.

achieve the first aim, statistical information is compiled, studies are conducted and contacts are maintained with the major shippers and shipping companies regarding their future wishes. To achieve the second aim, business trips abroad are undertaken and in the past the PHB even had a London office.

8.4. Nautical Management in Japan

All facets of nautical management in Japan are implemented by the *Maritime Safety Agency of the Ministry of Transport*. The country is split into eleven separate nautical management regions and a decentralised department of the Maritime Safety Agency has responsibility for each of these regions. These agencies are wholly financed by the Ministry of Transport. The port of Kobe is located in the fifth region which covers Osaka Bay and for this area the Maritime Agency provides the harbour master and the appropriate technical control instruments.

The tasks of the maritime agency comprise more than just guiding shipping traffic, carrying out port entry and exit procedures or allocating berths landside or at anchorage. Its responsibilities also include:

- guaranteeing the safety of maritime traffic by means of its own patrol boats (Japan has a total of 119 ships available for this);
- allocating berths;
- granting permission for the handling of dangerous substances;
- guaranteeing the safety of leisure activities;
- saving people and ships at sea;
- protecting the sea environment and preventing disasters;
- enforcing peace and order (policing tasks);
- conducting studies of the ocean and
- installing and developing navigational aids (buoys, beacons, lighthouses, vessel traffic system, etc. (Japanese Maritime Safety Agency, 1995).

These activities are implemented within the framework of the *Port Regulations Law*. The agency runs its own training courses and, together with the business community, holds conferences on safety measures. On a national scale, 12,000 people work for the Maritime Safety Agency and in 1995 the budget was 160 billion yen. Collecting tonnage and special tonnage dues is the responsibility of customs.

8.5. Port Planning

The advent of the steamship in the mid-1800s forced the Japanese to think about the development of ports. Due to the traditional role division, there was no coordinated policy-related government behaviour. At the beginning of the steamship age there were no wharves or berths so that all cargo from the larger steamships had to be transferred to smaller boats. Until the introduction of the Port Act in 1952, the Japanese harbours developed in a very ad hoc way. After this time, the national government became actively involved in port development, the level of its involvement varying depending on the importance of a port's economic function. Due to the regional function of the '*specially designated ports*', at present both the local and national level are closely involved in port planning. Before the method of coordination, the financing and the planning are addressed, the formal position of the national Ports and Harbours Bureau of the Ministry of Transport will be examined.

8.5.1. THE COORDINATING ROLE OF THE PORTS AND HARBOURS BUREAU OF THE MOT

The coordination of port planning takes place at national level through the Ports and Harbours Bureau of the Ministry of Transport (PHB/MOT). The Ministry of Transport has devised a number of concrete tasks for itself:
- formulating a national port development policy and establishing the necessary laws and regulations for port administration and development;
- making recommendations to port authorities regarding port administration and development;
- studying and coordinating the port plans of major ports;
- financing port works (including via loans);
- carrying out port works itself (generally major, highly technological projects in cooperation with port administrations);
- developing and maintaining channels outside port areas and
- establishing technical standards for port planning, design and construction (Japanese Ministry of Transport, 1995a:15).

The Ministry has its own building department with an office in every major port and its own research agency which keeps track of the technological development in the ports. For the coordination of projects between ports, PHB/MOT assesses the port plans in the light of the general requirements of the national development policy. The requirements which the national level imposes on the local plans are in the field of:
- assessing the plan in light of the national master plan laid down by the Ministry of Transport;
- economic assessment: what will the investment yield in terms of national profit;
- technical criteria: does the plan meet the standards which can be set as regards the content of the port projects.

Since the different aspects of projects are evaluated, The Ports and Harbours Bureau must consult with other departments of the Ministry of Transport, such as the building department and the research centre. The PHB/MOT is also responsible for coordinating the competition between the ports of Kobe and Osaka. This is not formally regulated but there are three ways in which it may be elaborated:

■ via inspection of the general master plan of Osaka Bay drawn up by the district office;
■ via inspection of the Kobe port plan and
■ via the granting of subsidies.

The PHB/MOT coordinates both the content and financial features. The PHB/MOT endeavours to offset the cost variations (e.g the dredging costs) between the two ports chiefly by means of their financial contributions. 'Thus Kobe, in its favourable position as a "natural" port, effectively helps cross-subsidise the dredging costs of its competitor Nagoya, which suffers massive silting from the eight rivers which converge at the head of the Ise Bay. (...). Thus the main port authorities, each entirely subservient to local community interests, and very mindful of the expectations and constraints of the national government's policy, do not appear to have different overall objectives. Only the degree of emphasis which each local authority places on its port's success seems to vary' (Containerisation International, 1986:63). Thus the development of the port of Kobe is designed on the basis of various formal objectives: '1) improving the port as the foundation of Kobe's economy, 2) conducting more intensified port promotion, 3) making the Port more people friendly and 4) enhancing port workers' welfare, while implementing the construction of a manmade island (Port Island 2nd stage), the redevelopment of existing piers, the improvement and development of waterfront green belts' (Port and Harbour Bureau Kobe City Government, 1994:1).

8.5.2. PORT PLANNING PROCESS

The planning process starts with the PHB in Kobe drafting it within the limiting conditions of the existing national port policy and local plans. Once the draft plan is ready, it is assessed at local level by the Local Ports Council. If it is approved by the local advisory council, the plan is sent to the Ports and Harbours Bureau of the Ministry of Transport which then considers it with the help of its own advisory board, the Ports and Harbours Council. A number of people participate in the national advisory council: representatives from various Ministries (Finance, Agriculture, Economic Affairs, Foreign Affairs, land affairs and building works), sundry professors from diverse universities, representatives from the prefectures, the fisheries, the dockworkers, the seamen and major companies. Normally, it is the national Ports and Harbours Council which makes recommendations to the Ministry, but since the earthquake an additional special (temporary) committee has been set up.

A striking detail is that the media are represented on both councils, because they have such a strong influence on social life/opinions (as an instrument of the civil servants). The special earthquake committee is regulated by the research institute of the Ministry of Transport and comprises: professors, press representatives, shipping companies, a bus corporation from Yokohama and two retired entrepreneurs. If the councils and the Ministry are satisfied, the plan can be approved. It is difficult to determine in advance how long such a process will take. It depends on how good the port director of Kobe and/or the mayor are at obtaining funds, i.e. whether they have the right people in the right places, and then it still depends on getting the right donations at the right time.

Once the plans have been approved, a project can be implemented. The way in which this takes place depends on the user's rights that will apply after completion of the project. Three different types of facilities are distinguished:

a) *basic facilities for general use* (water facilities, protective facilities, mooring facilities, harbour transport facilities) which are implemented as public works with national government subsidies, either by the city of Kobe or the Ministry of Transport;
b) *basic facilities for specific users and other facilities for general use* must be constructed by KPTC and the construction costs split in accordance with the division of shares within KPTC;
c) *private facilities for specific users* must be constructed by the users themselves.

Depending on the type of construction work, either PHB or the Ministry is responsible for the costs. Public terminals, for instance, are paid for by the Ministry and PHB, but PHB gets to net all the profits. In the case of a private lessee of a specific section of wharf, then it is the Kobe Port Development Corporation which is responsible and it nets the profits from this. Finally, the storage warehouses and container cranes located on the quay are entirely financed by PHB (cost of a container crane: 1 billion yen). The 'stacking cranes' (which stand inside the terminal) are also financed by private companies. The PHB gets its revenue from the leasing of public berths. These contracts have to be extended every 15 days. The Kobe Port Development Corporation gets its revenue from private lease via contracts of 10 years or more. In principle land is not sold.

Reclaiming land from the sea by constructing islands (Port and Rokko Island) takes place in consultation with PHB/MOT but is financed by the city (Chihaya, 1981). The construction of breakwaters is financed by the PHB/MOT. Dredging costs are divided between the local and national level and the same holds for public terminals. The private terminals are formally wholly financed by KPTC but as a result of the earthquake a legal exception has been made; the Ministry of Transport will finance 80% and provide loans to private companies at low interest rates. In principle, many development schematics are feasible depending on the type of construction work, the users and the owners, but the basic facilities for general use comprise the bulk of the port development projects (Japanese Ministry of Transport, 1995:21). Kobe, on the other hand, has 18 container berths (with 38 cranes) which were constructed by KPTC and 4 container berths (with 10 cranes) which were constructed by the municipality for public use. The precise overview of the division of costs is shown in the schematic on the structure of Japanese port development (see Table 8.4).

Any plans made by the local port authority undergo two formal assessment procedures. The first assessment is carried out by the district office of the Ministry of Transport in Kobe in light of the *master plan* for Osaka Bay (reviewed every 5 years). Then the local plan is assessed again by the Ministry of Transport, i.e. by the central Ports and Harbours Bureau in Tokyo. The district office thus functions as *interface* between the local and national level. As a result of the disaster, however, there is at present considerable direct communication between PHB/Kobe and PHB/MOT in Tokyo.

The national interest in the port of Kobe is still considerable and this is demonstrated among other things by the fact that hardly any transfer of powers exists and that PHB/Kobe always needs to have its plans approved by the Ministry. As a result of the earthquake this situation has become even clearer. Moreover, the port of Kobe was managed by the Ministry of Transport until 1950 so that some basis for the interference still exists.

Impact of the earthquake on port planning
In connection with the reconstruction of the port, the Chief Executive Officer of the PHB/Kobe made frequent visits to Tokyo to try and get money out of central government. A special body, the Committee on Kobe Port Reconstruction Plan, has been set up at the Ministry to deal with this problem. The mayor also plays an important role in such matters. The mayor is the Minister's direct communication partner which means that it is the mayor who must defend national interests at local level and incorporate them in policy. The earthquake cost the port 540 billion yen of which around 80% (390 billion yen) was reimbursed by the national level[3]. The national money was

---

[3] Compare this with the situation in the United States where the local authorities have to finance reconstruction costs themselves.

Owner	Type of Facilities	Use	Development Scheme	Responsible organization for construction	Share of Construction Cost		
					MOT	City of Kobe	KPTC
I Public or semi public sector	1. Water facilities	a. general use b. specific c. energy plant	a. general public work b. industry related project c. specific port facilities construction	a. City of Kobe/PHB or MOT b. MOT c. MOT	a. 2/3 or 5/10 b. ¼ c. 0.5-2.5/10	a. 1/3 or 5/10 b. ¼ c. 0.5-2.5/10	b. 5/10 c. 5/10-9/10
	2. Harbour facilities	general public	general public work	City of Kobe/PHB or MOT	5/10 or 2/3	1/3 or 5/10	
	3. Mooring facilities	a. general public b. general public (specialized facilities for bulk cargo) c. specific	a. general public work b. specialized port facilities construction work c1. foreign trade container wharf work c2. ferry wharf work c3. marina construction works	a. ditto b. MOT c1. KPTC c2. KPTC c3. marina company	a. 5/10 b. 4/10 c1-3 non interest loan 2/10	a. 5/10 b. 6/10 c. 1-3 non or low interest loan 2-5/10	b. 2/10 special charge c1. loan from user 3/10 c2. band loan 1/10 c3. band loan 7/10
	4. Supporting facilities (cargo handling, machine transit shed, etc)	general public	supporting facilities work	local government	provide low interest loan	loan-raising 10/10	charge
	5. Land reclamation	general public	land reclamation work	local government	provide low interest loan	loan-raising 10/10	rent-free loan price purchased
	6. Navigation aids	general public	navigation aids work	MOT	10/10	-	
	7. Others			City of Kobe/PHB		10/10	
II Private sector	All facilities	exclusive		private sector	-	-	10/10

Table 8.4. Division of Port Planning Responsabilities (Source: Japanese Ministry of Transport, 1995a:22)

desperately needed because the disaster had been so totally unexpected (the last was 400 years earlier; compare this with Tokyo, for instance, where an earthquake can happen at any time) and private companies were not insured against damage to their fixed assets (buildings, etc.).

As we mentioned earlier, it is difficult to recapture trade. In this situation of financial need, Kobe cannot offer attractive bonuses on a permanent basis. As long as reconstruction operations continue, there is a discount of 15%. It cannot do any more, partly because it is not legally possible. Moreover, an ordinance cannot be changed all at once since there are no fixed procedures or schedules. At the moment PHB/Kobe is running at a loss so there is also little to give away (they must borrow from the national level).

The Ministry of Transport's special committee, which provides recommendations on the reconstruction of the port, was appointed on 12 February 1995 (a month after the disaster) and in April 1995 came up with additions and instructions regarding the port of Kobe's existing development plan. In principle, the old development plan is maintained; the rubble from the buildings in the city will be used for a southern extension to Rokko Island. A prioritisation of the various essential projects has also been made. In the third place, the earthquake offers the opportunity to introduce innovations such as the early replacement of old conventional terminals and the expansion of the container terminals. In the reconstruction of the terminals, earthquake-proof techniques will be used. It is expected that all reconstruction activities will be completed by late March 1997 (Japanese Ministry of Transport, 1995b).

8.6. Port Services

The port service provision in Kobe is mainly taken care of by private companies. These companies need the approval of PHB for the construction of their own facilities. Certain clearly-delineated criteria must be met (safety, not causing disturbance, etc.) before a licence can be obtained. This approval is based on two separate regulations: the Port and Harbour Law and an ordinance aimed at the concrete construction in Kobe. The ordinance was drawn up by the municipality on the basis of delegated powers from the Ministry.

Stevedoring operations are carried out by private parties who may be allied to a shipowner. The work is seen as part of the freight handling; the stevedoring profession thus originated as an auxiliary to major trade societies or groups. The 'dedicated terminal concept' will thus not arise in Kobe because under the present regulations, terminals may not be leased to stevedores but only to shipping companies and for the latter one berth is often sufficient. 'Scale increase is only viable if shipping companies cooperate and jointly engage one stevedoring company. But because existing contracts with these operators cannot be terminated - each shipping company has its own historical ties with such a company, sometimes from the same parent company - this is not feasible. This is only possible at the public terminal operated by the municipality. But this operates on a "first come, first served" basis which is no use to major shipping companies working to tight sailing schedules' (Oosterwijk, 1994:20).

The pilots for the whole of Osaka Bay are all members of an association and for that reason have a monopoly position. The district office of the MOT supervises this activity. It also oversees the stevedoring, the work in the distribution sector and the ships' crews. Towing services in Kobe are offered by the PHB/Kobe as well as by private companies but there is no real competition: the PHB/Kobe owns only two of the 33 tugboats. The PHB/Kobe is responsible for supplying water and fuel. Kobe has a bunkering system on Port and Rokko Island so that refuelling can take place directly at the quay. Within the port, PHB/Kobe is also responsible for waste collection and preventing discharges, and consultations regarding such matters take place on a regular basis with the municipal environmental agency and the Maritime Safety Agency.

The police in the whole of the Kyoto prefecture are organised by the prefecture. There is thus no separate harbour police force. There are 10 branch offices for the whole area. The prefecture has no further control over the regulation of the port. The fire brigade is the concern of the city of Kobe.

8.7.   Port Management in Japan: Analysis and Questions

Japan has traditionally always been a seafaring nation, but it was only in the last century that Japan opened its doors to foreign trade. In order not to be overrun by the advancing West, crucial economic sectors, chiefly the mining, shipping and iron industry, were developed by the Meiji dynasty with the aid of government capital. Once these public enterprises had reached maturity they were sold to the private sector for a relatively low price. In order to achieve their objectives, nobles and industrialists became mutually dependent. When Japan had been defeated after the Second World War and the old guard of industrialists had to clear the field, the concept of mutual dependence between the political and economic powers did not change. The country was once more faced with the effort and challenge of matching itself against the wealth of the West and even surpassing it. Modern nobles (politicians and bureaucrats) and new industrialists confronted the task together and scratched each other's backs in the reconstruction of the country.

It was only after the Second World War that coordinated port development came about. The port of Kobe was one of Japan's major ports and would have to orient itself primarily towards foreign trade. Ports were seen as the instruments that would get the national economy, which was highly dependent on imported raw materials and the export of domestic products, to prosper. They formed part of the social capital of the country and thus did not have to meet numerous commercial criteria. Due to the shortage of land, the six ports in the Bays of Tokyo, Ise and Osaka in particular, became multifunctional, combining industry, residential areas, storage of raw materials or waste, recreation, etc.

Due to the multifunctional position which the 'specially designated ports' are assigned, numerous control relations can be expected in principle. After all, the primary function is not to make a profit for the port itself, but to provide a good infrastructure. The striking institutional element in the role division between public and private parties at the port is that the law expressly states that the port authority may not engage in private activities (read: activities pertaining to the freight). This is in stark contrast to the national intertwining between the political sphere, the bureaucracy and the industrial sector. In fact the division between public and private service provision in Kobe is also not as watertight as might be expected on the basis of the regulatory principle. The two clearest examples of this are the public terminals for general use and the KPTC terminals.

As regards nautical management and nautical service provision, however, there is a consistent implementation of the principle that national government is responsible for these public tasks. In the period since the earthquake, port planning has certainly become a national matter as attested to by the fact that the Ministry of Transport is financing no less than 80% of the reconstruction. But even in normal times, port planning is assessed at the national level and competition between ports is coordinated. Private sector input in port planning is ensured through the frequent talks and promotional activities organised by the PHB/Kobe, but also via their specific wishes regarding the KPTC terminals.

The main question is how the Japanese economy, and the ports along with it, will manage to develop in the coming decades now that the competition from other Asiatic countries is becoming tougher, the recent domestic trends towards emancipation are becoming more pronounced and the international pressure to open the Japanese market is being stepped up. Nevertheless, Japan's financial capability is enormous and it is still managing to keep its own market closed to international providers. In that sense, Japan remains an island. How long it can keep this up depends on the economic growth at home and the goodwill of the rest of the world to keep its own markets open to Japanese products.

If Japan really starts to join in with international competition, however, then the question is whether the present institutional situation in the port of Kobe will be capable of reacting to developments in time. Will port development, which is in government hands due to the shortage of space and the huge investments involved, be able to meet the new demands of the global market? Will the direct intertwining between politicians, bureaucrats and the business community lead to creative solutions even in times of declining economic growth? The most important question might be whether it will be possible to maintain the social character of the port or whether there will be a shift towards running it along more commercial lines.

8.8.   Summary

In January 1995 the port city of Kobe was largely destroyed by an earthquake. In view of Japan's idiosyncratic, economic group-driven regulation, the opportunity to conduct research there a year after the earthquake seemed an interesting addition to the existing case material. Two facts about Japanese port management stand out. In the first place, Japanese ports perform a number of economic functions simultaneously (working, living and recreation) and port management is not oriented towards making a profit. In the second place, the national Ministry of Transport plays an important policy-making and financial role in the reconstruction of the port and its further development. There is also government conditioning of the competition between the ports of Kobe and Osaka.

CHAPTER 9

The Port of Hong Kong

'Hong Kong is run by The Jockey Club, Jardine/Matheson, The Hong Kong and Shanghai Bank and the Governor - in that order' (Miners, 1995:46/47).

9.1.   Introduction

As a former British colony, Hong Kong has become famous chiefly for its impressive economic development in a 'laissez faire' environment. The natural harbour, which handles 90% of all trade with Hong Kong, is the main reason for this. Hong Kong is the largest container port in the world handling 14.5 million TEU in 1998. According to the prognoses, this growth is likely to continue. No less than 50% of the total cargo is shipped in and out in containers. The remaining transhipment consists of bulk goods such as oil, coal and cement. The port activities have always managed to thrive under the prevailing economic policy. The remarkable thing is that Hong Kong has never had a port authority (in the sense of land manager). That changed more or less in the late eighties when the development of the port reached its geographical and institutional limits. The developments in trade were taking place so rapidly that thinking and planning needed to occur on a much larger scale. This also required a different commitment from government and involved a departure from its laissez-faire attitude to some extent.

The port of Hong Kong forms a very interesting case from an institutional viewpoint. Under a permanently constant institutional structure a tremendous development of the whole colony has taken place. Moreover, this has been able to occur without the facilitating role of a port authority. A third argument is that a tremendous growth potential has been predicted for this port over the coming twenty years whereby the business community and authorities will (temporarily) need to embark on a different relationship than they have had in the past. A gradual change in the institutional structure was considered desirable by the government not so much due to China resuming control in 1997 but chiefly due to the required integration of policy areas in order to implement new development plans. In addition, economic development has ensured that in Hong Kong, too, other quality criteria than just that of economic growth have gradually become important. Finally, the reason for choosing the port of Hong Kong lay in the business community's tremendous financial involvement in the port which occurs on an internationally unprecedented scale.

This chapter is structured as follows. Section two provides a historical and institutional map of Hong Kong and the port. The third section deals with nautical management and the fourth section takes a look at the specific aspects of port planning in Hong Kong. Section five focuses briefly on port services while section six contains the analysis and a number of questions.

9.2.   Hong Kong: a Part of China

The former British Crown Colony of Hong Kong lies on the south coast of China on the east side of the Canton river estuary. The area comprises the island of Hong Kong and the New Territories which extend to the Chinese city of Shenzhen. The dry area covers a total of 1084 km^2 of which only a small part (161 km^2), due to the rest of the area's rocky terrain or other uses (nature areas, water reservoirs, etc.), is available for residential use, industry, transport, etc. (Daryanani, 1995:appendix 39). Until the first half of the nineteenth century, the area that is now Hong Kong consisted of an assemblage of various small villages and hamlets. The port thronged with masses of small fishing boats which also served as homes. Although it was an inhospitable area, the location had one important advantage: it offered pleasant anchorage sheltered from the dangerous storms in a region that lies on the edge of the tropics. For English ships sailing to the Far East in the eighteenth and nineteenth centuries this was the perfect spot to rest up and take shelter from typhoons.

In the early nineteenth century, Hong Kong gradually developed into an *entrepôt location* for English trade. Initially silver was traded for tea but later opium was chiefly traded. The opium trade was repugnant to the Chinese. They had banned this trade back in 1799 and finally came to blows with the English in 1840 but they were unable to drive them from Hong Kong. In 1841 the English planted the Union Jack on the rock of Hong Kong in order to secure a permanent location in this part of Asia[1]. Although a treaty was signed, the situation was highly unstable and in 1856 the Second Opium War broke out. This lasted until 1858 and was finally settled in 1860 by means of the Peking Convention. In 1895, other European powers of the time (Germany, France and Russia) requested concessions from China for their part in the war against Japan. Great Britain was able to settle this veiled expansion of power to its advantage by signing a 99-year lease for the neighbouring area, the New

---

[1] For an intriguing story on this subject, James Clavell's (1975) novel "Taipan" is recommended.

Territories in 1898. In this way Great Britain would be in a better position to defend its *rock*. It was this lease which terminated on 30 June 1997.

Against all the expectations of the liberal English, the number of Chinese in Hong Kong increased sharply (from 31,463 in 1851 to 878,425 in 1931 to 6.1 million in 1994). Due to its laissez faire policy, Hong Kong proved an excellent port for trade and contact with the Chinese community elsewhere in the world. The English residents insisted repeatedly on giving Hong Kong self-government but the democratically-inclined United Kingdom did not consider it essential for an English minority to govern a Chinese majority (Daryanani, 1995:456). At various times in its history, Hong Kong has proved a very welcome refuge for those Chinese who no longer wished to remain in their own country. This was the case, for example, when Japan invaded the province of Canton in 1938 but also in the sixties when many Chinese found themselves between a rock and a hard place as a result of the communist revolution in their country.

After the Second World War, Hong Kong went over to light industry, such as clothing and electronics, so it would not have to be solely dependent on the direct activities in the port. However, the port was still the major transport link in trade with China which had closed itself off from the world after the Second World War. In this regard, despite their totally different politico-economic structures, China and Hong Kong have remained dependent on each other. Hong Kong does not have its own raw and auxiliary materials (water, unrefined materials and cheap goods) and has to import these from China. On the other hand, Hong Kong provided its only gateway to the outside world in 1960 when Russia terminated its cooperation with China.

9.2.1. POLITICAL AND ADMINISTRATIVE CHANGES AFTER 1997

In 1984 the Prime Minister of Great Britain felt it might be useful to get together with China to consider the position of Hong Kong after 1997. After much arduous consultation, the Sino-British Declaration was established which clarified the position of Hong Kong after 1997. Hong Kong will become a Special Administrative Region of China with a high degree of autonomy apart from in the spheres of foreign policy and defence policy. The United Kingdom and China have agreed that during a fifty-year period after 1997, the companies located in Hong Kong will not have to pay taxes to China, that the local currency will remain freely exchangeable, that rights of ownership will be respected, that passenger traffic to and from Hong Kong will continue unchanged and that the seaport and airport will be freely accessible to all. This in fact creates two systems within China. After the 1984 declaration, a Sino-British Liaison Group was established which made preparations to ensure the effective implementation of the Joint Declaration in

1997. The group consisted of five members from each of the two countries and officially had no power because it only intended to advise the governments of Great Britain and China on numerous matters which would crop up during the transfer of power, and primarily in (potentially) discordant situations. It met at least once a year. The Land Commission was set up at the same time as the Joint Liaison Group to supervise any sale of land involving more than 50 hectares per year. With regard to port expansion, in particular, both advisory bodies proved to acquire considerable importance (see also section 9.4). In 1989, the cooperative mood of the Liaison group was seriously disturbed by China's reaction to the student protests. This led to a big protest march in Hong Kong. Negotiations between Great Britain and China ceased and many of Hong Kong's citizens became apprehensive. Many senior bureaucrats left Hong Kong and many a businessman transferred his possessions to Bermuda or the Cayman Islands. In 1990, the Chinese Congress adopted the Basic Law which has been in force for the Special Administrative Region since 1 July 1997. This provided some insight into China's formal intentions regarding Hong Kong and what Hong Kong may expect of the Chinese administration although it has not prevented the fact that 40,000 highly-qualified people per year are leaving the country and many others are having serious doubts about whether China will honour its promises.

In contrast to other British colonies, Hong Kong is not entirely independent: after all, it remains an administrative part of China. Explanations for this situation can be found in the fact that in the past no trend towards self-government or democratisation developed, but also in the fact that the New Territories were formally leased from China for 99 years. A guarantee that the politico-economic system will remain in its present form for the next fifty years is the best that the UK as its former administrator could manage to deliver in the negotiations with China. The publication of the Green and White Papers in 1984 and 1985, containing proposals for democratic reforms, were criticised at the time by China which stated that only reforms in accordance with the (as yet unwritten) Basic Law would be permitted. Moreover, despite Hong Kong's autonomous status, unilateral action in these areas was not required (read: patience is expected of Hong Kong).

In contrast to his predecessors, the last governor tried to make more arrangements for the people of Hong Kong without taking trade with Great Britain as the central focus. Great Britain wanted an honourable withdrawal in 1997, but despite its good intentions it was in fact unable to do much more for Hong Kong: China was already casting its shadow before it and governor Patten had a tendency to bring down China's wrath on his head rather than gain its silent assent. During the transition period, the poor relationship between the last governor and China had an adverse effect on people's faith in the bureaucracy. It was unclear to what extent civil servants were already promoting Chinese interests with a view to safeguarding their own position after July 1997.

From the Basic Law it can be deduced that the most important political-administrative changes will concern the position of the present governor and the EXCO (Executive Council) and Legco (Legislative Council). In the Special Administrative Region the governor will be replaced by a 'Chief Executive'. This person will be selected by:

- 100 people from the financial, commercial and industrial sectors,
- 100 people from the various trades,
- 100 employees from religious and other sectors and
- 100 former political figures and Hong Kong delegates of the Chinese National Congress.

All these people will be selected by a *Preparatory Committee* appointed by the Chinese Congress. The Chief Executive will be officially appointed by the Chinese government. His tasks and functions will be almost identical to those of the English governor. The chairman of the container shipping company OOCL has been selected as Chief Executive. This choice appears at any rate to safeguard the special position of Hong Kong within China.

The EXCO, whose members will all be chosen in future by the Chief Executive, retains its present function, but the Legco will acquire a stronger position. It can only be disbanded once per session and it can dissolve the EXCO in the event of dysfunction. Moreover, the Chief Executive must sign an Act if this has been adopted by a two-thirds majority in the Legco. In the new situation, double memberships of both the EXCO and Legco will not be permitted: this will reduce EXCO's power because it will no longer be able to submit private member's bills.

In 1997, all formal constitutional documents will be replaced by the Basic Law document. The only part of the British constitutional structure which will remain is the judicial system and the Common Law. The only point on which the Joint Declaration runs counter to the present system is that official positions (Chief Justice, Chief Judge of the High Court and members of EXCO and Legco) must be held by Chinese citizens who do not have dual nationality. On paper, Hong Kong with its autonomous status will have more freedom than a state in the United States, while China is a unitary state. Moreover, China will have less formal power than the British have now because China has no authority to reject an ordinance laid down by the Legco.

On the basis of the above it may be expected that for the port institutions and port management there will be little change after July 1997. The Marine Department will be able to operate more autonomously than is the case at present under the British. On 2 December 1990 Hong Kong became a flag state with its own register. This means that it employs its own quality standards with regard to the safety of ships. The fact that Hong Kong has an internationally recognised register with which it earns good money combined with the fact that the state of Hong Kong has its own flag and participates in all international forums (e.g. the International Maritime Organization) as a full partner, demonstrate not only the dramaturgical significance of its autonomy but also its international legitimacy. Since none of this poses a direct threat to China and given the fact that it can rely heavily on the far higher quality standards of Hong Kong, there will be little change in the government role in the port.

After 1997 the Chinese influence will formally come into force and a country with two systems will arise. For at least the next fifty years, Hong Kong will uphold an economy run along capitalist lines as agreed in the Sino-British Joint Declaration of 1984. In view of the many private construction works planned for the next ten years, the international private sector appears convinced of this too. The public/private role division will thus not change fundamentally after 1987.

9.2.2. ECONOMIC STRUCTURE

Traditionally, the economy of Hong Kong is strongly oriented towards trade, in particular (transit) trade with China. At present, 60% of trade is destined for or comes from China. This is the direct result of the lack of raw materials, the favourable trade location and the presence of a natural harbour. A second feature of Hong Kong's economy is the fact that the government actively pursues a laissez-faire policy. This is not entirely unrestrained, however, since there are many areas in which government intervention occurs, but it refrains from intervening in wages and labour relations issues and this is very important for economic relations in Hong Kong. In the fifties, for example, as a result of the international trade embargo on China and the substantial influx of immigrants, industrialisation got underway without any government intervention. It is unfair, however, to characterise Hong Kong as a 'bastion of laissez-faire politics' (Hart, 1985:20). The economic growth after the Second World War certainly prompted the government to increase its involvement in matters other than defence, justice, public order and public works. This gradually took shape in port planning and environmental legislation among other things.

The more or less essential emphasis on laissez faire lies in the absence of raw materials: the government can in fact do little to influence the relationship between costs and prices in a beneficial way. The only thing that the government owns is the land. In the past it granted or sold off land by means of private treaty grant but in recent years it has let it by auction on a 50-year lease (apart from types of land use which by nature cannot be competitive such as some forms of innovative industry, housing, hospitals, schools and such). Moreover, the government does a number of things that the private business community cannot do, such as providing cheap housing, enforcing export quotas, regulating trade in vegetables and fish, ensuring well-ordered monetary activities (which are furthermore entirely liberal), combating environmental pollution, operating the airport and the railways and providing welfare facilities, such as a health service for the more serious cases and education. Electricity, gas, telephones, buses, ferries, trams and the underground system are privatised services, however.

In the past, Hong Kong had mainly light, labour-intensive industry in which the wage component is a major cost item. The wages are high by Asiatic standards but since productivity is also high it is possible to remain competitive. Wages have hardly risen due to the abundant supply of labour, the minor influence of unions (only 15% of the working classes belong to a union) and the absence of large companies dominating the wages front. In recent years, however, part of the production in Hong Kong has moved to China or one of the other Asian tiger markets (Taiwan, Korea, Philippines, Malaysia, Thailand), because wages are lower there. The shift from a production economy to a service economy in the eighties is very telling in this context. The share of (small) production companies (clothing, electronics, clocks) in the total of the Gross Domestic Product (GDP) and employment dropped over a 25-year period from 47% in 1971 to 20% in 1994 (Daryanani, 1995:65-66). The tertiary sector (wholesale trade, import/export, financial and business service provision, etc.) accounts at present for 77% of the GDP. Just like Singapore, Hong Kong realises it will need to make the switch to high-quality industry which needs highly-trained people. The government will also play an important conditioning role in this.

9.2.3. SIGNIFICANCE OF THE PORT FOR HONG KONG

It may be stated categorically that the port of Hong Kong is the natural element that has made Hong Kong great. Hong Kong is first and foremost a maritime centre around which at a later stage other services (banks and insurance) have become established. Because of the port, Hong Kong has been able to grow into both a maritime and financial centre (Kagan, 1990). 'Hong Kong's eminence as a financial, communications and manufacturing centre is rooted in its pre-eminence as a port, presently one of the two busiest in the world. Its prime clients are the world's shippers, traders and manufacturers. If their needs are correctly anticipated and efficiently met, the city will thrive. If they are not, it will decline' (Lewis, 1992:2). In other words, the port forms the pivot of many activities which make the economy of Hong Kong thrive. Or in economic terms: with a goods transhipment of around 125 million tons, the port generates a total of 350,000 direct and indirect jobs which account for 20% of the GDP.

Hong Kong has a very open economy, which for the port means that there is a minimum presence of regulatory public bodies. 'The Hong Kong Government's trade policy seeks a free, open multilateral trading system and the government's role in this process is one of facilitation. No protection or subsidy is provided to any manufacturer and the very small domestic market means that the import of raw materials and export of finished products must be freely facilitated' (Hong Kong Marine Department, 1995a). The government can supply new ideas by making recommendations to trade and industry.

Graph 9.1. Port Performance Hong Kong 1988 - 1998

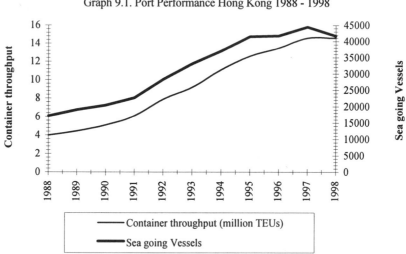

Regulation only takes place in accordance with the international provisions of the GATT, the Multi-Fibre Arrangement (MFA), the International Maritime Organisation and any bilateral treaties. Import and export taxes are only levied on spirits, tobacco, hydrocarbon oil and methyl alcohol. There are no further controls on export and re-export. It is striking that in Western Europe the term '*facilitating*' has exactly the opposite meaning. In Western Europe the concept is used to mean the government creating the most favourable conditions possible to facilitate the business community in a financial sense.

Until twenty years ago, the geographical section of the port only comprised the area called Victoria Harbour. With the advent of containerisation, however, the port of Hong Kong has shifted steadily to the west and it may be expected that the construction of new terminals on Lantau Island will continue the geographical separation of city and port. The development of the new Chek Lap Kok airport on Lantau Island may increase the integration of port

and airport which has not been the case up to now. In the new situation they
will both be served by the new road and rail infrastructure which runs from
Tsing Yi to Lantau Island via the imposing, two-kilometre long Tsing Ma
Bridge.

Due to its strategic position and natural depth, the port of Hong Kong is not
(as yet) troubled by competition. This is why it is the most expensive container
port in the world. The shipowners have to put their containers ashore here
because there are no good transport links elsewhere or because they want to
avoid the government corruption and political pressure of the Chinese.

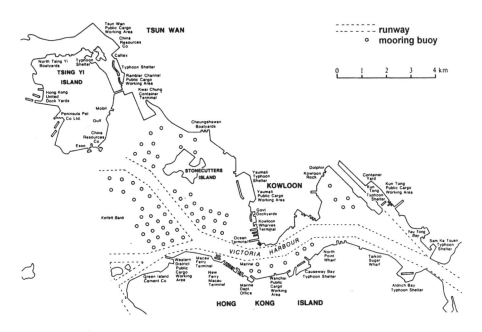

The Port of Hong Kong

In view of the technical complications, such as the difficult passage for
larger ships and the need for continuous dredging, there is not much point in
shipping companies constructing terminals elsewhere. This is the major
difference between the position of shipowners in the West and those in the
East. In the West, the port companies have to get down on their knees to
capture shipowners whereas in the East it is exactly the other way round[2].

---

[2] Hong Kong must not overdo it, though. With price increases of 58% in seven years it is
becoming increasingly attractive for shipowners to transport cargo to China via Kaoshiung in
Taiwan or via Chinese coastal ports (Yantian and Shekou) (Lloyd's List, 17 September 1996).
Moreover, in Hong Kong itself there is increasing interest in the development opportunities
presented by the Pearl River delta.

The strict division between public and private tasks in the port in the past ensured that there was no formal contact between the government and the business community. The business community could not count on any (financial) aid from the government because the bureaucracy itself had to break even. In the event of conflicts between actors from the business community these are not resolved through government intervention but by an independent body such as the Chamber of Commerce. The *Shipping Committee* of the General Chamber of Commerce provides an institutionalised forum for the various service providers. Participants in this committee comprise only service providers, however, (container shipping companies, transhipment companies, 'midstream operators', pilots' associations, etc.) and do not include the shippers. The aim of the committee is to look at specific problems. It only meets once a year, however. A recurrent theme at the end of each year concerns the wharfage and berth dues which are laid down by the government. In comparison with other Chamber of Commerce committees, the Shipping Committee is highly reactive to government proposals for changes in the port.

Table 9.1. Trade Partners in the Port of Hong Kong in Order of Importance (Source: Planning Department Hong Kong and Port Development Board, 1995:7).

Trade Partner	Share in Import and Export of Trade via the Port
China	14.5%
North America	12.5%
Singapore	11.0%
Japan	11.0%
Europe	9.8%
Taiwan	9.8%
Australia/New Zealand	6.0%
South Korea	5.3%
Rest of Asia	11.6%
Other	8.5%

Another subject that has caused a bit of an uproar is the increase in the handling tariffs by the container operators and shipping companies. The shippers, who are organised in the *Shippers' Council*, demonstrated forcefully against the increase. There had been no consultation with them. The shippers have no representation in the Chamber of Commerce although it was one of the many bodies which set up the Shippers' Council. By taking a neutral stance, the Chamber of Commerce has been able to mediate in this conflict. This was a new experience since usually the private partners are able to sort out their own problems.

The two most important public bodies in the port are the Marine Department and the Port Development Board. The Marine Department is above all the nautical manager of the port and also performs tasks in the framework of 'port state control'. The Port Development Board (PDB) is the advisory board to the governor on port planing and comprises on the one hand top civil servants whose policy domains include the port and, on the other, the most influential people from the port business community. These bodies are dealt with more fully in the sections on nautical management and port planning.

9.3.   Nautical Management in Hong Kong

Nautical management in the port of Hong Kong is certainly not an easy job. 'In 1994, The Marine Department recorded a total of 370,000 vessel arrivals and departures. This comprised 73,100 ocean-going vessels, 175,000 river trade cargo vessels and 122,000 river trade ferry vessels. In addition, there are 16,000 local crafts using the waters of Hong Kong on a daily basis' (Port Development Board brochure). The port of Hong Kong has no port authority in the traditional sense of the word but has always had a nautical manager who also developed a number of other activities. This function is performed by the Marine Department. In addition, a number of other public services have traditionally been directly involved in the port, some of which assist the Marine Department.

9.3.1. THE MARINE DEPARTMENT

The reason for the absence of a port authority as we know it in other parts of the world (combination wet and dry areas) is the fact that the government, afraid as it is of disturbing the economic process, wants absolutely no involvement in the port. The Marine Department performs an important role in ensuring the smooth-running of shipping traffic in an orderly and safe manner. The Marine Department is the government body that supervises shipping traffic, to which end it employs an advanced traffic guidance system among other things. This system, which covers 95% of the port, was constructed with the aid of a government loan and will be upgraded in 1997. The reason for choosing to use a government loan was partly because it was financially possible and partly because, as a service-providing component, it would become too expensive if private funding was sought. The government thus facilitated the business community. The shipping guidance system keeps a check on shipping traffic and the Marine Department can provide a captain with advice and directions should he so require.

The Marine Department is an executive department of the Economic Services Secretary and comes under the governor in the hierarchy. On the one hand it performs activities which fall to a traditional port authority and on the other it undertakes tasks which could also be undertaken by private companies. Its principal activities include:

- conducting maritime research;
- monitoring compliance with safety regulations (Port State Control);
- representing the International Maritime Organisation (IMO). The Marine Department is also responsible for the disaggregation and implementation of IMO conventions which relate to an increasing range of shipping areas and which are also becoming increasingly stringent (international law);
- keeping the register up to date (592 ships with a Gross Rate Tonnage of 8 million);
- coordinating search and rescue operations (140 incidents in 1994; 500 people brought to safety);
- running various training courses;
- supervision to prevent environmental pollution during bunkering and by oil companies;
- managing the government dockyards where all the government boats from the police, fire brigade, customs, harbour service, etc. are handled (approx. 500) and where around 700 people work (including the crews). In the dockyards, ship design and modifications are carried out although the implementation is put out to tender. This is always carefully supervised (Hong Kong Marine Department, Handbook, 1995a).

The Marine Department sees itself primarily as a *service provider to the shipowners* and endeavours to minimise the duration of its actions (stringent success standard). An attempt is made to constantly cut down on delivery times for various services, for example (Hong Kong Marine Department, 1995b). In the framework of the temporary presence of the English and the laissez-faire policy it is considered important that as a public service it should at least break even financially. It is useful, therefore, that in addition to the 'port, light and anchorage' revenue for the maintenance of the traffic guidance system (only ocean-going vessels) the Marine Department has developed a number of other activities which bring in a bit of cash, e.g.:

- the ownership of two ferry terminals which are managed independently of each other and which make a profit;
- the revenue from mooring charges;
- the operation of 20 passenger ships;
- the management and operation of 76 mooring buoys;
- the management and operation of the 'public cargo working areas' (public terminals) which also generate money, because the port's 2000 'lighters' and the trucks can load and unload in these areas at a charge;
- setting examinations which shipping companies pay for; and
- keeping the register up to date which shipping companies must pay for.

The port of Hong Kong is run on the market principle but this does not mean that public authorities refrain from the operation of private activities. The integration of nautical management and private activities is a more unique phenomenon than the fact that Hong Kong has no port authority in the traditional sense of the word. The shortage of income from the dry area of the port is offset to some extent by the revenue from other services. Although the Marine Department even has a good chance of running at a profit in the future, this needs to be contextualised. If we are talking about actual profits, then the office costs (depreciation of the building, office materials, etc.) ought to be included and this is not the case at present.

The fact that the Marine Department keeps its own register means it not only generates its own income but also safeguards partial autonomy. Until 1990, all ships were entered in the manner of, and in accordance with, the requirements of Great Britain. After 3 December 1990, Hong Kong became a flag state and it could impose its own requirements on the register. At first, the number of ships on the register decreased but has now increased once more. This is despite the stringent and highly expensive requirements which the ships and their crews must meet. Hong Kong keeps its own register primarily to protect itself as a maritime centre (ranging from financing and insurance to shipping docks, etc.). The register and its own flag stand for autonomy vis-à-vis China in the future and are even included in IMO negotiations. The register operates under local Hong Kong law. China has no problem with the register since the quality standards are much higher than those used in China.

The Marine Department, where in total 1,600 people work, comprises six separate divisions. First of all there is the *Planning and local services division* which is responsible for Hong Kong's less important harbours and the typhoon shelters. It works closely with the Port Development Board and the Planning Department. Then there is the *Port services division* which organises the management and maintenance of the traffic guidance system. It also operates the public terminals and is responsible for hydrography. Furthermore, it is concerned with mooring facilities and pollution control, and it forms the link with the pilots. The *Government fleet division* is responsible for the management of the government dockyards. The *Shipping division* is concerned with crew safety (exams), convention ships (certification of those ships which come under the IMO convention) and local activities (ship repairs and freight trading). The *Multilateral policy division* is concerned with international maritime legislation and implementation. And finally, the *Business services division* is responsible for generating resources and funds. A Trading Fund System will come into operation in 1996 to enable the provision of extra services to private companies and which will make the Marine Department even more financially independent.

The navigational markings and beacons are the tasks of the Marine Department. It has a number of means at its disposal to effectively regulate shipping traffic, e.g.: (re)positioning anchors (for dangerous cargo, e.g.) and shipping lanes, the traffic guidance system and other traffic regulatory procedures (ordinances for safety, dangerous substances and port control). At present, in view of the increasing congestion, a study is being conducted into a long-term strategy for shipping traffic policy. The Marine Department can impose fines and this frequently leads to legal proceedings. The fines are not high, however. In 1994, 1,942 fines were imposed and proceedings were instituted in 1,839 of these cases; the total sum involved was two million Hong Kong dollars. The Marine Department is also responsible for (household) refuse collection from ships (without charge), keeping the port clean (driftwood, etc.) and the cleanup of oil discharges (Pollution Control Unit). In 1993, around 5,000 tons of floating waste was removed from the port. The Pollution Control Unit works in close cooperation with the petrochemical industry in preventing oil pollution. The collection and processing of chemical waste substances from ships occurs on the 'polluter pays' principle. They can hand over their chemical waste to 'barges' (these are in fact lighters with a trough or tank) which then deliver it to the only private company which is designated by the environmental department for processing. The objective is to have all cleanup activities carried out under concession by a private company. In 1993 the port of Hong Kong, with its unique designating of a private company for the collection of chemical waste substances, became the first in Asia to have this type of treatment centre.

Position of the Marine Department director
The director of the Marine Department is *harbour master* of the port and has been given a number of discretionary powers on the basis of various port ordinances which have been approved by the governor. He has the final say, for example, in matters concerning the pilots, although all these ordinances contain provisions that objections to the director's decision may be made to the governor or the executive council. This occasionally happens. The Chief Executive Officer of the Marine Department is also directly accountable for the implementation of tasks and powers to the Legco panels which often contain no experts in the relevant area. The panels only assess the director on financial criteria.

Police and Fire Brigade
The port police and port fire brigade do not come under the authority of the Marine Department. They are divisions of the Royal Police Force and Royal Fire Brigade. However, the port police are busier catching crooks than providing assistance with shipping violations in the ports. In such a situation

the port division of the Marine Department must do the job itself. In the event of a disaster, due to the division of authority it is apparently not possible for the harbour master to issue direct orders to the fire brigade and police. This always has to be done via the directors concerned because it cannot be taken for granted that instructions from the Marine Department will be obeyed. It can happen, for example, that in the Marine Department's opinion a fire on a ship lying in the shipping lanes should not be extinguished because it might sink, which would cause a hindrance to other shipping traffic. The harbour master thus has to pass the problem on to his superiors because the fire brigade always want to extinguish first and ask questions later. The Marine Department has 25 boats of its own for patrolling the whole of Hong Kong's waters. Port dues are collected by the port service staff of the Marine Department and not by customs. In addition to the Marine Department there are a number of other departments involved in the port. The *Department of Civil Engineering* (Port Works Division), for example, is responsible for the non-commercial activities in the port, such as the construction of sea walls, public piers and breakwaters. The *Port Health Office* (Department of Health) is responsible for quarantine matters and the *Harbour Control Section* (Immigration Department) for ensuring that all passengers on board ships are entering Hong Kong legally. In addition to the control of imports and exports, the customs is responsible for the search for drugs and smuggling.

Coordination between Public Authorities
The Marine Department maintains relations with six of the fourteen policy departments of the Hong Kong government, i.e.: finance, economic affairs (controls 80% of the Marine Department activities), transport, safety (search and rescue), health and welfare, and spatial planning. The actual work is done by the Marine Department but these departments provide the financial means to do it. The Marine Department has signed a framework agreement with Economic Services in which requirements to be met and resources to be delivered are specified. Formally, the Chief Executive Officer of the Marine Department has a meeting with this policy secretary every quarter, but in fact this occurs every month. He meets with the other relevant policy departments once a year. It is the Marine Department that decides how much service the department will provide, but this generally take place in consultation.

In addition, the Marine Department maintains relations, through various disciplinary committees, with professional and trade organisations which are linked with local and international maritime communities (Port Operations Committee, Pilotage Advisory Committee, Shipping Consultative Committee, Dangerous Goods Standing Committee, Marine Ferry Terminals Standing Committee, etc.).

9.4.   Port Planning

The most extraordinary thing about Hong Kong is perhaps the fact that so many activities are carried out on such a small area of land. Many more are planned for the coming ten years: new terminals on Lantau Island, a new airport, rail and bridge links between the various islands, a river terminal, a feeder terminal, etc. Until a couple of years ago, because of its laissez-faire stance, the government had no direct control over port planning. 'Historically, port development in Hong Kong has come about through commercial interaction between private enterprise and interested government departments' (Containerisation International, 1991:71). It was the business community which approached the government with plans; nothing was either initiated or coordinated by the government. The construction of the first container terminals in the early seventies occurred by means of *private treaty grant* of an area by the government. The private sector then had to work out for itself how it would go about it. 'It has been up to the operators to develop the sites (usually involving land reclamation), construct buildings, install handling equipment and finally man/run the completed terminals without the benefit of government subsidies or sponsorship' (Containerisation International, 1991:71). The container operators, at least, were assured of a piece of land for a period of seventy years (contracts terminate in 2047).

Due to the fact that the initiative lay with the operators, the (expensive) development of container handling occurred in a fairly ad hoc way, thus increasing the chances of overcapacity. Moreover, the operators at that time were dependent on the government which, by means of the treaty grants, stipulated which sections were to be used for container handling. In 1984 it was determined via a legal agreement that: 'The go-ahead for a new terminal must be timed so as to enable that terminal's first berth to open as the port's total TEU demand reaches the total TEU capacity made available by the opening of that berth; neither sooner nor later' (Containerisation International, 1991:71). This agreement between government and container operators which is still in force is sometimes called the *Trigger Point Agreement*. The governor and EXCO give the go-ahead for building; it is Legco's task to provide funds. Even in liberal Hong Kong conditions are thus laid down for port planning.

During the eighties, the port became very prosperous and the agreement ensured that the construction of new terminals was always tightly planned. Around 1986, however, it became clear that once terminal nine (CT 9) came into use, there would need to be some changes in port planning. The opportunities for expansion in the present port would then be exhausted and the only option remaining would be to expand the port to the west onto Lantau Island. It was not only the port that had apparently reached its physical limits but Kai Tak airport, which had only a single runway, would have to consider moving to another location[3]. The joint *Port and Airport Development Study*, which was conducted for the government by external consultants, was a shot across the bows for the development of Lantau Island. The study was based on the figures for 1988 and was completed in 1989. There were such rapid developments in trade, however, that in late 1990 new projections were needed to keep a grip on demand and the planning for port facilities linked to it. Since then annual adjustments have been made, though these were often published too late for the planners. From the late eighties it became clear that the government would have to become more directly involved in spatial planning and in a different way than in the past. In April 1990, as an institutional answer to the development problem, the Port Development Board was established to coordinate port planning matters.

9.4.1. PORT DEVELOPMENT BOARD

The Port Development Board (PDB) was established from the government's need to channel and coordinate interests connected with the port. It functions as an advisory body for the governor and has no formal control. The governor appoints the board members and the chairman. The 16 members of the PDB range from top civil servants to the most important people from Hong Kong's business community. For the day-to-day activities, the board is manned by civil servants (around 25 in total) who are borrowed from the Marine Department among others. The board's task is to deliver more coordination with regard to public planning because infrastructural projects have become increasingly large-scale (links between Hong Kong and Landau Island, links between airport and port terminals, etc.) and there are more interests involved (e.g. more attention to environmental aspects). This also makes it the only coordinating body that the government has in regard to the port of Hong Kong. 'Terms of reference' of the PDB include:

- ascertaining the needs of port development against the backdrop of changing demand, port capacity, productivity and performance and the relative competition with other regional ports;
- devising and recommending strategies and port facilities;
- coordinating government bodies and private sector in the planning and development of the port;
- functioning as focal point for the interests of those participating in the world of the port or those who are influenced by it;
- forming specialist subgroups as and when needed and

---

[3] In 1994, Kai Tak handled an unbelievable 24 million passengers.

- undertaking any task relevant to the above as outlined by the government (PDB Annual Report, 1995).

Early on, the PDB was faced with the need to establish its status. The PDB had to advise the governor regarding the problems between the container operators on land and the lighters (simple, floating container cranes) on the water which on average handle 25% of the total containers. The lighters, which can collect containers in the water and then transfer them to another ship or set them ashore, are not faced with the high fixed costs of the landside operators. The container operators felt that the lighters were trying to ruin the competition and thought that the government ought to impose an extra levy on them. However, the shipping companies were keen to have alternatives to the terminals because they considered them too expensive and too powerful. 'Following the recommendations of its specialist committee, the PDB pronounced: "... it is neither necessary nor desirable to place [artificial] restraints on container handling in the stream so as to divert more vessels to the container terminals". Amen. The terminal operators were furious, but the PDB's pill was sugared by an assurance that, by 1995, natural economic and operational factors would reduce mid-stream container operations to a trickle. Gradually, the level of protest from the terminal operators subsided' (Containerisation International, 1991:71). Thus the PDB became an institutional fact which would have to be taken account of in the future.

The PDB has five separate committees which make recommendations about various port activities and which are chaired by individual members[4]. The Container handling committee, in particular, has a very responsible role in determining the exact moment when a new terminal will come into use. Due to the considerable financial interests involved, it is of great importance to the committee to time the opening as precisely as possible. In the early years, an extrapolation method (based on growth in the previous five years) was used to determine that moment but in the years 1992-1994 much higher growth percentages were realised so that the pressure on new terminals increased. Moreover, constantly improving techniques make it difficult to determine how quickly a terminal can be constructed. Terminal seven, for example, was completed two years early so that capacity was able to increase more rapidly. It was also the first terminal to be put out to public tender, something which later became the norm.

---

[4] The five committees are: the Container handling committee with the Lantau Port Liaison Group, Port land and transport committee, River trade activities committee, Ship repair and ancillary facilities committee and the Committee for strategic planning for Hong Kong Waters.

Every two years, the board publishes a prediction of the demand for land and port facilities based on developments in shipping and trade, development in the world economy, in China and the surrounding area and in other ports. This prediction goes to the Planning Department which turns it into a port development strategy. Due to the incredibly rapid developments, the strategies are in fact already out-of-date as they come off the press. Once it has been approved by the governor and the executive council, the plan goes to the legislative council which has to release the relevant financial resources. It does not concern itself with the content but merely looks at the funds required and offsets them against other public expenditure. It cannot make any policy changes.

Once it had established its status in the early nineties following its recommendations regarding the mid-stream activities, in the years that followed the PDB chiefly derived its status from the accuracy of its research results and the continuing participation of the major representatives from the port business community.

Now that the PDB has become institutionalised, minor changes can be traced in the approach to port planning. The Planning Department designates an area to be used for port expansion and the Land Department then issues a tender for it[5]. The money which the government receives via the tender from the highest bidder can be used to construct the necessary public infrastructure. The business community thus also indirectly finances the infrastructure. In this way, the government's contribution towards the total construction costs of the terminals is usually in the region of 40%[6]. The department of Civil Engineering has been very directly involved in terminals 10 and 11 right from the design stage. The government's plans and those of the private financier are generally implemented by the same building consortium. This is in contrast to previous terminals whose planning, implementation and integration into the environment were entirely financed by private developers/operators. Once a private investor has paid his premium, he does not subsequently have to pay any rent or other annual monies to the government. The new lease is not renegotiated for another 50 years.

---

[5] This had not happened with CT 9 and China used this as an argument for reassessing the choice of an operator whereby the construction was delayed.

[6] The construction of CT 10, for example, cost 16.2 billion HK$ of which the government paid 6.9 billion for the infrastructure by means of the premium which it receives for that area. The business community paid another 9.3 billion to equip it.

Flexibility and Adaptability

In the late eighties, there were many complaints from shipping companies that port development was not happening fast enough. The desired progress ran up against institutional boundaries; although the Hong Kong government had interfered very little in the economy, it became apparent that the business community could no longer cope with the challenge of keeping the port city in its totality attractive to both shipowners and residents. The further development of the port seemed to entail the problems of a 'common pool' good; the water was rapidly becoming polluted, the roads were becoming increasingly congested and the terminal operators of the time could not be directly compelled to commence the construction of new terminals. Kagan (1990) blames the government's procrastination chiefly on the fact that they, together with the terminal operators, were shackled by the Trigger Point Agreement. The container operators were able to bid ridiculously high sums for a new tender because they would be able to keep it profitable by using the profits from the other terminals. This scared off potential competitors and the container operators were able to plan their own initiatives for new terminals. Due to the legal agreement, the government could do nothing until the terminals really had reached their absolute maximum.

It is not surprising that a government in such an institutional setting was not abreast of events. First of all, it was neither permitted nor used to coordinating between private parties. Up until the projected construction of terminal nine, port development was an established private affair which progressed incrementally. Moreover, the developments in the container sector advanced far more rapidly than anyone in the private sector had thought possible. And finally, the influence of China in the approval of terminal nine proved to be greater than anticipated.

The Environmental Issue

For many years, Hong Kong's only success standard was economic growth. This is also the theme of the major projects which are now being implemented. It is a slow process for environmental aspects to actually be turned into policy. The *Environmental Protection Agency* has been in operation since 1983, but it was not taken seriously until the late eighties (Kagan, 1990). At present a large-scale sewage programme for the city is being implemented and a joint fund from the government and business community is enabling attention to be devoted to environmental matters (information campaigns and education). The government also has its own research laboratory and the water quality is regularly monitored and licences checked. Moreover, Shell in Hong Kong has its own vessel for locating oil pollution.

In the Hong Kong Yearbook (Daryanani, 1995) extensive consideration is given to environmental matters (more extensive than that given to the large-scale spatial projects) although a few reservations are in order. Environmental aims are still stated in very general and vague terms and it remains to be seen whether measures will be taken for the sake of the environment or for economic reasons (Kagan, 1990). An example of this concerns the relocation of the planned terminals 12 and 13 from the north side of Lantau Island to the east, partly due to the concern about noise nuisance for the local residents and the white dolphin. There was insufficient research data and ultimately it turned out to be more attractive for economic reasons to place the terminals elsewhere. Moreover, it is quite usual to buy off certain interests, e.g. those of the fishermen. Finally, the government has a policy of not punishing violations but of tackling problems jointly with the business community. Thus the government does not take the lead but makes itself dependent (also in a financial sense) on the goodwill of the business community and the options which they come up with.

9.4.2. INFLUENCE OF CHINA

In approving the private treaty grants for terminal nine, there was no way round the Land Commission and the influence of China. China made an objection to the operator who was to run terminal nine, the famous firm of Jardine/Matheson. This firm was not popular with China for various reasons. It had been, for example, the direct instigator of the opium wars with China in the previous century. But the final straw for Peking was when a reform proposal from the present governor was approved, in spite of strong opposition from China, partly because of this firm's influence via its representation on Exco. According to Peking, Jardine/Matheson got terminal nine as a thank you for abstaining from the vote.

All in all, the construction of terminal nine was considerably delayed because China threatened to institute legal proceedings if Jardine/Matheson was permitted to operate the terminal. This made it impossible for the firm to take out a loan and the existing terminals came under even more pressure. Ultimately, the firm withdrew. The PDB could not actually do anything about China's objections. An attempt was made to find a solution to this affair through diplomatic channels and in the Joint Liaison Group between the Chinese and the British (the Hong Kong government was deliberately excluded, but certainly had indirect influence). The reason that this group was involved lies in the agreement that the Sino-British Land Commission will always approve projects which exceed the annual volume of 50 ha. land grant, which was the case with CT 9. But despite the approval of the commission, the Chinese dragged their feet. This commission was established among other things to prevent money from the leasing of land disappearing to the United Kingdom. This is not an inconceivable situation: the sale of land is the only substantial revenue for the government because taxation is low, China was thus already exerting influence on Hong Kong investments, and in particular private investments, which meant that the government lost direct control. The 2.2 kilometre-long bridge from Lantau Island to the mainland was financed with public money and then sold to a private entrepreneur for further operation. The new airport behind Lantau Island was financed with public money and will be managed by a private operator. Such constructions are approved by China because the government retains control via the capital invested in it. This is not the case with container terminals.

9.5.   Port Services

Shipping services, such as water and fuel provision, towing services, etc. are all provided by private companies under competition. Due to the many mid-stream activities, there are 350 tugboats in operation which belong for the most part to three major tugboat companies. The pilots come under the pilotage ordinance which is supervised by the director of the Marine Department in connection with compulsory pilotage (over 5,000 GRT and, with dangerous substances, over 1,000 GRT) and the pilotage tariffs. 'The Director of Marine is the Pilotage Authority and he is advised on all pertinent matters by the Pilotage Advisory Committee' (Hong Kong Marine Department, 1995a). In fact, the pilots with their private association form a monopoly (75 licensed pilots in total). Around 3,000 ships are piloted per month (a continuous operation). Following a study in 1993, the conclusion was reached that in the near future compulsory pilotage will be slimmed down and a closer organisational cooperation will need to be introduced between the pilots and the traffic guidance system.

The port has four container operators, of which HIT and MTL are the largest. In addition, Sea Land has its own terminal and it can hire capacity from MTL. For terminal 8, HIT has set up a joint venture with the major Chinese shipowner Cosco. Following the PDB's decision not to interfere in mid-stream activities, HIT and MTL have bought up a large number of lighter enterprises and are now in a position to control the container market. They also own the major feeder companies which sail from Hong Kong to other Chinese ports and have a number of terminals in China. In fact, the whole container handling is run by two container operators. And because trade is booming they can demand exorbitant prices. This appears to confirm Kagan's (1990) conclusions. The development capacity of the port is in the hands of the present terminal operators and it is not in their interest that new terminals should be constructed in the near future.

The major problem for the operators in Hong Kong, however, is that in the short term there is too little land available for all the necessary container activities so that extra time is required for handling activities. The older terminals have only very limited space for storing containers. And with the delay in the construction of CT 9, the existing operators have been forced to make super-efficient use of the space they have[7]. For example, in 1994 Sea Land handled 800,000 containers at one berth. The lighter operators are also short of space because in many places land is being reclaimed so that some public working areas are no longer, or not easily, accessible.

9.6.   Port Management in Hong Kong: Analysis and Questions

The port plays a crucial role in Hong Kong's open economy; for many years, it has been not only the principal international trade gateway to China but it also generates myriad activities which are directly and indirectly connected with Hong Kong as service economy. Hong Kong has thus become not only a *maritime* but also a *financial* centre for Asia.

A wind of change will blow through Hong Kong after the transfer of the area to China in 1997. The value of the agreement that Hong Kong's social and economic freedom and capitalist lifestyle will be preserved for 50 years will only really be known after June 1997. In any case, it may be expected that under Chinese rule the business community will not have as much influence in the national political arena as it does in the present situation. At the moment there is no direct reason to assume that after June 1997 fundamental changes in port management and port planning will occur, so that the present institutional set-up serves as a basis for questions.

In the past, the British were forced to refrain from government intervention in Hong Kong and this ultimately led to the city being able to develop into the foremost free trade centre of Asia. On the basis of this liberal economic regulation it may be expected that the division between imperium and dominium will be very strict. After all, the threshold for government action is very high and there is felt to be no direct need for regulating all kinds of matters.

The empiricism shows, however, that although Hong Kong is labelled a very liberal port it does not entirely live up to its name. As a government agency, for example, the Marine Department takes care not only of the ins and outs of shipping traffic and international legislation but also various private

---

[7] The shortage of space in Hong Kong is the most intensive in the whole world. In 1988, Hong Kong handled 29,432 TEUs per hectare, Singapore 17,078, Kobe 12,896 and Rotterdam 'merely' 8,430 (Kagan, 1990:86).

activities, such as the management of ferry terminals, passenger boats, public terminals, etc. The Marine Department sees itself as service provider to the business community and tries to disaggregate this primarily into various time-saving measures and objectives. How does this success standard relate in the long term to the procedural quality requirements which are linked with its monitoring and supervisory responsibilities? How are the various quality criteria guaranteed?

Another example which contextualises the liberal economic regulation in Hong Kong is the existence of the Port Development Board as advisory co-arrangement. Although it started out as a fairly minor political body, it has now acquired considerable authority since the dispute regarding the role of the lighters and based on its various studies. In the coming years, to what extent will Hong Kong tend towards an oriented market economy due to its need to solve 'common pool' problems? What will be the status of the Port Development Board: a purely advisory body as it is at present or a body tailored to a broader supervision of the port with far-reaching powers?

Another important difference between the normative point of departure and the empirical situation is the existence of a capacity-planning measure such as the 'Trigger Point Agreement'. The measure was intended to safeguard private investments in new container terminals and has resulted in a situation where the further development of the port is in the hands of the container stevedores. How does such a measure relate to the actual need for coordination between the various authorities and the business community in the construction and planning of new terminals and infrastructure? How does the measure relate to safeguarding competition, the cardinal principle of Hong Kong's economy?

Port service provision also has a number of hybrid forms between the public and private sector. The pilots, for example, form a private association, but their market is emphatically regulated by the director of the Marine Department (compulsory pilotage and tariffs). The private chemical waste collection company is designated by the Environmental Department as the sole authorised body. And the prevention of oil pollution is tackled jointly by the government and the business community. The question relating to the Marine Department also applies here: how will the various individual quality standards be safeguarded in the long term? How can the service provided by the pilots be guaranteed in relation to competitive prices? How is the control and supervision of the collection and processing of chemical waste regulated? What safeguards are in place to prevent the government and the business community blaming each other regarding oil pollution because after years of cooperation they have now become too dependent on each other?

9.7.   Summary

The port of Hong Kong is the largest container port in the world. In the last ten years, as a result of the economic development of South-east Asia, tremendous growth figures have been achieved. This trend appears to be continuing. Although the English have withdrawn from Hong Kong, agreement has been reached with the Chinese that in this Special Administrative Region of China rights of ownership will be respected for a period of fifty years.

Hong Kong is well-known for its avoidance of government intervention and the tremendous investment of private capital. This also holds true for port management up to a point. Hong Kong has no traditional port authority in the sense of a simultaneous nautical and land manager. Dry areas and wet areas (literally: water) are sold to the business community for an extended period and are subsequently developed with private capital. In the granting of concessions, market share agreements are made to avoid overcapacity. In the increasingly busy port, however, nautical management remains an exclusively governmental task in which the emphasis lies on optimising the efficiency of implementation. Due to the shortage of space and the tremendous growth figures over the last five years, the call for coordination has now been clearly heard. A distinctive feature of Hong Kong's approach are the committees in which government departments and the business community cooperate on an equal footing.

CHAPTER 10

The Port of Singapore

'A democracy, with its full panoply of institutions and processes, that does not produce the greatest well-being of the greatest number is obviously not functioning according to its design and can be considered as little more than a facade, a sham' (Vasil, 1992:xi).

10.1.   Introduction

This chapter addresses the institutional map of the port of Singapore. The city-state of Singapore has three million inhabitants and lies at the southern tip of the Malay Peninsula. Its location on the Strait of Malacca forms the gateway for east-west trade. The port of Singapore is the busiest in the world and its logistical function can be compared with that of the container hub and port city of Hong Kong[1]. As far as port administration is concerned, however, there are substantial differences between these two eastern giants. Whereas in Hong Kong nearly all port activities are in private hands, the port of Singapore is entirely managed and regulated by a public corporation.

The situation in Singapore will be described and analysed in the following way. The first section describes the economy and sketches the significance of the port for the country. The next three sections provide an account of the institutional structure of the port, broken down into the formal position of the port authority, the method of port planning and the organisation of the various port services. Finally, the descriptions are analysed and a number of questions posed.

10.2   The Economy and Significance of the Port

In no other country in the world are people so aware as in Singapore that to earn a living you need to work hard and be smart. The fact is that the country, which covers an area of 636 square kilometres, has no natural resources and is entirely dependent on trade opportunities and foreign investment. The natural harbour, the labour force, the predictable and stable government and the good infrastructure are the only resources which the country has (Vasil, 1992:165).

---

[1] At any time of day, over 800 ships will be found in the port of Singapore. In 1996, 117,723 ships with a registered dead weight tonnage of 768.5 million tons put at the port.

Singapore has a market-based economy. This means that the government 'frames policy directions, invests in infrastructure, provides housing, education and health services, and ensures a conducive business environment for the private sector to expand and upgrade. In a nutshell, the government's role was to make Singapore one of the easiest places in the world to do business' (Economic Development Board, 1993:4). On page 9 (ibid.) it summarises: 'The government's role in economic development and management in Singapore is to set policy directions, provide the institutional and regulatory framework and develop efficient infrastructure'.

Location of Singapore in Southeast Asia

Although formally speaking Singapore has a market-based economy, the division between the public and private spheres is paper-thin. The bureaucracy is small; policy preparation is carried out by the various ministries but policy implementation is realised by the public corporations which manage and operate private activities or enter into participation with private (foreign) corporations. In 1995, the Singapore government participated in over 3,000 multinational corporations via its statutory boards (Economic Development Board, 1995:2).

Another economic principle which is followed is that of linking Singapore's economy with the global economy by (promoting) foreign investment and implementing a free trade policy (free port) so that Singapore's goods compete on an international market. A third economic principle that exists is the system of *meritocracy*; a person is rewarded in accordance with his/her contribution to the total, with the guarantee of equal opportunities for all. The principle of meritocracy has made the city-state of Singapore a real corporation; in 1994, the revenue of the Singapore government amounted to over 23 billion S$ compared to a 9.6 billion S$ expenditure (Ministry of Information and the Arts, 1995: appendix 26).

When it became clear in the sixties that Singapore's economy rested entirely on the transit trade without any value being added to it or any items actually being produced, the PAP leaders had to admit that this was not a healthy situation from which to try and transcend its Third World status. With the aid of various studies conducted under the guidance of the United States (Winsemius report) and financial resources from the World Bank, a new economic perspective was offered. Attempts would be made to attract industries which would thrive on the low wages and reliable labour force. The Economic Development Board (EDB) played an important role in capturing companies through their permanent acquisition offices in the world's major industrial cities.

Companies were attracted to Singapore by means of its internationally open market economy (profits may be pumped back into parent companies), low tax levies and financial inducements. This cut both ways; on the one hand, multinational manufacturing companies could safeguard their profits and on the other, employment increased significantly in Singapore and people learnt a trade. The major industries at present are the electronics and chemical industries, oil refining, mechanical engineering and such things as metal products, electrical machinery and applications, transport equipment, etc. Singapore's three most important trade partners are Japan, Malaysia and the United States in that order. Due to the open character of the economy, economic fluctuations in these countries have a major impact on Singapore. The following Table shows the import and export values.

Until the mid-eighties, the economy of Singapore grew steadily. After that time, Singapore became too expensive for labour-intensive companies and the political leadership was forced to rethink its economic policy. There was a shift to attracting high-quality industrial labour and Singapore developed into the most important (petro)chemical location in Southeast Asia.

Table 10.1.Value of Imports and Exports in 1994 in million Singapore dollars
(Source: Singapore Ministry of Information and the Arts, 1995:296-297).

Product	Import	Export
Food	5,217.9	3,542.4
Beverages and tobacco	2,141.3	2,447.5
Raw materials	1,938.5	2,193.2
Mineral fuels	13,787.8	14,074.8
Animal and vegetable oils	640.9	574.1
Chemicals	10,113.6	8,418.0
Industrial goods	16,523.4	8,855.8
Machines and equipment	88,306.3	94,198.7
Various industrial products	15,427.7	11,185.2
Other	2,298.4	1,837.6
Total value	156,395.8	147,327.2

After the brief recession of 1985/86, which was the result of declining world trade, over-hasty wage increases and other price increases, the Singapore economy revived remarkably quickly. At present, growth percentages of around 10% are being achieved. A problem for Singapore's economy in the near future will be getting well-trained personnel. Foreigners are only allowed limited contracts and the government has already exhorted the population to increase the birth rate. Moreover, the physical space for new initiatives is decreasing.

For the coming years, the government has put regionalisation high on the politico-economic agenda. 'Singapore's Next Lap in national development aims at achieving a balanced economy and a developed country status and standard of living. Investing in the region is part of Singapore's long-term strategy to move up the economic ladder. By going regional, Singapore can participate in the region's growth by interlinking the regional economies with own domestic economy, especially the manufacturing sector. Up to 30% of the Republic's reserves will be gradually invested in regional economies to build up this external economy to contribute and participate in the region's growth' (Economic Development Board, 1995:3). Now that there is a threatened shortage of people and that by regional standards huge financial reserves have been created, there is a desire to earn money long-term with the aid of the regional dimension. Together with the business community, projects have already been initiated in China[2], India, Indonesia, Vietnam, South Korea, Hong Kong and Italy (Port of Singapore Authority, 1996). Whether or not in

---

[2] Singapore's investment in a China which is becoming increasingly open economically should come as no surprise. The Chinese businessmen have always been an important background partner in the economic construction of Singapore (Vasil, 1992:164). Moreover, over 80% of the population of Singapore is Chinese.

cooperation with the business community, Singapore buys into activities with good commercial expectations. This is its only option for profiting long-term from the economic growth. A picture of a new kind of colonialism is beginning to emerge, however.

As has already been mentioned, the only natural asset that Singapore has is the port. Due to its limited geographic size there is not only an economic but also a strong physical link between the city and the port. In the last ten years, the port of Singapore has experienced unprecedented growth. The graph

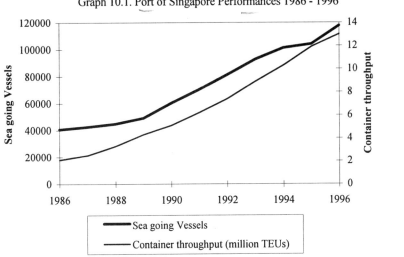

Graph 10.1. Port of Singapore Performances 1986 - 1996

illustrates this. In 1996, 242.5 million metric tons were handled and no less than 12.9 million containers were loaded and unloaded[3].

At present, the transport sector as a whole provides direct and indirect jobs for 175,000 people in Singapore. The port of Singapore is well-known principally for container handling and performs a hub function for the region (Vietnam, Malaysia, Indonesia). In contrast to the more hazardous (petro)chemical industry, the container terminals and the distribution centres are located on the outskirts of the city, while the former are located on a separate group of islands outside the port. Almost all the activities in the port are run by the *Port of Singapore Authority* which has two distribution parks and six terminals where every type of ship can be handled.

---

[3] In 1994 at one point it looked as if Singapore was going to take over Rotterdam's top position in the world but a study conducted by the Rotterdam Municipal Port Management department made it clear that Singapore was using different standards.

10.3.   The Position of the Singapore Port Authority

The port is managed by the Port of Singapore Authority (PSA). This port authority, which employs 7,500 people, is nautical manager, environmental enforcer, port planner and port service provider rolled into one. This explains the large number of staff employed. The PSA was established by law in 1964, in conformity with the government's general desire to control economic activities and its specific desire to develop port infrastructure. As a public corporation it has to perform a variety of functions and tasks. Article 9 of the Port of Singapore Authority Act gives an almost exhaustive account of these tasks; 'it shall be the duty of the Authority -

- to provide and maintain adequate and efficient port services and facilities in the port;
- to regulate and control navigation within the limits of the port and the approaches to the port and to provide pilotage services;
- to promote the use, improvement and development of the port;
- to provide and maintain adequate and efficient lighthouses, beacons, buoys and other aids to navigation in the territorial waters of Singapore and the approaches thereto (...);
- to disseminate navigational information;
- to provide, where expedient, a ferry service for the transportation of passengers, vehicles or goods within the territorial waters of Singapore; and
- to carry out such other duties as are imposed upon the Authority by this Act and any other written law'.

In addition to these tasks, the PSA has the responsibility for the functions described in the second section of the Act (Port of Singapore Act, pp. 85-86). Twenty of these functions are exclusively listed and range from the provision of fuel and water for ships, organising the fire brigade and housing for port employees, imposing and enforcing regulations pertaining to the prevention of pollution of territorial waters and the actual cleanup operations. From this extensive list of activities it is clear that the PSA has complete legal control over the port and the territorial waters. Even if a private company is in the port for a particular port activity, it is wholly dependent on the PSA to grant it a licence.

The port is of major economic significance to the city-state of Singapore. Its strategic location is the principal reason for its success. When it became an independent republic, the government chose to consolidate and further develop the country by means of full control over the most crucial economic sectors. The port, which attracts freight and many allied activities (storage, distribution, processing, etc.), was and is the prime government instrument to help Singapore become an international power. This is why in the past it was chosen to let the port authority operate as a *statutory board* (public corporation) with both private, and far-reaching public, powers. The Port of Singapore Authority is a monopolist in many areas of port activities.

It is worthwhile to look at how the institutional structure of such a statutory board serves to make it actively accountable for renewal and good service provision. The PSA's legal format also gives it both political trust and broad freedom of action. As a public corporation, political management only occurs along very general lines. Via various mechanisms it has, in so far as it can actually be separated from the bureaucracy, the possibility of safeguarding its control. The following section will examine the special nature and the specific structural elements of a 'statutory board'.

10.3.1. PORT OF SINGAPORE AUTHORITY AS 'STATUTORY BOARD'

In Singapore it is usual for the task of policy preparation to take place within small ministries and for the task of implementation to be carried out by 'statutory boards' which can be seen as a specific form of *public corporations*. Numerous public corporations were set up in Singapore when it became a self-governing state and the People's Action Party was founded in June 1959. Their specific aim was to promote *national development*. Three reasons can be mentioned for setting up various public corporations (Quah, Heng Chee and Che Meow, 1985:125):

- a public corporations model was chosen because the emphasis lay on maximising *efficiency* in task implementation without being hampered by bureaucratic restrictions (e.g. strict regulation and a lack of flexibility). The civil service in Singapore is located at the Ministries and is responsible for the regulatory and routine tasks. The government's philosophy has always been that the public corporations should break even. Making a profit is not a priority since the main thing is functional success and survival. The Port of Singapore Authority group (including direct and indirect participation in other companies) is highly successful and made a net profit of 717 million Singapore dollars in 1994[4];
- public corporations were set up to reduce the workload for the civil service and

---

[4] In 1995, the tax payments for the whole PSA group comprised 9.2 million S$. The PSA as port manager is exempted from paying income tax although the companies related to it are not. In 1995, PSA had to contribute S$ 173 million to the Government's Consolidated Fund. The size of the net profit in 1995 could not be deduced from the annual report.

- they provided the opportunity to attract competent people from both the public and private sphere who would be able to participate in the various development programmes by being part of the highest administrative body.

The Singapore public corporations are corporate bodies which are set up by means of special Parliamentary acts or statutes. These acts specify their functions, their various commitments, responsibilities and powers, the management composition and the relationship with the minister responsible (for PSA this is the Minister of *Communications*). Moreover, the arrangements with regard to the appointing of staff, the financial provisions, accounting methods and audit are all laid down by law. The public corporation has an independent legal status, separate from the public service, and has legal personality. However, it does not enjoy the legal privileges and immunity which the departments have. On the other hand, the 'statutory boards' have a greater degree of autonomy and flexibility in their day-to-day affairs than the government departments. The latter have no individual legal personality, they obtain their funds via annual approved government budgets and consist of civil servants. The specific institutional characteristics of 'statutory boards' are intended to enable them to react more quickly to changing situations and to tackle problems more effectively.

Generally speaking, the legally independent public corporations in Singapore have the following characteristics (Nguyen-Truong, 1976:172-173). In the first place, each 'statutory board' is created by means of an ordinance or Act of Parliament which establishes its legally independent status. This status gives it the right to autonomy regarding entering into commitments and reducing or increasing its property. Moreover, the Singapore public corporation has a typically individual organisation chart which comprises various layers. The first layer is that of the *Board of Directors* whose chairman is appointed by the minister for two years. The second layer in the hierarchy is formed by the management team that normally consists of a Chief Executive Officer, a secretary and the heads of the various departments. The management is appointed by the Board of Directors and the Chief Executive Officer is usually also a member of the Board. Finally there are the support staff which comprise the administrative personnel, executive personnel and administrative employees all of whom are appointed by the Board of Directors.

The size of the Board of Directors ranges from between 6 and 17 members; at present, PSA has 12 part-time members. The chairman of the Board is usually a Member of Parliament or top civil servant from the ministry concerned. The other Board members are generally top civil servants, entrepreneurs, businessmen, representatives of unions or professional groups, or from the academic world. It is not unusual for a top civil servant to be a member of various Boards simultaneously. This is probably due to the shortage of highly-qualified personnel and the fact that top civil servants are accustomed to assuming quasi-political responsibility. In this phase of economic progress, the chances of a concentration of power in '*old boys*' network is very high.

(4) A fourth institutional feature of the public corporation in Singapore is that the employees do not have civil servant status because they are not recruited and appointed by the *Public Service Commission*. The Board of Directors employs its own salary scales, fringe benefits, opportunities for promotion and internal control instruments. These payments are usually similar to those of the public service but there are differences to express the specific functions of the Board. A fifth institutional feature is the fact that the public corporations aimed at providing infrastructure are expected to cover their expenses with their own income (self-sufficient). The Boards of Directors of these public corporations are allowed to make investments which are not directly required and the revenue from the various activities may be reserved. For PSA these are the reserves made for the benefit of the pension fund. The opportunity to participate in many different port-related activities makes the (financial) audit of public corporations even more complex.

Finally, a last institutional feature is that the accounts of the public corporations are audited by the *Auditor General* (Ministry of Finance) or by any other person who is charged with the task and certified by the Minister. The Board of Directors' annual budget estimates must be approved by the relevant minister. Moreover, the Boards must submit their financial reports and annual reports to the minister so that he can send them to Parliament.

10.3.2. POLITICAL CONTROL OF 'STATUTORY BOARDS'

The relevant ministers' power and authority over the various types of public corporations varies. In the case of the 'statutory boards', the control options are laid down in the various incorporation Acts and statutes. Normally, these allow for the provision of general directions with regard to policy implementation, the exercise of powers and for the Board of Directors to delineate its own function. For the Port of Singapore Authority, the Minister of Communications is also empowered to give *specific directions* with the aim of improving the functioning of the Board. He may only give specific directions, however, after consulting the Board of Directors of the PSA (art. 11, PSA Act).

The Minister of Communications appoints the members of the Board of Directors and, together with the Public Service Commission, is consulted regarding the appointment of the PSA Chief Executive Officer. The Minister also has to approve this appointment (art. 29, PSA Act) and can discharge members of the Board of Directors before they have completed their full term. Via the appointment of various top civil servants on the Board of Directors the Minister can clearly demonstrate his relative power and coordination between the Ministry and the PSA is safeguarded to some extent. It is not the intention, however, that the appointed civil servants should interfere too much in the day-to-day running of the organisation because this would represent an incursion on the desired principle of autonomy. The Minister of Finance is empowered to appraise and approve or reject financial matters relating to the various public corporations. These might involve such things as increasing loans from non-government funds, creating and issuing acknowledgements of debt, cashing in reserves and deciding terms and conditions under which loans will be repaid, etc. The authority of the Minister to approve or reject matters also extends to the appointing of the private auditor.

The most important policy decisions with regard to the PSA are taken by the Minister of Communications and after consultation with the other members of the Cabinet. The Board of Directors of the PSA is responsible for the tactical policy level, such as appointing the various departmental heads, budgetary control, investment of profits and public relations. In addition to these tasks, the Board of Directors of the PSA is responsible on occasion for making available to the Minister information requested by him.

The chain of decision making within the PSA is organised as follows. The management under the Chief Executive Officer is responsible for the day-to-day administration within the framework of established policy. In order to achieve an effective delegation of powers and responsibilities and an efficient coordination of information, the Chief Executive Officer is a member of the Board of Directors. In contrast to the relationship between the Minister and the Board, it is not usual for the relationship between the Board and the management to be tied to legal rules. It has been suggested that the best Board is one which has *de jure* powers to determine policy but is a *de facto* advisor to the Chief Executive Officer. In order to provide legal embedding for this practice and desire, the Chief Executive Officer should also be directly appointed by the Minister. Singapore has not chosen to do this, however. The Chief Executive Officer is appointed by the Board of Directors and in some cases, as with the PSA, the appointment needs the approval of the Minister. In practice, the degree of autonomy of a public corporation varies according to the national political interest that attaches to the organisation, the nature of its functions and the political power and personalities of the Minister and the chairman of the Board of Directors. Such findings are also applicable to the relationship between the Board of Directors and the management. The need for Ministerial approval of the Chief Executive Officer reflects the major importance of the port to the city-state. 'In addition to the specific powers given to Ministers, it was discovered that Ministers, in fact, exercised significant influence over the policy of nationalized industries by their frequent and informal discussions with the Chairmen of the Boards. More influence was found to be generated not through the power to give directions, but through the power to appoint and reappoint members of the governing body without external influence on his selection' (Nalliah Pillai, 1983:100).

Parliament is excluded to some extent from control over the public corporations. There has been no political opposition for thirty years and the only institutional instrument that might possibly serve as a psychological deterrent is the *Public Accounts Committee* which examines whether public monies really are being used for the objectives in question (Chung Pui Hoong, 1984:62). After an Act has been passed sanctioning the creation of a 'statutory board', the organisation becomes an institutionalised fact and parliament only obtains information regarding its day-to-day operations indirectly via the Minister. Parliamentary questions concern only the strategic issues which are far removed from the problems of daily management. It is no wonder that Von Alten (1995:209) concludes that the public corporations in Singapore are managed and controlled by an exclusive network of technocrats.

The institutional control at the PSA works on four levels. On the first level, there is the powerful and influential *Directorship and Consultancy Appointments Council* which comprises seven important ministers and top civil servants. The Council makes proposals for the appointment and removal of members of the Boards of Directors of the larger 'statutory boards'. The second level of PSA is that of the Minister of Communications who also has responsibility for a variety of other public corporations and participation in other companies. The relations between the Minister and the third level (the members of the Board of Directors) are very close and direct which makes it easy to pass information on to the first level and the prime minister. The final control level is that of the chairman of the Board of Directors of the PSA. Most Board members are appointed for political reasons or, to be more precise, they are appointed thanks to their relations with the inner circle surrounding the ex-prime minister (Von Alten, 1995:210). So it should come as no surprise to learn that there are many relations between the various control levels and within each individual control level.

10.3.3. PRIVATISATION OF PSA

Following the recession of 1985/86 there was increased criticism of the economic structure of Singapore. Until then, this structure had always had a solid twin base, i.e. the multinationals on the one hand and the Singapore government-related companies on the other. As mentioned above, the public corporations in their turn are linked to a variety of other private companies. Until 1985/86 the national private corporations did not play a major role. As long as the economy continued to prosper, they played the part of suppliers for industry. However, when the multinationals transferred their production lines to other, cheaper countries these private 'subcontractors' suffered huge losses. They started to complain to the government about the cutthroat and unfair competition, two aspects to which the Singapore government is highly sensitive.

The government dismissed the claims, however, and emphasised that there was no evidence of the public corporations receiving preferential treatment. The list of privileges for public corporations is a long one, however, and attests to a rather different picture. They enjoy a special position either by means of licences or concessions to undertake activities or special agreements with the government to acquire numerous products at a cheap rate, etc. (Von Alten, 1995:213). It is equally understandable that the Singapore government wants to keep a tight rein on the economy because otherwise it would not be possible to transform it from an entrepôt economy to a more diversified one.

In the general discussion on privatisation in Singapore, there is a high degree of congruence 'between the aims of the PAP and conservative-liberalism. Changes are coming in the form of allowing the private sector to have more economic liberalization with privatization, which is inevitable in view of the changing socio-economic environment. But this is not being translated into political emancipation, because conservatism as exemplified in Western Europe, clearly distinguishes politics and economics. This reinforces the one-party dominance that the government in Singapore seeks to perpetuate. Also, it enables political control of economic power so that even as the private sector becomes the engine of economic growth and development, the state will continue to be its driver and director' (Low, 1987:87). The recession of 1985/86 made it clear that there could be no wage increases because the country was still stuck in a position between First and Second World. After 1986, a tremendous 'technology push' was given in a whole range of economic sectors to improve the economy and in order to achieve the status of a First World country within a number of years (Singapore Ministry of Trade and Industry, 1991). A principal reason which guarantees the survival of the Singapore public corporations is the fact that in that case the taxes can remain low and the government does not depend on its own people for its revenue.

In the framework of Singapore's economic developments (large reserves owned by government companies, limited national labour force and decreasing amount of available land) opportunities were sought to earn money via projects in foreign countries with links with Singapore. This regionalisation tendency (see section 2) also had consequences for the position of PSA. First of all, a '*corporatisation*' of PSA was initiated for an indeterminate period (Containerisation International, 1995:79) which involved an organisational division between private and public tasks but with ownership for the time being remaining entirely in government hands. Since 1 February 1996, the former PSA has been split into two separate public corporations: the *Marine Port Authority* which is responsible for nautical management and the maritime service provision. The PSA retains responsibility only for goods handling. According to the PSA, the reasons for this split are threefold:

- to become more independent of the government and to be able to react more quickly to developments in the market;
- to be able to invest independently in the Southeast Asia region because otherwise it might smack of neo-colonialism (cf. Griffiths, 1995) and
- to increase awareness of market forces among its own staff.

The ultimate goal is to float PSA on the stock exchange. After the period of 'corporatisation' it is felt that PSA will be ready for the market and be able to operate as a real market party with private money and private risks. Until that time, the division remains merely an organisational split. The privatisation process is progressing slowly and carefully. After all, it is the Singapore government and parliament which will have to take the decision. An important reason for its slow progress couold be the politico-economic philosophy of the PAP which is still based on the fact that the government rules economic affairs (cf. Low, 1987). 'Although the PAP has decided to reduce its excessive participation in Singapore's economy through privatisation and although a more flexible wage system was developed in 1987, the shift to greater liberalisation has not been very strong. Despite the realised measures the government is still not willing to limit its direct control of economic activities in Singapore. After almost thirty years of growth - enforced by the economic policy of the PAP - the government is not convinced that the private sector can do better when left on its own' (Von Alten, 1995:38).

The major shipping companies in particular are keen on a privatised PSA and are gambling on a future share in the container handling company. For the time being, PSA wants nothing to do with 'dedicated terminals' in any shape or form because this would mean the loss of the successful '*multi-user principle*' and PSA would lose control of the terminals. It thus remains to be seen whether the Singapore government will want to surrender its control over the coordination of the terminals and the profits involved. For the moment it may be concluded that the function of the PSA as 'statutory board' has changed: from a means for national economic development in its first thirty years to a means for guaranteeing economic expansion and revenue for an independent Singapore. In other words, an attempt is being made nowadays to effect its transformation to *regional hub coordinator* (cf. Griffiths, 1995).

10.3.4. PSA'S ACCOUNTABILITY

The Singapore public corporation is institutional, 'based on the theory that a full measure of accountability can be imposed on a public authority without requiring it to be subject to ministerial control in its managerial decisions and multitudinous routine activities, or liable to comprehensive Parliamentary scrutiny of its day-to-day workings. The theory assumes that policy, in major matters at least, can be distinguished from management or administration; and that a successful combination of political control and managerial freedom can be achieved by reserving certain powers of decision in matters of major importance to Ministers answerable to Parliament whilst leaving everything else to the discretion of the public corporation acting within its legal competence. The government is further endowed with the residual powers of direction and appointment which marks its unquestionable authority' (Nalliah Pillai, 1983:99). PSA is formally under the control and supervision of the Minister concerned and indirectly under that of Parliament. 'Essentially it is this autonomy-with-control feature that makes statutory boards the most popular form of apparatus for implementing development policies and providing complementary infrastructure. But the proper balance between autonomy and control is in fact the most difficult task facing the government which opts to use this type of organisation' (Nguyen-Truong, 1976:171).

The constant search for the most appropriate form of distance also makes it extremely awkward to discover the exact control mechanisms which are intended to induce the PSA (pre and post split) to be accountable for the policy it pursues. It is primarily the very direct intertwining of public and private tasks with those of politicians and top civil servants which make it almost impossible to discover what the control mechanisms of this institutional structure are.

Neither is the position of shipping companies and other ports in the area such that they either have or offer alternatives respectively. Due to its unique location, Singapore has shipping traffic and trade in abundance. PSA does not undergo any direct competition from Malaysia or Indonesia because either the ports in those countries are not sufficiently developed or they have permanent dredging problems. Due to its legal monopoly position, the survival of PSA is not at stake and there still appears to be room for growth although this can only be achieved in the medium term (construction of new terminals in its own port and abroad). With the huge profits that it clears year after year, the PSA seems increasingly to be becoming a state within a state. This impression is bolstered by the 'corporatisation' and the creation of the opportunity to make autonomous regional investments. The main question remains of whether PSA will indeed be floated on the stock exchange.

10.4.   Port Planning

The port area of Singapore can be split into two geographical parts: the city section and a combination of islands which will be turned into one large island over the next few years by means of the Jurong project. Adjoining the city section there are the container handling facilities, the distribution centres and the ship repair yards, while in the Jurong area the petrochemical industry is flourishing. The main distinction between the two areas lies in the management of the property: the city section is managed and operated by PSA and the Jurong area is leased by the Jurong Town Corporation to private companies (Shell, Van Ommeren Tank Terminal, etc.). The Jurong Town Corporation is a 'statutory board' like the PSA. As regards container handling, all infra- and superstructure is financed by PSA which passes the costs on to the users. The Jurong Town Corporation is responsible for industrial locations and the business community has to cough up the necessary funds for the superstructure unless PSA or the Economic Development Board has a part interest in the location.

Decision Making on Expansion Plans
Major expansion plans (such as e.g. the construction of the new Pasir Panjang terminal) are prepared by PSA itself and the temporary approval is given by the Board. The plan then goes to the *Master Plan Committee* which assesses it from the perspective of other disciplines. The Committee contains representatives from various Ministries (finance, trade, spatial planning, environment) which apply their own quality criteria to the plan and pass judgement on it. The Committee has substantial control and determines whether the plan can go ahead. PSA has no representation on the Committee but can exert indirect influence on the decision making via its own Board of Directors.

A project also needs to have cabinet approval if it covers an area in excess of four hectares, in which relatively large amounts of money are involved, which plays a role for the country as a whole and which is generally a long-term undertaking. PSA develops all plans which are related to the port area while the *Urban Redevelopment Authority* is responsible for developments on the landside (land which has no port function). The Urban Redevelopment Authority can also voice its reactions to PSA plans to the Master Plan Committee. In the construction of the first container terminals, the decision making went through different channels and there was a more direct line to the cabinet via Board members. The Board had considerably more control in those days than it does now because the link with the cabinet was more of a formality. Now that the Master Plan Committee has come between them this direct influence is no longer possible (Lim Kay Hwan, 1976).

Singapore Port Terminals

The construction work on the new Pasir Panjang terminal is being carried out under tender to private companies. The terminal itself has been designed entirely by PSA, however, which is also supervising its implementation. Major engineering projects are usually carried out by the *Public Works* division of the *National Development Department* but because of the specific requirements attaching to the construction of this particular terminal, they returned the contract.

Another major project which will take place within the port area over the next ten years is the consolidation of various small islands into one big island (Jurong). This will provide sites for the (petro)chemical industry. This project does not come under the responsibility of the PSA, however, but under that of the Jurong Town Corporation. As a public corporation it is responsible for the development of industrial sites. The Maritime Port Authority sees to the supervision of shipping traffic in this area, however. In theory, if PSA wants to develop an initiative in this area, it must buy a site from JTC. The administration of the port is now divided into specific sectors which was not yet the case fifteen years ago. This trend is likely to continue in the future. Each public corporation has full responsibility for its own task. Whether this will improve coordination remains to be seen.

Commercial Leasing of Sites

In principle, PSA maintains no contacts with industrial site lessees because it needs all the land for its own activities (terminals and distribution centres). In the early eighties, however, it entered into a lease with Van Ommeren Tankopslag which wanted to construct a storage and transhipment installation for the petrochemical industry on one of the islands (at present this power rests with Jurong Town Corporation). A 30-year lease was signed and the commercially-minded PSA stipulated that Van Ommeren had to take an option on the whole island and that the conditions of the lease would be renegotiated at the termination of the lease period. For PSA, this renegotiation meant that after thirty years it would regain control of the whole island including all the installations paid for by the business community.

Ten years into the lease, when Van Ommeren decided to expand over the rest of the island, it appeared that under the current contract they would not have sufficient time to write off the new investments. They wanted to renegotiate and only then discovered PSA's crafty construction for the expiry of all rights of ownership. In order to get out of this construction, Van Ommeren had to pay a premium to ultimately gain a fifteen year extension on the contract. Due to the tremendous economic growth in this area and Singapore's prime position as a trade and industry partner, the Singapore government is in a position to take a very hard line in negotiations at present. It pursues a highly selective and commercial acquisition and leasing policy. All potential activities are chiefly assessed according to commercial criteria. Environmental considerations scarcely play any role since they are still in their infancy and are principally controlled by international standards. Moreover, multinational corporations work with installations that have been devised in developed countries and meet higher standards than are customary in Singapore. Due to the Singapore government's emphasis on commercial criteria, it is not until private companies actually threaten to leave that the government is willing to renegotiate earlier agreements. On the one hand, such a hard line provides companies with security because they know exactly where they stand. On the other hand, such inflexibility can lead to irritation and a bureaucratic image.

Another planning tactic employed by the Singapore government is that when a new commercial enterprise becomes established in the area, one of the government companies participates in it financially. This cuts both ways: the company remains up to date on the most recent political developments and the Singapore government has a share in the profits from economic growth. As soon as an initiative reaches full maturity, the Singapore government pulls out and participates in a new project. In this way many co-arrangements are forged.

10.5.   Port Services

As far as the institutional structure of the port services is concerned, we can be brief. As the summary of PSA activities in section three made clear, there is no competition in port services because PSA controls all the activities. With the new 'corporatisation' and the ensuing privatisation, the Marine Port Authority will have total control of the nautical service provision. Until 1999, the PSA will remain a monopolist in the highly profitable container handling sector. To a lesser extent, the absence of competition also applies to tugboat services because in addition to the Marine Port Authority there are potentially two other companies. These must meet the requirements laid down by the Marine Port Authority (subject to licensing procedures). Moreover, they are only used when the towing services required exceed the Marine Port Authority's capacity. The private tugboat companies hope that a more extensive privatisation will mean that the authority to issue licences will be transferred from the PSA to the Ministry of Communications which really will open up the towing market. At present, there is a fairly clearly-delineated area for each of the towing companies and the other two can report to the Marine Port Authority at the appropriate time. The organisation itself owns a number of boats and leases the remaining tugboat capacity from the private providers. The Marine Port Authority also has six water tankers and five smaller ships for collecting domestic and other waste. The berthing of ships is also taken care of by its own tugboat companies.

Pilotage of ships is a component of the Marine Port Authority. In certain areas of the port, ships are required to use a pilot. The electronic information system, Portnet, enables the pilotage and towing services to be directly coordinated. The Marine Port Authority has its own police force which comprises 374 police and safety officers. All vehicles carrying freight are checked at the various gates to ensure that the correct documents and cargo are being transported.

The fire services are taken care of by 120 Marine Port Authority staff members who have seven pumps at their disposal. They not only come into action if there is a fire but also in the event of oil spills, and they regularly check the fire equipment installed in the buildings within the port area. The harbour service monitors the water quality and cleans up oil spills. The Marine Port Authority also sees to the collection of ship's waste (domestic and chemical) and supervises the handling of dangerous substances.

Relationship with Employees
The unions have played an important part in the history of Singapore's independence. 'Unions played a vital role in the formation and later growth of the PAP. During the early years, the party had presented itself as a party for

the workers and was viewed as such by others. Although union leaders had been substantially responsible for the electoral successes of the party during its critical formative years, the pre-eminent leaders of the party never included any unionists. The PAP was not modelled on the British Labour Party. It did not allow direct representation of unions at different levels of its organisation' (Vasil, 1992). The PSA, with its 7,500-strong staff and high level of union membership (over 70%) is dependent on the wishes of the unions.

Two unions are active in the port: the *Port Officers Union* for management personnel and the *Singapore Port Workers Union* (SPWU) for the port workers. Collective negotiations only take place with the SPWU; the managers negotiate separately with PSA. Until now, wages and fringe benefits have always been discussed in effective consultation. 'The working relation between the SPWU and PSA is a strong one, based on frank and open communications. There is without doubt a spirit of trust and co-operation between the two parties. Both recognise that workers and management depend on each other for the well-being of the workers and the efficiency of the port' (Singapore Port Workers Union, 1986:7). Joint projects for the well-being of the employees have been undertaken, such as the provision of cheap meals and housing. In the past, PSA and the union were regularly involved with each other financially on a project basis. At present, PSA runs training courses for its own staff and for foreign interested parties (17,000 in total in 1995) via the Singapore Port Institute. State of the art simulator techniques are used among other things.

In the early eighties in particular, the union played an important role in disaggregating the necessary technological innovations to the PSA employees. PSA and the union were once again closely involved in this. 'In 1981, a joint union-management steering committee was formed to promote Quality Control Circles (QCCs). The QCC was introduced to further improve working conditions, work methods, productivity, and labour-management co-operation' (SPWU, 1986:54). In 1985, the SPWU arranged that in future there would be a monthly rather than a weekly wage which finally brought an end to the double negotiations of the previous thirty years. Today, the union is very involved in the strategic management of PSA and consultation takes place on a monthly basis (Port of Singapore Authority, 1996).

10.6.   Port Management in Singapore: Analysis and Questions

The port of Singapore, as the country's only natural asset, forms the heart of the Singapore economy. The port has played a crucial role in the economic development of Singapore. Initially as a transhipment location between west and east, at a later stage as the basis for the construction of its own industry, and in the present period increasingly as a coordinator of trade to other parts of Asia. In order to be able to develop into a first world country, in the sixties the government opted for total control over the crucial economic sectors. It was aware that the country was operating in an international environment and the products and services needed to meet international standards. By opting for an open economy all port activities, albeit under government ownership, were run right from the outset as private activities which had to attract as much cargo, and later industry, as possible. Finally the government managed to run the whole country as one big enterprise with a profit motive.

Principally due to the recently-introduced economic regionalisation policy, Singapore has become a national state operating in a world market. The combination of total government control and operating the port as a company was found in the 'statutory board' construction. As a public corporation, the Port of Singapore Authority was able to take on both the public nautical management and the commercial management of the handling activities. Moreover, over the years it played an increasing role in numerous private (industrial) corporations which had some link with the port.

It proved very awkward to identify the institutional control mechanisms of the Port of Singapore Authority. On the one hand there is the mixing of public powers and private activities (towing services and container handling). On the other hand there is the close intertwining of politicians and top civil servants: an 'old boys network' with its concomitant mutual back-scratching. The unique location of the port, the rapid economic growth and the way political and economic power are both held in the same hands have made the country and the port what it is today. advantages

It will be interesting to see what happens in the near future when the PSA is privatised. The question then will be how much control will the government want to keep? How will PSA's relationship with other private companies change? Which activities will the Marine Port Authority continue to undertake itself and what will be its relationship with the Ministry of Communications? Assuming that the privatisation of PSA really will go ahead, the profit motive (growth) will become the first matter of importance. The survival of the port of Singapore would seem assured for the present. It need expect little competition from ports in Malaysia, Indonesia and the Philippines over the next decade since:

- these ports are less strategically sited;
- they are not natural harbours, so that suitable access can only be achieved by incurring tremendous dredging costs;
- Singapore has highly modern handling techniques and relatively highly-qualified people which make port operations considerably more efficient;
- Singapore is already engaged in constructing good hinterland connections to Malaysia and

■ the port as corporation is given the organisational option to invest in ports in other countries.

It is far more likely that the state of Singapore as corporation will encounter competition from the major container shipping companies which want to decide for themselves which ports in the region they sail to. In other words, the question is whether Singapore will acquire a position as hub coordinator in the future or whether the major shipping companies will claim that function for themselves. This will depend on what financial capability Singapore is willing to invest in such an effort. Moreover, the question is whether in the long term the governments of those countries might not want to profit from the growth in container traffic themselves. And finally, a lot will depend on the developments in world trade in this area; the positive prognoses still have to be actualised.

If the PAP really does opt to change its politico-economic principles in favour of private initiative, there will be no getting round the fact that the competition principle will have to be applied to more port activities. In the coming years it remains to be seen whether the PAP, having made a complete u-turn, will be able to formulate an equally consistent long-term vision of economic role division as it has had over the last thirty years.

10.7. Summary

Over the last ten years the port of Singapore, just as the container hub Hong Kong, has undergone a period of unprecedented growth. The most striking difference between the two, however, is that in Singapore the various port activities have all flourished under total government coordination. As a government corporation, the Port of Singapore Authority has kept a tight and firm rein on port development. It is only very recently that the position of this all-embracing port authority, with its participation in myriad private port companies, has been influenced by the separation of the nautical management and container handling activities. Whether this will really clear the way for competition within the port remains to be seen. For the time being, the city-state is making a very decent profit from the port and the intertwining of public and private activities seems to be continuing unabated.

CHAPTER 11

South Africa: the Port of Durban

'A port is the lifeblood of a city and also acts as an economic barometer. (...) A port is also a kaleidoscope of events, characters and legends - a richer history one cannot find' (Port Manager, Port of Durban, in: Pearson, 1995).

11.1.   Introduction

This chapter explores the institutional format of port management in South Africa. In early 1996, I visited the head office of the national port organisation of South Africa, *Portnet*, in Johannesburg and also travelled to the ports of Durban on the east coast and Cape Town on the south west coast. The text concentrates primarily on the port of Durban because it is the most important port in South Africa and no notable differences were found between Durban and Cape Town in terms of the management of the various port activities.

Visiting South Africa in early 1996 was a very appropriate time to conduct research. This was nearly two years after the political upheaval and the country had just made a start on the more equitable and balanced construction and regulation of the economy and the legal system. The formulation of central and decentral government tasks was very topical at that time, but as regards conducting research, the most important thing was that the port staff were allowed to speak freely and to provide information. This freedom and openness was unknown under apartheid.

The chapter is constructed as follows. The following section addresses national economic regulation and the function of the ports. The third section looks at the formal organisation of the port authority and then, by focusing on the organisation of nautical management, port planning and port services, the division of responsibility between government and private actors is examined. Finally, the more descriptive information is compiled and questions asked from the perspective of the theoretical framework.

11.2.   Economy and Significance of Ports

South Africa is traditionally famous for the variety and wealth of its raw materials and minerals (chromium, coal, diamonds, manganese, platinum, uranium, vanadium and gold). Due to apartheid and the international economic sanctions, it was not easy to trade with South Africa or invest in the country

277

until the early nineties[1]. The industrial manufacturing sector (steel, cars, chemicals, foodstuffs, beverages and tobacco products) is the largest economic sector in the country. It makes up about forty percent of the gross domestic product and is concentrated in four areas: Transvaal, Western Cape, Durban-Pinetown and Port Elizabeth-Uitenhage. These areas together provide over 75 percent of the net industrial production and employment.

Another major national sector is that of the mining industry in which a boom is expected over the next five years. Coal and uranium from the mines are also principally used to provide energy for the home market; South Africa has the lowest energy tariffs in the world. 'For matters such as clothing, transport requisites and chemicals the country is (still) dependent on the import of raw materials or semi-finished products from abroad' (Stichting International Contract Research Centre, 1996:44). The role of agriculture is decreasing. This is due not only to the development of industry and the service provision sector but also to the higher productivity resulting from improved agricultural techniques.

KwaZulu-Natal is the province on the east coast, covering a total area of 91,481 km^2 and with a population of over 8.5 million people, about 4 million of whom live in Durban. Historically, KwaZulu-Natal has the highest population concentration (80%) of inhabitants who originally came from India. The contribution of the industrial manufacturing sector to the gross regional product in this region comes principally from footwear (12.9%), textiles (8.9%), paper (8.7%) and industrial chemicals (6.2%). The beautiful beaches along the coast of this province mean that many people earn a living from the tourist trade. The service sector is also well developed (financial sector, trade and transport). This fact, together with the battle against the high rate of illiteracy in this region, is perhaps the reason why the province is seen as a trailblazer in the field of education in South Africa.

KwaZulu-Natal contains the two most important export ports in the country: Durban and Richards Bay. Durban is the most modern (container) port in the country while Richards Bay has developed itself into a bulk port (chiefly coal transhipment). By means of the KwaZulu-Natal Marketing Initiative, a coordinated marketing bureau between the government and the business community, an attempt is made to provide new investors with the necessary information and contacts and to promote the further industrial development of the region[2].

---

[1] The transhipment figures for chemicals in the port, for instance, have only been made public since 1994.

[2] The initiators have industrial and infrastructural interests in the province. The port of Durban is one of the participants.

At present, reconstruction and development form the most important government objectives, and a recent addition to these is an emphasis on economic growth. At national level the *Reconstruction and Development Plan* (RDP) has been developed which must first of all tackle unemployment and the housing problem. South Africa has a total population of over 40 million people of whom 15 million are economically active. In 1992, no less than 20% of the latter group were unemployed. Unemployment is seen as the greatest threat to the political stability of the country (Stichting International Contract Research Centre, 1996:43).

The national development plan also addresses the design and construction of the water supply (water is a scarce good in the country due to the irregular and scant rainfall), the electricity supply, infrastructure, public health and the raising of education to an acceptable level within a short time. In the first year the RDP took up 3% of the national budget but in later years this will rise sharply to around 45% of the budget[3]. The government's primary financial objective is to control inflation and the national debt, so that a tight rein will be kept on other government spending and taxation will increase as little as possible.

As regards economic regulation, the South African government has opted for an oriented market economy (KwaZulu-Natal Marketing Initiative, 1996:32). Until 1990 the government was only concerned with enforcing apartheid structures and there was no labour policy. Since it was partly thanks to the unions that the government came to power in 1994, things are likely to change on this point. Twenty-five percent of the economically active population belongs to one of South Africa's many unions[4]. A tripartite consultation system has already been established in which government, employees and employers negotiate with each other on equal terms. The Ministry of Employment has composed a five-year programme in which attention is devoted to training, human resource development, job creation, international relations, etc. The new labour legislation has a political character in order to show the electorate that they have not been forgotten. The new Labour Act contains mechanisms for settling disputes and unrest concerning strike situations and the dismissal of employees.

---

[3] In concrete terms this involves creating 2.5 million jobs over ten years, the redistribution of 30% of the agricultural land over five years, the construction of five million dwellings over five years and connecting 2.5 million dwellings to the electricity grid over five years (Stichting International Contract Research Centre, 1996:49).

[4] There are about two hundred unions registered with the Ministry of Employment (Stichting International Contract Research Centre, 1996:77).

Moreover, in the past, public administration equalled white supremacy. The government administration was the apartheid instrument par excellence and was not directed towards providing a service to citizens and companies. There was no cooperation between the various layers of government and transparency was absent. Civil servants were judged in terms of their adherence to procedures and not on their effectiveness and efficiency. Inefficiency was also the result of understaffing, a shortage of skilled personnel and a lack of awareness regarding training and refresher courses. The relationship between managers and the rank and file civil servants was hostile rather than cooperative: this tended to encourage a lack of dedication and fostered corruption (International Contract Research Centre, 1996). Since the political upheaval this situation has been changing rapidly. Education is high on the political agenda and as a result of the multiracial elections there is a strong impulse towards openness.

The Reconstruction and Development Plan is not a universal panacea and the programme has been severely criticised (naturally) by the white population. It is anticipated that the plan will start to bear fruit from 1996. The Government of National Unity is keen to make multilateral and bilateral agreements and to seek regional and international links that are aimed at promoting the development of the country. At present, South Africa is one of the parties to the international GATT treaty that arranges agreements on tariffs and trade and efforts have already been undertaken to increase the competitive power of local industry. For example, in 1991 a regional industry programme was set up in which use is made of financial stimuli to attract new industry.

Around 60% of the South African population has had no education. Some people have never even heard of electricity. In many people's minds the idea of new technology is coupled with the fear of job losses. Many people are thus not interested in new technology and reason from a communal viewpoint: if one person loses his job it will have financial consequences for those around him. In addition to this cultural element, South Africa also has a tribal system in which the role and perception of the 'chiefs' towards technology is very important because they ultimately decide whether a change will be good for the people. This implies that new techniques and technology are finding their way little by little into South African society. The political parties also have an important role to play by constantly explaining why innovations mean better working conditions.

Role and Significance of Ports
The ports of South Africa traditionally form gateways to the country. Three periods in the national significance of the South African ports can be distinguished: the period from 1833-1908, from 1909-1990 and from 1990 to the present time (Portnet, 1995b). The first period is characterised by the

initial legislation on the construction of various port facilities and the creation of individual port authorities. 'Before the formation of the Union in 1910, the harbours were viewed as entities wholly financially separate from the various colonial railways, and all revenue generated within harbours accrued to harbour administrations, who also bore responsibility for all harbour assets and all expenditure incurred. Each harbour, on its own, tried to obtain the largest share of imports and exports. Attempts were made to set common port tariffs but to no avail' (Mors, 1986:23).

Due to the uncoordinated character of port development and through the close cooperation with the railways, in the period 1909-1990 the ports were placed under the railway administration which also switched from private to public ownership: *South African Railways and Harbours* and later the *South African Transport Services* (SATS). The SATS operated as umbrella organisation for the various ports which retained their decentral administration. The ports were run as individual companies and the SATS were of a non-profit making nature.

In the present period (from 1990) the management and administration of the ports has been brought under *Portnet* which operates as a separate divisional activity of the national transport and infrastructure organisation *Transnet*. This has caused no fundamental change in the institutional management of the ports; Transnet is still 100% public corporation. However, in contrast to SATS it operates financially autonomous of the South African government and may make a modest profit. Cross-subsidisation between various Transnet activities is no longer possible. It may be expected, therefore, that the South African ports will be run increasingly as commercial activities.

The most important function assigned to the ports is the promotion of regional and national trade and economic growth by enabling import and export at competitive prices (cf. Nyawo, 1995). Economic growth and opportunities for the future are very closely linked with the capacity to boost the international trade position. South Africa only has a couple of natural harbours so that sizeable investments in existing port infrastructure remain essential. The ports offer the economic opportunity to link reconstruction and development. Maritime development will thus form an integral component of national development. In this period of reconstruction and strengthening, in particular, developing effective and efficient ports is a national concern. In other words, one of the conditions for the success of the Reconstruction and Development Plan is economic growth and the South African ports form one of the major conditions for achieving this. At present, job creation is of course another important objective for the ports.

As yet, the national government still lacks a clear national port policy so that expectations between the port business community, port authority and other layers of government are not explicit. This stems from the absence of a clear legal framework so that areas of authority are not clearly delineated and many uncertainties exist in the maritime industry. In the current period, the key issue is to seek the desired delineation of authority between the central and decentral level but also between the public and private sector. The role and position of the South African ports is principally interpreted in a political way. Central to this is their instrumental function for the economic construction of the country.

The port of Durban is the largest port in Southern Africa. Due to the good infrastructural facilities (road and rail) and reliable service provision, it also functions as a container hub for countries such as Mozambique and Zimbabwe which border on South Africa to the north. The port has a total surface area of 1,854 ha. of which 892 is water (high tide). In 1998, Durban handled 31.2 million tons (including 75% general cargo) part of which was transported in the one million containers (containerisation level over 60%). The port generates on average around 64% of the total income from all South African ports (Portnet, 1995a) and around 4500 ships per year put in at the port.

Graph 11.1 Port Performance Durban 1986 - 1996

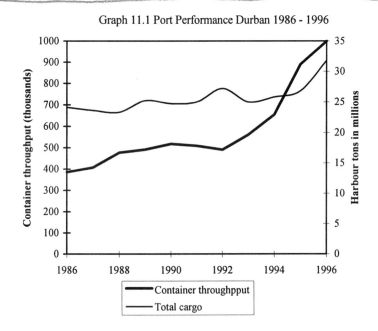

11.3.   The Organisation of Port Administration: Transnet and Portnet

Portnet, a division of *Transnet* Ltd., administers and manages all seven commercial ports along the South African coast - Durban, Richards Bay, East London, Port Elizabeth, Mossel Bay, Saldanha Bay and Cape Town. Transnet was formed in 1990 as a public enterprise following eighty years of direct intervention by government and parliament. The present corporation is financially autonomous and represents various large-scale transport networks including Portnet, Autonet (road transport), Petronet (pipelines), PX (parcel services), Spoornet (national railways) and the national airline (South African Airways). Companies related to Transnet and its subsidiaries are involved in the chemical industry, tourism, data processing and communications technology, house building and real estate development and management.

The transformation of SATS into Transnet is regulated under the 'Legal Succession to the South African Transport Services Act 1989'. In accordance with legislation, Transnet will provide services in the field of public transport (regulation 15)[5]. More specifically: Transnet's most important success indicators are 'financial viability, a focused and capable player in all sectors of the transport industry and its related environment, a highly client orientation, a valued partner in big and small business, a caring employer concerned with the development of all its people to the fullness of their potential and a positive influence in the broader community through active participation in the Reconstruction and Development Programme' (Transnet, 1995:83).

Transnet is managed by a *Board of Directors* who are appointed by the *Minister for Economic Co-ordination and Public Enterprises*. This Minister is responsible for the management of the 'parastatals'. He is able to issue directions which serve the strategic and economic interest of South Africa. Parliament is apprised of Transnet's annual financial reports by the Minister within two weeks of their being submitted. Differences of opinion between Transnet and the Minister regarding, e.g, contracts are settled by mutual agreement through an arbitrator. The arbitrator may be instructed by the Minister.

Organisation of Transnet
Transnet used to be a part of the public sector but seven years ago it formally became a private corporation which is under full government ownership as described in the transitional legislation to turn SATS into Transnet. The Board of Directors comprises a variety of prominent figures, chiefly businessmen. No-one at this level is a clear expert in the field of ports. The day-to-day operations are run by the *Managing Director* who is supervised by the *Group Managing Board*. This board meets once a month and determines Transnet's general strategy. The Group Managing Board prepares proposals for the Board

---

[5] Transnet's transitional legislation contains no fundamental arrangements for the institutional format of South African port management. It only stipulates that pilotage in the ports is compulsory.

of Directors whose final approval is necessary. The Group Managing Board is also responsible for the strategic policy which comprises the principal conditions for the various subsidiaries and/or divisions. The provision that the divisions must have two black employees for every white employee, for instance, is the policy of the joint divisions.

The coordination between the various transport activities of the individual Transnet divisions used to be better than it is now. This is partly due to the fact that the financial administration systems of the individual divisions are no longer interlinked. Cross-subsidies used to be possible but since the split that option is no longer available. For the development and construction of terminals, for example, Portnet may hire the expertise of other Transnet divisions by means of direct contracts but is no longer obliged to do so. In Cape Town, for example, the management has opted to put the job out to tender instead of automatically doing business with other Transnet divisions. The smaller ports in particular do not necessarily have to use the knowledge and expertise of other Transnet divisions.

Centrally-administered Ports and the Coordination between Transnet and Portnet

Portnet is the national organisation responsible for the day-to-day running of South Africa's seven seaports. Various arguments are used to justify the centrally-organised port management. In the first place, policy between the various ports is coordinated whereas they would otherwise compete with each other which would involve extra costs for infrastructure. In addition, ports are seen as instruments to serve the national economy. In the third place, South Africa has no natural harbours so that substantial capital is needed for port construction and development (chiefly dredging costs and engineering works). This capital can only be raised at the national level. In the fourth place, central decision making does not exclude the delegation of various decision-making powers to the individual ports.

The Table below shows a brief profile and comparison of the South African ports. It is striking that only Richard Bay has undergone considerable growth while the port of East London appears to be on a downward slide. The negative impact of the apartheid regime on economic growth can be clearly seen.

The various South African ports are not in direct competition with each other due to their specialised terminals, national tariffs and their different hinterlands. In the past, competition only occurred on the basis of service. Due to the economic growth, this is gradually disaggregating to other areas, too.

Table 11.1. Profile of Goods handling in the South African Ports in 1985, 1990 and 1998 (Source: H. Stevens based on Institute of Shipping Economics and Logistics, 1991:372 and 1999. Amounts in million harbour tons).

Ports	Goods type (proportion general cargo versus bulk)	Total 1985	goods 1990	handled 1998
Richards Bay	coal, steel, wet bulk (5% - 95%)	46.7	52.4	86.1
Durban	conatiner, grain, repairs, sugar (75% - 25%	24.4	24.7	31.2
East London	container, grain, combi-terminal (99% - 1%)	1.7	1.9	0.63
Port Elizabeth	container, ore, fruit (65% - 35%)	6.5	4.1	5.7
Cape Town	container, fruit, fish, grain (94% - 6%)	4.5	5.1	6.8
Saldanha	iron ore, oil (1% - 99%)	10.9	17.5	22.6

The hinterland behind Richards Bay is increasingly overlapping that of Durban, for example. Moreover, as a result of the standardisation of train tariffs to Johannesburg, Durban is also encountering competition from the East London container segment. The policy freedom of the individual ports is relatively small, however. The national tariffs mean that there is no opportunity to negotiate independently with clients. In the present period of high interest and inflation, in particular, the absence of these powers is perceived as a deficiency.

Portnet is managed by two Chief Executive Officers, one for the *port authority* activities and one for the activities which come under the heading of *port operations*. Activities which fall under the port authority include:

- **the promotion of trade**:
  guaranteeing the quality of the service level by supervising and measuring port performance, the handling of goods and the duration of a ship's stay in port. In addition, cooperating with port partners to 'sell' the port abroad and to promote trade;

- **strategic planning and spatial planning**:
  guaranteeing the necessary facilities for future trade by means of port development for the long term and strategic 'master plans';

- **the construction and maintenance of basal infrastructure**:
  guaranteeing that breakwaters, channels, turning basins, quay walls and roads are constructed and maintained;

- **optimising the present land use**:
  guaranteeing that sites are leased in the most appropriate way, and offering incentives for freight transfer and for promoting improved land use;

- **safety for shipping**:
  guaranteeing a safe passage for ships within the port by the installation and maintenance of lights, buoys and other navigational aids, dredging channels and removing wrecks;
- **protecting the environment**:
  maintaining the environmental quality in the port.

The component of Portnet that is concerned with port operations, on the other hand, is responsible for:

- **the construction and maintenance of superstructure**:
  guaranteeing that the equipment used for goods transhipment (e.g. quay cranes, forklift trucks, handling equipment, etc.) and the fixed assets for e.g. the storage of goods, are offered and maintained under lease conditions;
- **the transhipment**;
  guaranteeing the actual implementation as port operator by maintaining guaranteed cargo throughput and service standards, in accordance with the conditions laid down in the implementation contract with the port management.

Both Portnet managers are appointed by the director of Transnet. The distinction between the port authority and port operations was made in accordance with the policies laid down by national government which are aimed at making Portnet more efficient and are intended as an initial step towards complete privatisation of the 'port operations' activities. Due to the power of the unions, this final phase has not yet been reached (Seatrade Review, 1995). The most important reason for this is Transnet's dependence on Portnet's revenue. In 1995, Portnet made a net profit (after financing costs) of 1.1 million Rand which also made it the largest contributor to the Transnet total, i.e. 57.5% of Transnet's total net profit (Transnet, 1995). On an annual basis, Portnet makes a substantial contribution to Transnet (in 1995: 48% of all expenditure) to offset the deficit in the pension fund. It seems likely that Portnet's 'port authority' activities will be transferred away from Transnet because these are seen as public tasks. It may be anticipated that this transformation will take place in the next few years (Nyawo, 1995). One could even argue that instead of the 'port operations' activities being part of Transnet it might be preferable to sell them to the private sector. In the current situation (1996) both components are still part of Transnet although they function under separate financial-administrative systems.

The two Chief Executive Officers of Portnet in Johannesburg are in charge of around ninety people. Each month they consult with the *Portnet Managing Board* which consists of the managers of the various ports in the country and the directors of the various divisions in Johannesburg. The Chief Executive Officers and the directors of the divisions in Johannesburg together form the *Portnet Executive Board*. These people are jointly responsible for national port development but not for the financial issues and the more important appointments in the ports because these fall under Transnet's exclusive authority.

The day-to-day operations are the responsibility of the port managers and head office in Johannesburg. The relationship between Transnet and Portnet is primarily of a financial nature. The annual budget, port tariffs and major investments need to be approved by the Transnet Board of Directors. Portnet has various sources of revenue, the most important of which is wharfage. Wharfage is levied as a percentage of the value of the cargo and makes up about 60% of the total Portnet revenue. This form of port dues is the subject of stringent criticism in the international community (GATT) and is detested by the shipping companies. Portnet's other financial sources include the handling revenue from the container and break bulk terminals, managing the stock of trucks (about 70 million Rand per year), charges from port services such as pilotage and towing services, income from repairs to ships (dry dock) and the leasing of sites.

Portnet's budget is approved in March each year by Transnet. The financial year runs from 1 April to March 31. The various ports have their draft budgets prepared in September and negotiations with Transnet get underway. The financial-administrative system is not run on a cash basis but on a commitment basis so that each potential flow of funds is registered. Via national computer links, the financial director in Johannesburg is able to gain an up-to-date overview of the financial accounts of the individual ports although this is not as detailed as the accounts kept in the ports themselves. Transnet has prepared its own guidelines for the depreciation of fixed and floating assets and for the profit requirements based on prognoses. Each month, the financial manager of the port of Durban sends the financial situation to Portnet in Johannesburg and tries to clarify any major discrepancies between the budget and the actual expenditure.

In Transnet's current financial policy it is not possible for the individual Transnet divisions to negotiate a separate loan. The aim is to obtain the necessary funds through depreciation and positive results. Portnet defies this policy by entering into private cooperative links with private actors by selling services or contracting them out. The fact that it cannot make investments without Transnet's approval is Portnet's main reason for desiring privatisation. Despite the fact that Portnet provides the lion's share of Transnet's revenue, Portnet's investment plans are treated in the same way as those of other Transnet divisions. The result of this is that the port management cannot make the necessary investments directly. It is only in the last two years, since the economic liberalisation of South Africa, that it has become possible to make major and by this time highly necessary investments. Much of the superstructure had been written off years before and the existing infrastructure was beginning to show serious capacity problems.

At the present time, Portnet has a number of options for privatisation under strict conditions as formulated by a joint commission from Transnet and Portnet. A first condition is that the aim of the privatisation must be to serve the expansion of the port, given that a port is not an end in itself but a booster to trade. A second condition is that the national government must be clear about the aims, methods and costs of the privatisation. There is a fear that unclear definitions and unrealistic expectations will result in a port crisis. In the short term, Portnet is very aware of the fact that it will have to take account of the wishes of the port shareholders. This is why the division between 'port authority' and 'port operations' seemed the most important thing that could be done to attract trade. However, a year on from this commercialisation, the shippers and shipping companies began to complain about the close interaction between Portnet as port authority and Portnet in the role of private provider. 'In most South African ports, Portnet acts as both authority and operator. Since the initiative to commercialise its activities, while at the same time awarding operating concessions to private companies, Portnet's position has become increasingly antagonistic' (Griffiths, 1993:33).

Negotiations with the unions take place at central level between the port managers of Portnet and union representatives. One of the largest union organisations is the *South African Railway and Harbour Workers Union* (SARWHU). The union looks at Transnet as a whole and, while the ports continue to contribute so much money to the pension fund, privatisation is out of the question. The union is able to maintain this strong political stance because it, together with the other unions, made it possible for the Mandela government to come to power. Portnet is only now realising the fact that it has no direct contact with the Minister of Public Enterprises which makes it almost impossible to generate any recognition of its own wishes at the national political level. Moreover, the union plays an important role in the disaggregation of new technical developments to the employees. In the coming years, education will be the most important thing. Portnet hopes to improve productivity by means of bonuses.

Coordination Between the Central and Decentral Level
An important responsibility for Portnet in Johannesburg is the coordination of the developments in the various ports. For example, if at any time the number of containers wishing to enter Durban exceed its handling capacity, it is essential that these should enter the country via another South African port without incurring extra costs. Portnet has thus arranged with the ports of

Durban and Port Elizabeth, and with the Transnet division Spoornet, that the containers can be transported by rail via Port Elizabeth, instead of via Durban, at the same costs. The result of this coordination is that ships no longer need to wait in Durban and the freight transfer still occurs via South Africa.

The port manager in Durban is the person responsible for the day-to-day operations in the port, where about 5,250 people work for Portnet. The port manager is assisted by a number of divisions which either come under 'port authority' or 'port operations' activities. 'Port operations' can be broken down into three divisions: breakbulk, container, and goods vehicle management. In concrete terms, the 'port authority' activities cover the following divisions: the harbour master (pilotage and shipping inspectorate), the marine manager (dry dock and tugboat services), the port engineer, the real estate manager, marketing manager, public relations, port secretariat, personnel manager, port planning, finance and computerisation.

Portnet Johannesburg draws up the main lines of policy within which the port manager must operate. Normally, these directions function as guidelines which means that the port managers are free to provide customer satisfaction within the policy margins. They need to be aware that clients are likely to try and play one division off against another. The port managers present and justify their port development plans in Johannesburg. At present, Durban has a plan to construct a new container terminal because the current capacity of one million TEU has now been exceeded.

11.4.   Nautical Management

The nautical management is a component of the 'port authority' activities and comes under the responsibility of the harbour master. The *Harbour Regulations* provide the formal, general framework for the powers of the harbour master. These regulations cover a great many matters relating to order and safety and also specify the various port areas of the seven South African ports within which the individual harbour masters may use their powers (Portnet, 1995b).

Towards the end of the eighties, Portnet tried to reformulate and minimise the number of regulations but they have not managed this yet. The regulations thus function as *guidelines* for the nautical responsibilities of the port management. The port of Durban has the simplest form of traffic guidance, i.e. permanent radio contact between ship, pilot and tugboat.

The harbour master is not responsible for the safety of shipping traffic outside the port. The reason for this is that it is impossible to guarantee the safety of ships outside the port during storms. There is a separate, non-profit aid organisation for ships in distress at sea (the South Africa National Sea Rescue Institute). More general rules for shipping traffic and the prevention of pollution by ocean-going vessels are laid down in national statutes. In the event of oil pollution, the Ministry of Transport's *Anti Pollution Services* are alerted. They determine whether the polluter should be fined and how much. This responsibility has been very deliberately imposed by the Ministry of Transport because the water in the sea is also the water in the port. Pollution of the water from the landside is regulated by the *Water Act* which is implemented by the Ministry of Water Management. The fines imposed are not substantial, however.

Although the port does have a crisis contingency plan, it is doubtful whether the plan would actually work in a crisis because too few drills are carried out. Along the coast, the national crisis plan is in operation. The harbour master has ultimate responsibility for emergency situations in the port. In cooperation with the petrochemical industry, Portnet has compiled a list of available equipment. In an emergency, the harbour master has direct contact with the well-equipped city fire brigade and with the captain. Portnet has its own staff which constantly monitor the fire safety of buildings and port equipment.

The Portnet security staff also have a training problem. In the normal course of events, Portnet has to rely on its own staff because the South African police will only provide security and carry out checks on the port area on a contract basis. Now that the port is expanding rapidly, criminality is becoming increasingly widespread; containers are regularly broken into and sometimes disappear completely.

An important problem for the harbour master is that of laying an embargo on a ship which has an outstanding debt to a bank or other company. These legal procedures are sometimes very protracted which means the ship occupies a valuable berth for an unnecessarily lengthy period. Portnet is also responsible for the maintenance of the lighthouses. It is a member of the *International Association of Lighthouse Authorities* which has its own management requirements. Portnet has a seat on the administrative body of this organisation as the representative of the central and southern African regions.

In 1988, Portnet formed an independent *Dredging Services Centre* which was charged with managing the dredging maintenance work in the seven South African ports. The centre acts as '*in-house contractor*' for the various ports and is stationed in Durban. It has its own fleet of floating dredges. Most dredging work takes place in the east coast ports (Richards Bay, Durban and East London) where the sediment from the northerly current must be permanently intercepted in order to prevent the entrance channels silting up. Cape Town, with its low level of silty deposit, has its own dredging equipment.

11.5.   Port Planning

For years there was no strategic port planning. Until the international liberalisation of the South African market, the existing capacity was adequate and economic growth was very modest so that new developments could be anticipated in plenty of time. Since the economic liberalisation this is far from being the case and a start has been made on replacing old handling equipment and preparing plans for the expansion of sites. At present, there is a drastic shortage of space in Durban and the shipping companies are complaining that new developments have not been adequately anticipated, so it looks as if the South African ports are becoming victims of their own success (Seatrade Review, 1995). Waiting times of up to three days in Durban and six days in Cape Town are the result. The fact that the problem is so acute at present is not only attributable to Portnet but also to the shipowners themselves. The private sector had also not expected such rapid growth in the short term and had been supplying Portnet with relativised growth figures for years which meant Portnet was working with incomplete data.

The economic liberalisation of South Africa has thus also had a major impact on the way in which port planning is given institutional form in the future. Soon after the creation of Transnet, the first joint marketing initiatives between Portnet and private port users were developed (Philips, 1991). But Portnet retained control of e.g. the container facilities and the physical layout and so could determine for itself when replacement investments and expansion would take place. Thus a situation arose in which straddle carriers remained in service longer than the planned period and there was no question of interaction between the port and the city (neither in Durban nor in Cape Town). Moreover, for the financing of major projects Portnet is tied to the deliberations of the Transnet Board of Directors and it was some time before Portnet was able to get the desired budget. The first steps on the way to more integrated port planning have now been taken. It is still too early to talk of an institutionalised situation but the initial outlines are becoming visible (Nyawo, 1995).

New Forms of Port Planning in the Port of Durban
'Historically, Port Planning has been a "sacrosanct refugee camp" for exclusive technical marine and engineering sciences. The reason for this approach to port planning and development can be traced from the British maritime planning philosophy among others. The major emphasis of this old paradigm has been on technical and physical elements of ports (...) Port master planning process and management remained undemocratic, exclusive and often racist and sexist (depending on local and regional population dynamics)' (Nyawo, 1995).

It is interesting to see how the planning philosophy in Durban is changing. In developing plans, the South African ports can no longer ignore the fundamental elements of the business world and the requirements of society. It is becoming increasingly clear to the port of Durban, for example, that it is closely linked to the city of Durban and that this requires a broader perspective than the old-style planning in which technical norms applied as sole success criteria.

As explained in section two above, until recently there was no coordination between the various layers of government. Tasks were undertaken without any outside professional help. The city had its own plans, for instance, and built a number of offices in the immediate vicinity of the port which are now obstructing the further development of the port. The relations between Portnet and the city of Durban are a fairly recent occurrence and have lately become far more open. A number of joint commissions have been set up in which the two parties disaggregate the port developments to the citizens and in which there is more coordination of land use.

The nautical management area is clearly defined in the Harbour Regulations but there is a lack of clarity regarding the exact status of the geographical area of the port adjoining the city. Such a distinction will be essential to the further development of the port and will form a key subject for commissions both at local and national level. At present, Transnet pays a sum of money to the city for the use of such rights as energy and the use of roads. Here, once again, the uniform division of tasks is not always clear: some roads are constructed by the city and some by Portnet. Separate negotiations are needed each time for the construction of new roads. The rail links to the terminal are the responsibility of Spoornet. A regional planning procedure is not yet available. Private lessees in the port must pay municipal taxes to the city of Durban via Portnet. This is seen as an unhealthy situation because the private lessee frequently thinks that he can them claim rights from Portnet which is not the case.

The current port planning and development paradigm places the accent on 'integrated port planning'. Integration emphasises an interdisciplinary as well as a regional approach to port planning; economic, managerial, physical, technical, social, political and cultural factors are jointly considered. This restructuring of port planning also changes the success standard within which port planning takes shape: the process of port planning determines not so much the content of the plan but rather the success of the plan (Nyawo, 1995). In South Africa in particular, where First and Third World exist side by side, port planners have a lot of explaining to do. The planning for the use of new straddle carriers, for example, had to take into consideration that the Japanese high-quality technological straddle carriers were too vulnerable in the South African context compared to a solid German machine. Again, for the layout of the new container terminal in Durban (number three) lengthy explanations and talks were required with the environmental movement regarding the possibilities (or lack of them) of preserving the unique microcosm in the area.

For the port organisation, the philosophy of integrated planning requires the presence of highly-trained staff who can discuss the options for integration and deliberate with each other from various disciplinary viewpoints. The long-term planning function is seen as an activity which comes under the 'port authority' and the layout of terminals is seen as a 'port operations' activity. Each activity is responsible for the development of plans for the medium and long term. The port authority is responsible for the general long-term plan for the port but this is not a collection of long-term plans from the port operations division. The port authority's long-term plan provides the basis for the long-term plans of the port operations division. The intention is that each port operations activity should have its own planning team which cooperates with the port authority planning team on a number of problem areas. Portnet Johannesburg has also formed a *Planning Board*. The integrated approach to port planning is a fundamentally different way of thinking from the old paradigm, in which external consultants were appointed to develop master plans and in which the Portnet staff might be given specific tasks without being shown the big picture.

11.6.   Port Services

The control of the port services in Durban is broken down into various categories. In the first place, the *nautical services* (pilots, tugboats, etc) are controlled by Portnet. The pilotage of ships is compulsory by law and only the harbour master can grant a discretionary exemption. In Durban, Portnet has six tugboats which usually only operate inside the port and also look after the linesmen's work. The training of both the pilots and the tugboat captains is carried out in cooperation with private South African shipping companies (Unicorn and Safmarine).

In the second place, the *goods handling* is carried out by both Portnet and private providers depending on the type of good. The port has 57 berths and 14 terminals of which seven are operated by private companies. The latter comprise the terminals for the handling of coal, paper, sugar, citrus, liquid bulk, dry bulk and granite. These are goods which have always been entirely in private ownership. The remaining seven terminals are for the handling of passengers, containers, ro-ro, breakbulk, grain and wood and these are operated by Portnet.

In 1977 Portnet began operating the container terminal. At that time, it was a method of transport that still had to prove itself and substantial initial investments were necessary. From the perspective of the familiar operational uncertainty, it is logical that Portnet was concerned about the goods handling in this area. It is also understandable that the *stevedoring* in Durban has always been a private affair. Traditionally, Portnet employees do not board the ships, although these days Portnet does provide the automated, ready-to-use work schedules for loading the ships.

In the third place, the *port-related transport* by rail or freight vehicle is also conducted by a public corporation. As a colleague division of Portnet within Transnet, Spoornet is responsible for the construction of rail links to the port, and Portnet itself operates 200 trucks and 1,200 trailers. A working agreement has been made with the private sector that the Portnet trailers will look after the short-haul road transport up to 110 kilometres, and beyond that distance any freight will be transported by private provider. This agreement springs from the fact that Portnet does not want any private truckers at the terminal. It requires tight planning, particularly during peak times in the supply of truckers with cargo (Crichton, 1991). In the future, the truckers company will function as a subsidiary of Portnet Port Operations. Finally, Portnet operates the port services aimed at the ships themselves. It is the owner of the Prince Edward Graving Dock and also operates the floating dock. At once point it looked as if these ship repair facilities would be transferred to private ownership but that plan has been shelved for the time being.

Principally due to its combining of the two functions (port authority and port operations) there is considerable criticism of Portnet's behaviour from the private sector. Portnet insists that it functions just as effectively and efficiently as the private sector in other countries. Concrete benchmarking results are not available as yet. Portnet is not only kept in an iron grip by the union but it also fears that monopoly situations would arise if port operation activities were to be sold. Portnet wants to promote competition inside the port but it is also convinced that there is no point in privatisation without real competition (Jaques, 1993). In an interview, a respondent made the following comment: What we don't want is for all the transport activities in the port to fall into the hands of the 'big green R' (referring to the private company Rennies). This company, with the aid of its subsidiaries, already acts as a chain manager and is a keen advocate of the link between investing and operating (Philips, 1991). It is thus a proponent of the 'dedicated terminal concept' which it has already embarked on for the bulk sector. Portnet, on the other hand, would rather stick to the 'common user principle' under which, as handling company, it retains complete control of the terminals.

After years of negotiation, Portnet has entered into a cooperative agreement with the Japanese supplier on the maintenance of the new straddle carriers during the depreciation period[6]. As far as technology is concerned, Portnet

---

[6] The type of straddle carrier was selected in consultation with the union.

does not want to be a trendsetter but has opted for reliability. Since specific knowledge and specialist components are not available at home, a choice has been made to pay a little over the odds for the straddle carriers but with the assurance of reliable service. Such matters are not put out for tender because the lowest price is not the cardinal factor. Portnet calls this an effort '*to Africanise technology*': the use of modern technology in a less developed country. The agreement does not cost the supplier of the straddle carriers any money but he is risking his good name. Portnet is confident that it has the guarantee of rapid delivery of spare parts and the transfer of know-how to its own people.

Portnet Durban has regular meetings with shipowners and shipping agents to keep each other updated on a whole range of matters. There is also a Port Commission on which the majority of port users are represented and which can inform the port management about the situation in the port and any 'moans and groans'. The meetings take place six times a year. Since the economic liberalisation of South Africa, joint marketing initiatives have also been developed.

11.7.　Port Management in South Africa: Analysis and Questions

The end of the apartheid regime has heralded the dawn of a new era for the South African ports. Not only has the economic liberalisation of the country brought an unprecedented opportunity for economic development but above all it implies a complete rethink of existing institutions. Existing divisions of tasks and powers are critically reviewed and new ideas are launched. This struggle and discussion are still relatively new and by no means fully crystallised. In the description of the various port activities, the expectations of further change surfaced time and again. In the light of the as yet uncrystallised discussions, it is difficult as a researcher to theorise about the existing institutional situation although an indication can be given of where problems are likely to arise in the long term if attention is not devoted to particular matters.

Before the economic liberalisation of South Africa, the ports played a minor economic role. The opportunities for international trade were limited and the ports were not expected to make a profit. Any profits they did make were skimmed and used for other transport infrastructures (mainly the railways). The formation of a nationally coordinated port management served to prevent overcapacity being created with scarce national funds and to preclude the various ports from competing with each other. Moreover, the various ports serve separate hinterlands and the tariffs are laid down nationally. These were not consciously deployed as a national economic development instrument, however, since the country was too dependent on foreign trade to take such a measure.

The creation of the public company Transnet, in 1990, prompted a number of changes in the institutional format of South African port management, in particular the relationship between the Ministry of Transport and the former South African Transport Services. Transnet will henceforth come under the Minister of Economic Co-ordination and Public Enterprise and be financially autonomous of national government. The Minister retains only the power to appoint the Board of Directors, to appoint an arbitrator in the event of a dispute, to present the annual accounts to Parliament and to issue general directions regarding economic policy. The more substantive influence of the Ministry of Transport has thus disappeared and management can only take place along very general strategic lines by a non-specific minister.

The creation of Transnet has meant that ports tend now to be judged on their service level, effectiveness and efficiency and a modest profit may be made. Despite the financial arrears in pension provisions, replacement investments and expansion investments, however, Portnet rapidly started to make substantial profits. The role and function of the ports are defined by Portnet as being the engine for the development of trade and as generators of employment. It is also felt that the ports should be run on a commercial basis. Given the role and function allocated to the South African ports in formal documents, it might be expected that the public port activities will be strictly separated from the private ones and that whenever possible, private activities will be offered under competition.

The descriptions provide a different picture, however, namely that of a public company that takes care, virtually autonomously, of both the nautical management (navigational marking, allocating berths, dredging, etc.) as well as undertaking and monopolistically controlling numerous private services (pilotage, tugboat services, ship repairs, training, container handling, etc.). It will come as no surprise that the expected control relations in Durban do not always coincide with the set of control relations actually found. In the first place, the South African ports lack a formal national port policy and statutory framework. Moreover, it remains unclear how the various ports can contribute to the success of the national economic policy as it is incorporated in the Reconstruction and Development Programme. In the third place, the discussion on the desired relationship between Portnet and Transnet in particular, and between the 'port authority' activities and the 'port operations' activities has still to take place. Finally, choices regarding the above matters can only be made once the chosen relationship between the national government and the unions is actually given a concrete shape which excludes the possibility of becoming too dependent on each other regarding specific objectives.

Starting from the principle that South African ports are the success conditions for the achievement of economic growth and thus the Reconstruction and Development Programme, it is certainly surprising that port management is situated at such a distance from national and regional politics. It makes no difference to the formal options for control whether one chooses to pursue a socialist or a liberal economic policy. As a division of Transnet, Portnet has considerable policy freedom and can determine for itself whether components are sold or contracted out. It remains to be seen whether it will be possible to realise the desire to reach a more integrated planning methodology within the existing institutional relationships. After all, the dependency relations with Portnet are likely to hinder this integral approach.

Due to the transformation process that South Africa is going through at present, many questions remain. For example, how will the institutional position of Portnet manifest itself in the long term? Will the intertwining of imperium and dominium foster the construction and development of the ports (cf. Singapore)? How will Portnet resolve the present complaints from shippers and shipowners about the mixing of public and private activities? How will the relationship between the national and provincial levels be regulated and how will national and/or provincial ministries form a counterbalance to Portnet in the various ports?

The questions are simply endless and can only be answered in the long term. In the next and last chapter an attempt will be made to reveal a small part of the future.

11.8. Summary

Now that the apartheid regime has been abolished, the South African ports can actually perform the role intended for them: promoting trade. The economic liberalisation of the country has entailed such rapid developments, however, that the nationally organised port authority has had the utmost difficulty keeping up with the consequences. Although there are considerable variations between the individual South African ports, they are all managed by the national public corporation Portnet. This organisation, to the great annoyance of the private business community, is responsible both for the nautical management and for a large part of the port services. There is no chance at present of a separation of public and private activities because the powerful union will not permit it. In any event, the highly profitable Portnet can be used to elicit extra facilities in exchange for voter's support which is something that would not be likely to occur under privatisation. A close look was taken at the port of Durban as an example of South African port management.

CHAPTER 12

The Learning Capacity of Seaports

'Are the ports faithful to us?' (adapted from Slauerhoff, 1996)

12.1.   Introduction

In the previous chapters, the position, function and management of each of the various ports was analysed. In each case, the analysis was structured along fixed lines. First, a look was taken at the national constitution and (inter)national economic regulation within which the port operates and the socio-economic significance assigned to the port. Then the position of the port authority was described on the basis of various indicators (formal structure, relations with layers of government, sources of revenue, etc.). Finally, the institutional format of a range of public and private port activities was subjected to a more detailed examination. At the conclusion of each chapter, the question was asked of to what extent the port regulatory principles, whether explicitly formulated or not, were actually reflected in the format of various concrete port activities. Where this was not the case, a number of questions were then asked. These were compiled with the intention of discussing the learning capacity of the port. The aim of this final chapter is to reflect on those questions by comparing and contrasting the ports and to draw some general conclusions.

12.2.   International Comparison

The intention of this section is to clearly catalogue the differences found between the ports. On the one hand, important *quantitative* differences between the ports can be found and on the other, there are *qualitative* differences in the institutional format which can provide useful insights into the functioning of an institution. The purpose of the comparison is not to judge whether one port is more successful than another. That angle would be more likely to raise questions regarding the *comparability* of ports. If the geopolitical context of the port were taken as starting point, then conclusions would either be too generalised (e.g. 'ports always function under competition') or there would be no clear-cut success criterion. After all, it has proved extremely problematic to even achieve an internationally uniform quantitative performance criterion such as 'tonnage' let alone factors such as productivity, service level and so forth. As was already explained in chapter one, this study is intended to provide a *qualitative* assessment of institutional port management. Tentative expectations regarding the learning capacity of ports are reserved for the next section.

The seaports studied vary widely in terms of their institutional format but also in size, cargo quantities and impact on employment. The table shown below provides an overview of a number of these differences.

Table 12.1. Quantitative Comparative Overview of Ports Studied
(Source: H. Stevens based on Annual Reports; websites; Rotterdam Municipal Port
Management, 1997; Institute of Shipping Economics and Logistics, 1996).

Port	Total area (ha.)	No. of ocean-going vessels 1996	Total cargo (million metric tons 1998)	Total containers (million TEUs 1998)	Employment (direct and indirect) (1996)
Rotterdam	10,500	33.701	315	n.a.	338,000
Antwerp	14,055	15.417	n.a.	3.27	65,000 (direct)
Hamburg	8,700	11.500	75.8	3.55	140,000
New York	1,462 (dry)	4.636	56	2.5	140,000
Seattle	396 (dry)	1.117	13.54	1.54	26,000
Los Angeles	3,035	2.608	68.6 (revenue ton)	3.2	247,000 (+Long Beach)
Vancouver	6,565	2.785	72.0	0.84	17,300
Kobe ('94)	n.a.	10.836	171.0	2.7	n.a.
Hong Kong	n.a.	41.760	127	14.58	350,000
Singapore ('96)	n.a.	117.223	242.5 ('96)	14.1 ('97)	175,000
Durban	1,854	4.550	31.2	n.a.	30,000

The ports studied cannot be compared on the basis of quantitative differences. What it boils down to is that the roles that these ports play in the global economy are essentially different. Since the emphasis of this study is on institutional comparison, the table will not be further discussed.

In accordance with the concept underlying the various chapters, it is useful here to present a three-part overview and further refine the analysis. The first part concerns the function and socio-economic significance of the ports. The second part of the qualitative comparison comprises the role and position of the port authority and will address both intra- as well as interorganisational aspects (section 12.3). The final part contains conclusions with regard to the various control relations for the different port activities (section 12.4).

Differences in the Socio-economic Significance of Seaports
In the comparison of the seaports studied, major differences between the continents are evident as well as the more minor variations between ports on the same continent. In Western Europe, seaports were studied which typify the

Hanseatic tradition. These are locally-administered ports which traditionally have a high degree of autonomy and which are run as far as possible in accordance with commercial criteria. In Europe, ports are seen primarily as *components of the transport infrastructure* which should be managed as efficiently as possible on behalf of shippers and carriers. The aim is to reduce the costs of the total logistics chain. This is why, in the projected European seaport policy, Brussels wanted and still wants to make the rules for competition between the ports as equitable as possible although this has not been managed so far. Moreover, the ports of Rotterdam, Antwerp and Hamburg are important *instruments of industrial policy* (cheap land prices) the primary aim of which is to maximise the share of 'captured' freight. European port management is still a component of the local administration (municipal or city-state) of the individual member states, where the port performs a regional (Antwerp and Hamburg) or even a national (Rotterdam) function.

Despite the general European idea and norm that ports should play as efficient a role as possible in various logistic chains, this does not imply that essential differences in the concrete interpretation and meaning given to the Western European ports do not exist. The port of Rotterdam has two distinct de facto regulatory principles with respect to the division of responsibility between public and private actors. First and foremost, there is the rule that the business community must stand on its own two feet. In addition, there is the principle that the government is responsible for the infrastructure and that the costs of this will be passed on as far as possible to the user. Revenue from infrastructure should subsequently be used for port improvements. The port of Rotterdam has abandoned the objective of creating employment and emphasises instead the promotion of its transit function. The port of Hamburg, on the other hand, continues to uphold the employment objective and the generation of city revenue. In Hamburg the government feels itself called upon to build the infrastructure and there is a strict division between this and the superstructure. In Germany, costs of infrastructure are paid by the state (with the exception of a negligible amount for quay walls). Finally, of the three Western European ports studied, the port of Antwerp has the least clear regulatory principles. In principle, the privatised Port of Antwerp can apply itself to any activities which further the management and operation of the port.

The ports of North America cannot be grouped under one socio-economic heading. In the United States, the significance of a port varies according to the state or city in which the port lies, whereas in Canada there is national coordination and ports are expected to be self-sufficient. There is no federal statutory regime for port management in the United States and it has no direct national interest. Where it can, the federal level tries to create conditions for individual ports which are as favourable as possible without actively intervening. This allows considerable freedom for individual state or municipal interpretation, ranging from performing a regional logistics function to providing employment and commercial enterprise for the local residents. The port of New York is a component of the transport infrastructure, whereas the port of Seattle is an important employment instrument and the port of Los Angeles integrates the two elements. In Canada on the other hand, on the grounds of spreading the risks, considerable value is attached to the federally embedded port administration legislation. Although, in the past, frequent attempts were made to use this to set up a national coordination of port policy, this objective has not been proved effective so that examples can be found of regional ports which compete with federal ports under different rules. Canada's federal ports play an important role in national trade. They are judged by their service provision to the port users.

Now that the United States and Canada, together with Mexico, form a single power bloc (North American Free Trade Agreement) it seems to me that the differences in their regulatory principles have become more important. After all, if you were the port of Seattle what would your reaction be if you saw the port of Vancouver in times of economic recession falling back on federal funds or developing a new terminal with the aid of a low-interest loan (by means of the federal guarantee)? Or if you were the port of Vancouver, what would your reaction be if the Seattle port authority can rely on its tax revenues at a crucial moment?

A comparison between the ports of Western Europe and the United States makes it clear that the crucial difference in expectations regarding the functioning of ports lies in the (political) significance which is assigned to ports and which in turn is linked with the relationship between the member states and the federal level. The European Commission in Western Europe is still trying to get to grips with the *conditions* under which the ports compete. This means carrying on a ceaseless discussion with the member states about the division of control, whereas the federal government of the United States traditionally has responsibility for nautical management. The horizontal competition between the member states functions as a control mechanism for port planning. A comparison of port management in Western Europe with that of Canada, moreover, demonstrates that only a very limited role should be assigned to the federal coordination of port planning and that traditional defensive considerations in this context are outmoded.

In Asia, the ports form the economic instrument par excellence *to promote national trade* (attracting new commercial enterprise and cargo). Ports are seen as an essential component of the infrastructure over which a government rules by definition. One of the strongest motivations for this is the fact that these countries do not possess an adequate or sufficiently varied supply of raw materials and minerals. Competition between the national ports (Japan) is appreciated up to a point but the main thing is to coordinate the domestic competition in a centrally dirigiste way. This coordination is possible because the bulk of the infrastructural funds can be released at central level and the decision-making procedures are laid down by law. Finally, the institutional format of port management is regulated under national legislation (Japan and Singapore).

There are considerable differences between Japan, Singapore and Hong Kong, however, in the way they address the trade theme. Whereas Japan openly asserts that ports have no profit motive and should also function as locations where living, recreation and work take place in a coordinated fashion, the port of Singapore is administered and operated as one of the country's most important national assets. In Hong Kong, too, the main aim is to operate the port in the most profitable way possible. Moreover, the individual regulatory principles also differ. The port of Kobe has two regulatory principles concerning the public-private division of roles. In the first place, as in Hamburg, the principle of infrastructure versus superstructure is used. In the second place, a distinction is made between the control over, and division of finances among, the terminals for national versus international trade. In Singapore, on the other hand, nearly all port activities are run on the principle of a centrally-controlled economy. In that sense, the port has the function not only of promoting trade but is also and primarily an investment object with which the city-state obtains considerable revenue. In very recent times, Singapore has been trying to let the port function as a regional coordinator of international trade flows for reasons of spreading the risks and opening up new markets. In Hong Kong, finally, the cardinal regulatory principle is that new initiatives should be supported, financed and managed by the market. The government is mainly responsibility for rights of ownership and is now attempting to bring about some coordination between private actors and various policy areas.

Due to the economic strengthening of the country, the South African ports are expected both to promote foreign trade and to stimulate employment. At least, this combination appears to be the chief reason for keeping port management wholly in government hands.

A general comparison demonstrates that in principle two functions are assigned to seaports although not necessarily simultaneously. A port can be an economic instrument to promote trade and attract transport flows but it can also have particular significance as a socio-economic instrument for the region. Placing the emphasis on the trade function does not necessarily imply the promotion of employment. Due to working conditions or other locational factors, a considerable amount of cargo can be transported direct to the hinterland. The port then functions primarily as a transit point which cannot be used to directly pursue employment policy.

The direct trade function of the port is also doubtful, however. The attempt to promote trade via one's own port is a long-term matter. In the first place, a change in the choice of port will have to produce enough surplus value in time and/or money to make a change in traditional sailing schedules worthwhile. The port of Kobe, for example, must bend over backwards to try and recapture the remaining thirty percent of freight from Osaka, even though their former contacts are still fresh in the memory. Moreover, it is apparent that ports are part of a whole range of goods flows and that it is thus not only the port costs but also the transportation costs and the labour climate that count. Finally, the attracting of industry or other port-bound activities which mean 'captured' freight is a long-term exercise whose outcomes are often highly uncertain. In those cases, the port in its totality functions as a locational factor and competes with other ports. The way in which industry settles on a choice of port is still incomprehensible to many port authorities. The port representatives which function as initial trading posts in the port's major trade countries can often do no more than maintain an extensive network of contacts and put port authorities in touch with potential clients.

In other words, the fact that the ports are instruments for promoting trade and that they are in a position to create employment are both realistic thoughts for the long-term perspective. But in the short and medium term these two considerations are not sufficient motivation to induce port authorities to be accountable for their actions. In a more general sense, the defining of *ports as instruments* for short-term economic aims is highly debatable. Consider, in this context, the ideas underlying the economic dispersion policy in the Netherlands in the sixties and seventies. In my view, it is essentially the productive coexistence of port administrations and port business communities which must be safeguarded in the normalisation of concrete port activities.

Following on from this, it must be said that it is often not clear on which regulatory principles the position of the port has been embedded in national economic regulation and what the regulatory principles within the port itself are. In the long term, this can result in a lack of clarity regarding the function(s) of the port or in policy initiatives being developed which no longer tie in with the existing regulatory principles and inadvertently obstruct them. The first case is exemplified by the port of Seattle which in the late eighties explicitly reformulated its rationale and had to gain the trust of the local residents. Another more recent example is the port authority of Los Angeles which seeks to redefine activities in cost considerations even though these are directly linked with the political significance which is assigned to the port. The second case is exemplified by the behaviour of the Rotterdam Municipal Port Management with regard to ECT which in 1996 suddenly found itself faced with competition.

12.3. The Position of the Port Authority

In the analysis of the ports in each chapter, the position of the port authority was examined closely. The port authority is the principal subject of this study not only because in many ports the port authority carries out a whole range of activities but also because in a broad sense it is the actor directly responsible for the day-to-day running of the port. In the account of its position various elements were reviewed. First of all the formal structure; what is the nature of the activities undertaken by the port authority, how is the port authority organised, how is the political control regulated and where does the port authority acquire its revenue? Moreover, in the sections on nautical management and port planning, attention was devoted each time to the relationship with layers of government and the business community. Finally, the port authority's options for influencing the implementation of, or responsibility for, various port activities were explored.

From the various accounts it appears that all port authorities are in hybrid positions. The extent to which a port authority operates in a hybrid way, and which potential outcomes this might have for the port as a whole, can only be addressed when various port activities are examined. The following sections will deal with this point. In this section, a general comparison is made between the position of various port authorities.

Based on the internal organisation and formal external responsibility relationship, a distinction can be made between *territorial* and *functional* forms of port management and between *directly* and *indirectly (appointed), democratically* elected political representatives. If this framework is placed over the various port authorities the following picture emerges.

Table 12.2. Autonomy versus Political Control of Port Authorities

	Directly elected representatives	Politically appointed representatives
Territorial administration	Rotterdam Hamburg Kobe	Los Angeles Hong Kong
Functional administration	Antwerp Seattle	New York Singapore Vancouver Durban

The table shows that the ports and port authorities of Rotterdam, Hamburg and Kobe have the lowest level of autonomy while those of New York, Singapore, Vancouver and South Africa have the highest. After all, in the first category the port authority is in the service of a greater bureaucratic whole and the direct responsibility for port policy lies with a directly elected representative (committee chairman or mayor). In the second category of ports listed, the port authority has a separate legal status, it has complete financial independence (the buck stops here) and the members of the board of directors are appointed for political reasons or on the grounds of expertise. The appointed members do not have direct political responsibility which leaves room to pursue a long-term policy.

This finding needs to be contextualised. In the first place, the Rotterdam Municipal Port Management is a municipal agency at arm's length (it has its own budget) and various powers are mandated to the Chief Executive Officer. It should be added, moreover, that the committee chairman's control is limited since as a rule he only serves for a couple of years and the information asymmetry with the municipal agency is actually very considerable. This is also the case in Kobe. A second contextualisation of the conclusion is that the port of Hamburg does not have a separate government agency which is responsible for the port. This means that the control over various port matters is highly diffuse. Of the three ports, the port of Hamburg also has the greatest degree of direct political management. Finally, it is striking to see that the functionally-managed Seattle and Durban pursue a strong (cross-sector) employment policy whereas the territorially-managed port of Rotterdam, for example, chiefly concentrates on the transit function.

In Los Angeles and Hong Kong, the port authorities are a part of the municipal administration and the Chinese regional administration respectively. The port authorities have their own financial budget which is approved each year by the politically appointed representatives. Within the budget, the two port authorities have a relatively large degree of freedom; on the basis of their expertise in port matters they have considerable authority. In both ports the political control clearly functions as a back-up circuit. The city council can revoke decisions made by the Harbour Commissioners within a certain period and the director of the Marine Department in Hong Kong can be directly asked for information on the implementation of policy by a commission from the Legislative Council (bureaucratic responsibility).

The autonomy of the port authorities of Antwerp and Seattle principally lies in the specific legal position which has been created for the port management. In both cases, the directly elected political representatives can keep a policy finger firmly on the pulse. In Seattle, political control can be far more stringently applied due to the substantially smaller and specifically elected group of port commissioners. In Antwerp, there is a good chance that general municipal council discussions will carry over into those of the privatised port authority. However, in both cases the elected port commissioners have a dual responsibility: on the one hand, to the concrete, port authority issues and on the other, to the electorate. As the history of port administration in Seattle has demonstrated, it is quite conceivable that the two may conflict and this can lead to indecisiveness.

What Justifies the Autonomous Status of a Port Authority?
Most ports are managed by port authorities who often hold an autonomous position. In general, it may be asked what justifies this autonomous position. The anecdote from chapter 4 (footnote 1) makes it clear that a port authority has a variety of relations with other actors (business community and layers of government). The underlying thought is that the emphasis on the private element might be at the expense of public confidence in the port authority and vice versa. Apparently, the need for independence forms the justification for the autonomous position. As the port of Hamburg demonstrates, however, the independence argument is not *sufficient* reason for a specific, autonomous position. After all, the port authority has no separate (legal) position within the public administration of the city-state; its independence is apparently sufficiently encapsulated in the public ownership of and the direct political control over the port.

Another motive which is often given for the autonomy of the port authority is so that it can have the opportunity to specify costs and revenue (own bookkeeping) but if the port runs at a loss year after year this does not constitute a reason for autonomy. Compare in this context the situation in the port of New York, for example, and Vancouver's wish to remain part of the national system.

It is thus not essentially the demand for port authority autonomy that is at issue. That avenue leads only to answers to the question of how the port authority can realise its own objectives (expressed in client's needs) more effectively than at present. The relevant question in fact is to what extent the institutional format induces the port authority to furnish accountability.

12.3.1. THE POSITION OF THE PORT AUTHORITY AND EXTERNAL EXPECTATIONS

Although the table provides a neat schematic view of the formal position of the port authority, it only shows a limited part of the material examined. In this study, it is the port authority's various relationships with (other) layers of government and the business community that are the central issue. These diverse relations give an insight into the different ways in which a port is administered, what the external expectations of the port authority are and what the port authority itself may lay claim to being.

Depending on the individual port authority's package of activities, certain behaviours are expected of it by the public authorities, shipowners and industry (shippers). The port authority of Hong Kong, for example, is primarily a nautical manager. Although a number of other activities are developed, this is only to try and balance costs and revenue. Since the safety and training standards it employs are very high by Asiatic standards, the Marine Department is actually autonomous. As a rule, the aim is to minimise administrative expenses so that they do not turn into substantial transaction costs and disturb trade.

The port authorities of Antwerp and Rotterdam are responsible for both nautical management and port planning. The competition between the two ensures that the shipowners and international industry have a strong negotiating position. This means that both ports must offer the most competitive prices they can (port dues, leases and rents) and that the costs of the port authority itself and other public expenditure (for its own administration and licences) must be driven down as much as possible. Rotterdam has more 'natural' advantages than Antwerp, however, to offer the largest ships so that by providing more 'quality' (good access) it can ask higher prices. Rotterdam as port authority has a tendency to concentrate on offering the best possible service which can manifest itself in any number of ways (ranging from a hypermodern traffic guidance system to a new manager's vehicle). Although these investments undoubtedly benefit the access to and safety in the port and the know-how might be sold abroad, it is ultimately the total initial costs which determine the attractiveness of a port. Rotterdam needs to consider each time whether an increase in the service provision can be borne by the market. Especially now that Antwerp can offer more or less the same advantages and level of service since the deepening of the Scheldt, the market links tariff increases in Rotterdam to the administrative expenses of the Rotterdam port authority. Service provision thus has its limits.

The port authority in Kobe (Port and Harbour Bureau) is primarily a port planner. Due to the combination of functions which the port of Kobe performs on a relatively small scale (living, working and recreation) and the fact that ports do not have to make a profit, the port authority is chiefly concerned with the coordination of various (often conflicting) objectives. The external expectations of the port authority's function in Kobe are mainly fostered by the local residents and industry and to a far lesser extent by the shipowners. The latter are more concerned with the public-private cooperation in the Kobe Terminal Corporation in which they participate financially.

The task of the port authorities of Vancouver, Seattle and Los Angeles is a combination of port planning and port service provision. The port service provision in Seattle and Los Angeles is emphasised more strongly than in Vancouver and New York because the port authority itself carries out activities which could also be done by the market. Vancouver and New York actively intervene in port service provision to the extent that they construct terminals themselves and select a private manager for them. The external expectations with regard to the functioning of the port authorities do not differ. As the conference held in Vancouver in September 1996 illustrated, the shipowners feel the ports along the west coast should not worry so much about the growth of the port as such but should make more efficient use of the available capacity. Whereas the port authorities worry about attracting new investors and/or freight (as far as this is in their power) and devising new port expansion plans, the shipowners and other port users are far more interested in the fast transfer of goods at existing terminals and the coordination of transport flows.

Finally, the control over both nautical management and port planning as well as port services in Singapore, Hamburg and South Africa is incorporated in one or more public organisations. For Hamburg and Singapore, this combination of activities does not apply as strictly as in South Africa because in the first two ports the port services are placed under a separate management organisation. This does not really alter the fact that ownership is totally public and that strategic decisions are taken by democratically elected representatives. The external expectations with regard to the functioning of the port authorities vary widely. As a result of the boom in trade in Singapore, the emphasis there is on generating as much capacity as possible whereas in South Africa the shipowners agitate against Portnet's omnipotent coordinating position and argue in favour of a distribution of tasks and powers. What they really expect is low prices for goods handling and more control. The shipowners and users in Hamburg, on the other hand, are concerned with the functional success of the port (deepening of the Elbe and reliable passage) because the competition with Rotterdam, Bremen and Antwerp keeps the prices competitive.

12.3.2. THE PORT AUTHORITY'S PERCEPTION OF ITS OWN POSITION

The role which a port authority can play is determined to a great extent by the position which is assigned to it in the port by others. Problems can arise when the perception of one's own position and the position ascribed to one by others are not congruent. The chances of this are not inconceivable given the autonomous position which port authorities often hold. The perception of one's own position and role achieve clear expression in the policy plans which a port authority publishes. After all, numerous objectives are announced in it, some of which will be realised in cooperation with others. They mirror, as it were, the port authority's perception of its relationship with public and private sector. The real confrontation between perception and the actual, sustainable relationship only comes in the realisation phase. The final part of this section will take a brief look at the position that the port authority sees itself as having.

The Rotterdam Municipal Port Authority's Corporate Plan 1997-2000 makes it clear, for instance, that the port authority sees itself as the conditioner, but also primarily as the *facilitator* of market activities. For the latter task, the Rotterdam port authority sets itself up as a partner for joint ventures with the market. This idea was also encountered in the new statutes of the privatised Port of Antwerp. In both ports, the port authorities try to shift the traditionally dominant supply policy onto the shoulders of the market. In Hamburg they still espouse the exclusive public supply policy.

It is only in the port of New York that the port authority knows that all it can do is create the conditions for market activity. In the odd isolated case it will act as mediator between various port actors but its main task is to maximise the cost-effective leasing of sites and, to a lesser extent, cranes. The port authorities of Vancouver and Los Angeles, on the other hand, have the idea that they primarily function as service providers towards (potential) clients. The acquisition of revenue is not under discussion in these ports due to the growth of trade with Asia. In contrast to the other North American ports in this study, the port authority of Seattle has the perception that as a functional organisation it must concern itself with the water and air transport on the one hand and the welfare of the local residents of the region on the other. It sees itself as a partner of every other public or private actor, which suggests that it functions in a purely cooperative system. The emphasis lies very strongly on the creation of public goodwill.

In Singapore and Durban, the port authorities are well aware of the division between public and private responsibilities but they are not yet convinced that the market would be able to undertake the operation of private activities in a more efficient way. Moreover, both ports are strongly attached to the healthy flow of revenue deriving from the handling activities. In the third place, the public character of the ports is referred to (multi-user principle). Finally, Durban justifies its public monopoly position for private activities with the excuse that otherwise a private monopolist would arise. In short, in these public systems one's own position is seen as inextricably bound up with the total port management.

The port authority of Hong Kong sees itself primarily as a nautical service provider the costs of which are passed on to the users.

12.3.3. AUTONOMY AND ACCOUNTABILITY

Due to the often autonomous and financially independent position of the port authorities, it is not inconceivable that external expectations and the internal perception of the functioning of the port authority might differ. For example, it remains to be seen whether all sections of the Rotterdam business community will recognise the projected facilitating position of the Rotterdam Municipal Port Administration or whether individual companies will decide for strategic reasons not to tie themselves. A similar question may be raised regarding the privatised Port of Antwerp which may formally obtain ownership of private activities. How will shipowners and private handling companies react when, over time, the Port of Antwerp obtains ownership of an increasing number of port activities?

For the service-oriented ports of Vancouver and Los Angeles, the question is to what extent will the supply-driven services ultimately be borne by the market. Is it not quite conceivable that these port authorities will assign themselves tasks which in effect contain an element of luxury and superfluity and which in economically less prosperous times will vanish again? The confrontation between their own perception and the external expectations of the port authorities' positions in Singapore and Durban prompted the question of how long they can continue to hide behind the multi-user principle.

A general question which applies to all the ports is: how much merit is there in the assertion that competition between ports will keep the port authority on its toes? The issue of the corrective effect of competition between ports is a difficult one which has been vexing port economists for years. After all, it depends on which market segment is undergoing competition (e.g. service provision, prices in the case of comparable quality, conditions for transportation, etc.). Not all market segments can be directly influenced by the port authority. In basing expectations about the functioning of ports on the port's significance, the corrective effect of the competition between ports needs to be contextualised. For instance, the fact that the port of Rotterdam performs a national economic function for the Netherlands and that under the mainport policy a substantial amount of money can be released for hinterland connections (Betuwe line) and port infrastructure (projected second Maasvlakte) has structural significance within Europe. In fact, this does not involve the competition between ports but the competition between the various member states. For political reasons, a country may opt to increase capacity but also offer the lowest prices. The potential commercial profit on the investment is secondary and difficult to determine due to the broad timeframe and the macro effects. The competition between ports thus always appears to involve a 'mix' of locational factors whereby some 'natural' advantages often amply compensate for other factors. The scale increase in shipping, for example, has in many cases forced shipowners to opt for one port.

In nearly every port at present, a good deal of time and energy is being devoted to discussions on the privatisation (also termed the repositioning) of the port authority. These discussions mainly focus on the formal position of the port authority. Due to the constant desire for increased performance, it is assumed that the port authority will benefit from a private law status or other means which could increase autonomy (its own staffing and financial policy, etc.). In view of the fact that a port authority's function always unites public and private responsibilities, privatisation always has direct consequences for the way in which the port authority carries out its activities. These still have to be conducted under more normative relations than the private activities. Privatisation of the port authority can lead to the uncoupling of public and private activities in order to keep relationships clear. The more the port authority pushes and pulls to engineer its autonomous position, the stricter the conditions laid down by the imperium. One wonders whether port authorities are aware of this before they start up the privatisation discussions.

From the analysis in the various chapters, it must be concluded that the position of a port authority always comprises a mix of public and private activities and powers. Whether it concerns the exclusive land management in the port of New York or the profit principle-based nautical service provision in Hong Kong, it always involves a combination of various (exclusive) port activities. Section 2.4 has already addressed the expected consequences of such hybrid constructions and how they are dealt with. This does not alter the fact that extensive empirical research into the functioning of port authorities can definitely bring more insights and aspects to light. It should be remembered that the outcomes of behaviour will not become visible for a number of years and that essential differences can exist in the operationalisation of the learning capacity (cf. the definition of the concept of norms by Argyris and Schön).

This study not only explored the activities of the port authority but also focused on a broad range of port activities. In the institutional analysis used here, the actual position of the port authority can only be defined in terms of other port actors and thus of other port activities. The purpose of the study is to get to grips with the learning capacity of ports and not with the individual port authorities. In order to achieve this, the normative line of approach was used: clear relations and straightforward checks and balances. In the next sections, expectations will be formulated regarding the conditions needed to activate the learning capacity of the various port activities studied.

12.4.   The Significance of State-Market Relations

This study addresses the institutional-theoretical issue of the way in which state and society interact. The international seaports function as an empirical particularisation but comprise highly diverse activities which must meet various conditions. They proved to be an appropriate subject for reaching a clearer definition of the theoretically formulated control relations. The intention of this section is to analyse the theoretical-normative basis and the typology of interventions with the aid of the port activities studied. The purpose of the distinction between state-market relations was not only to try to get to grips with the multiplicity of port activities, but was intended primarily to gain an insight into the type of state intervention used. The general theoretical-normative starting point of clear relations formed the cornerstone in trying to gain an insight into the minimum conditions for the control relationship. The presentation of the analysis does not provide tables which show at a glance the difference between expected and actual institutional formats. The chief reason for this lies in the instrumentalist and static character of such a presentation which moreover does not always serve to simplify interpretation.

The conditions which diverse port activities must fulfil do not just appear out of thin air. They generally have a clear link with certain values which are safeguarded in the institutional relationship. The difference between one port and another lies principally in the fact that the conditions which port activities must fulfil vary because they have acquired a different value (public or private). A researcher is not in a position to judge the (political) value but the institutional effects of this value can be judged. The universal theoretical-normative principle of this study postulated that as soon as agreement has been reached on the value of an activity (e.g. primarily oriented towards profit, or expertise, or safety, etc.) it must then be clear to which corresponding institutional conditions the port activity is linked. After all, these conditions encapsulate the sustainable external expectations with regard to the functioning of the activity.

External expectations regarding the behaviour of government and social actors vary according to the relations between state and society. Four unique state-market relations can be distinguished depending on the position of the state. In a *government monopoly*, a government has exclusive rights over the production and arranging of goods and services. The external expectations of social actors in this case are based on the functioning of law and democracy. In the case of *government conditioning*, production by the market exists but the conditions of that production are set exclusively by the government. The government determines whether the expectations have been satisfied (government arranges). In *co-arrangements*, government and consumer are both in a position to arrange external expectations of production by the market. Finally, in the case of *market regulation*, the government expects that the competition between providers and the freedom of choice for the consumer will keep the providers on their toes and that the government will abstain from market intervention.

The theoretical-normative stance of this study reasons from the idea that concrete conditions for port activities must be based on a workable programme of normative basic principles. If it is clear what is expected of actors then they can steer their concrete behaviour accordingly. In other words, if relations between actors are transparent and checks and balances are uniform then clear points of reference exist on the basis of which information can be interpreted and converted into concrete behaviour. In this way, the learning capacity of actors is permanently activated and innovations can develop. Moreover, a clear normalisation of responsibilities can prevent the generation of negative motivations or to the neglecting of responsibilities. As already mentioned, an academic analysis cannot go into the value of a port activity but by using the political value as a fixed point of departure, a useful contribution can be made by providing an appraisal of the conditions for the learning capacity of port activities. In other words, whether the port authority opts for a public procedure or decides on a private mode of operation for a particular activity, the appropriate control relations must be observed to prevent it carrying out the activity off its own bat. In the next subsections, various port activities are examined to find out to what extent the value corresponds to the institutional format of the activity. In short, if a private solution has been chosen for a particular port activity, the institutional format must make it clear how the private value will be made responsible and guaranteed.

12.4.1. PERCEPTIONS DERIVED FROM NAUTICAL MANAGEMENT

Nautical management comprises those port activities with a public character (allocating berths, shipping guidance, buoys and beacons, dredging operations, recording water depths, inspection of ships, etc.). All these activities are aimed at providing a safe passage for ships, offering protection to ships and increasing safety on ships. The public character of the activities finds expression in the general application of the rules and in the fact that a government is responsible for (a part of) the activities. The control over the nautical management activities involves the delineation of public powers, (standards, supervision and enforcement) and the adequate delineation of standards (e.g. for the environment and spatial planning) because conflicts of interest can occur between various public tasks and powers, and between public and private responsibilities.

In the ports examined in the United States, Canada, Japan and Hamburg these requirements are met. In these cases, nautical management of the ports is separate from the port planning and the port services and there is a clear distinction between national and local responsibilities. The separation of responsibility for nautical management and port planning sometimes means that decision making can be a long time coming, as was the case for example with the dredging issue in New York and Los Angeles and is still the case in Hamburg. But the dependence on federal endorsement and funds does justice to the character of the political decision making and calls into question the routine nature of port investments. Moreover, such a separation prevents the dual interests of shipping safety and fast transfer from becoming entwined in an unclear way.

In the other ports, a choice has been made to incorporate nautical management powers in a single public organisation (Hong Kong) which may also devote itself to port planning (Rotterdam and Antwerp) and/or port services (Singapore and Durban). It is not always clear what the reasons are or were for the autonomous (local) responsibility and/or the integration with other port activities. It often seems to be a matter of tradition which is no longer even questioned.

The Marine Department in Hong Kong, for example, is traditionally responsible for order and safety in the port but sees itself primarily as a service provider to the shipping traffic which wants to work as efficiently as possible. Although nautical management is seen as a public task, services are not always provided free of charge from public funds. The management of the unique traffic guidance system, for instance, is funded directly from the port dues. A relevant question here is why a government monopoly offers such a service when the emphasis is on efficiency and service provision to individual ships. Would the learning capacity not be more effectively served if this activity were offered by means of government conditioning? After all, a choice could then be made to put it out to tender once every so many years. Or the total port area could be split up into separate regions and subcontracts negotiated for particular years which would provide benchmarking.

Due to the relatively high safety standards in Hong Kong (international guidelines) compared to the rest of the region, the enforcement assumes a high degree of autonomy for the Marine Department. After all, there are no competing ports of any significance and on an international scale the tax climate is very favourable. The emphasis on service provision and autonomy means that the Marine Department can decide for itself which additional activities it will undertake. In other words, because economic rationality takes absolute precedence in Hong Kong and the Marine Department manages itself, those public interests which tend to be more conflicting, such as e.g. the care for the water in the port, are not directly seen to. If these activities take place at all, it is on a small scale and they have no direct priority.

In the port of Rotterdam, the Rotterdam Municipal Port Management's *expertise motif*, the *unity of maritime policy* and the *single counter principle* were the reasons underlying the local nautical responsibility and the integrating of nautical management and port planning. As far as shipping guidance is concerned, this means that the Rotterdam Municipal Port Management is running the show. The Ministry of Transport, Public Works and Water Management has delegated all powers to the national harbour master and only performs a back-up function. The national delegation of responsibilities for this important activity means that the Rotterdam port authority has to activate its own learning capacity. It is not inconceivable that, with a constantly increasing flow of revenue, this autonomy will lead to technical dilettantism (e.g. the latest gadgets in the field of traffic guidance and other service provision).

Although the public and private powers of the Rotterdam port authority come under separate directorates, it has recently become the practice to introduce a greater separation of various public powers on environmental grounds. This is the case, for instance, with ship to shore transhipment and floating transhipment which is enforced by the DCMR Rijnmond Environmental Protection Agency. If we take the case of floating transhipment, it is clear that the distinction between wet and dry infrastructure is not an essential regulatory principle for the division of responsibilities between the Rotterdam Municipal Port Management and the DCMR.

The fact that the distinction between wet and dry infrastructure is not a regulatory principle is also illustrated by the delineation of powers between the port police and the municipal police in Antwerp. The port police of the Port of Antwerp have sole responsibility for the total *port area*. Although this delineation of powers is unmistakably clear, the question is to what extent will other public and non-specific port-linked interests, such as environment and spatial planning, have proper attention devoted to them? The exclusive control of the privatised Port of Antwerp over the port area would not appear to automatically offer scope for these things.

The example of the Rotterdam inspection agency referred to above demonstrates that it is not taken for granted that the port authority should also be in charge of controlling safety standards. It still remains unclear which more sustainable principles are used as a basis for regulating the division of responsibilities. Up to now, in the division of tasks between the city of Rotterdam and central government, only ad hoc and pragmatic solutions, aimed at reducing the number of inspections, have been produced. Surely the question of the division of various public interests is also a key issue here? After a time, when circumstances and situations have changed and new staff have arrived, surely the same competency problems will reemerge because there is no fixed strategy?

In response to both these examples, the question may be asked of whether the public division of responsibility is about safeguarding the expertise of the Rotterdam Municipal Port Management, the unity of maritime policy and the single counter principle per se or whether it is essentially about the division of public quality requirements which certainly does not imply the exclusion of the first objectives. Moreover, the unique public control over nautical management would appear to lead initially to monopoly by the port authority, but after a time components of this are reclaimed by other layers of government because they demand control over other public interests for themselves. If the delineation of responsibilities is only thought through on the level of goal-rational wishes and possibilities, they often generate unnecessary and prolonged negative energy.

Nautical management activities are not always undertaken by public authorities or public enterprise but are sometimes contracted out to private companies. In such cases, it is extremely important that the public authorities concerned make it clear which (auditable) quality standards the implementation must meet and which financial commitments will be borne by these companies. The example of TCR in Rotterdam shows that the contracting out of a public task to a private company can be a perilous undertaking. Not all public tasks can be converted to market terms because there is often no directly designated market. Moreover, there are limits to the range of political and bureaucratic controls. The abundance of political enforcers does not necessarily enhance the effectiveness of the control in such cases and requires very substantial control efforts (long-term deployment of personnel). It is remarkable that in setting up the new installation for TCR's successor, little had been learnt from the previous institutional problems; the institutional relations had not changed and it was still assumed that bureaucratic controls (pecuniary penalty) would be adequate to keep the company in check. Is the collection of ship's waste at this stage really commercially feasible or, due to the plethora of additional criteria, is it in essence a public activity which must be provided by the port authority and indirectly expressed in port funds?

In Singapore and Durban, the ports play an important role in the promotion of trade and the economic construction of the country. In the past, the port authority's justification for the strict organisational integration of nautical management with the other port activities was the necessity for central coordination. Nowadays, a choice has been made to create an organisational and financial division between nautical management and port planning versus port services. In Singapore, a separate state enterprise has been set up for this purpose whereas in Durban both types of activity still come under the same organisation but keep separate accounts. It remains to be seen whether such organisational divisions will suffice in the long term to keep the division between public responsibilities and private opportunities clear. After all, for a number of years the South African port authority has been obstructed by the union from implementing more far-reaching privatisation which gives the division an artificial character. Moreover, in both countries environmental standards are still in their infancy and there are no checks and balances. In view of the autonomous character of the government enterprises and the permanent, seemingly endless line of ships waiting to enter the ports of both

Singapore and Durban, a rapid introduction and enforcement of public quality standards other than the fast transit of ships looks doubtful.

The following table outlines areas of attention for nautical management:

Table 12.3. Areas of Attention for Nautical Management

Port	Expected Control Relation Based on Current Norms and Values	Actual Control Relations	Areas of Attention and Questions
Hong Kong	government conditioning	government monopoly	■ how is the consideration of other public interests in the port organised? ■ isn't the chance of introversion highly likely?
Rotterdam	government monopoly with division of various public interests	government monopoly with limited division of public interests or government conditioning	■ who prevents technical dilettantism? ■ isn't it conceivable that with the absence of a clear-cut normalisation of responsibilities and opting for ad hoc agreements, the chance of disputes regarding powers will reappear? ■ won't unnecessary conflicts arise because the court reverses RMPM's monopoly (e.g of floating transhipment)? ■ won't government conditioning cause a lack of transparency in enforcement and direct control (cf. the TCR case)? ■ won't government conditioning produce limited public law controls and difficulties in furnishing proof?
Antwerp	government conditioning	government monopoly	■ how can other public and non-specific port-linked interests (environment and spatial planning) be given adequate attention? ■ won't ad hoc policy and introversion get the upper hand?
Singapore and Durban	government monopoly	government monopoly	■ is the organisational division adequate to keep public responsibilities and private opportunities clear? ■ will public quality requirements be rapidly implemented and enforced? In other words, which public authority looks after which public interest?

12.4.2. PERCEPTIONS DERIVED FROM PORT PLANNING

On the one hand, the activities which come under port planning have a general economic conditioning character. This concerns the construction of quays and sites for attracting cargo and industry. On the other hand, there is the more specific construction and financing of terminals which may have various uses (multi versus single user). Port planning activities often involve huge investments on which the costs will only be recovered in the long term (20-year period). As a rule, the business community is not in a position to make long-term investments on this scale which accounts for government behaviour here. Port authorities are also interested in such investments, however, because the revenue can be used to realise other policy objectives. Depending on the definition of the port authority's task, it can lease sites or superstructure or both.

The land, the port area, is the most important possession that a port authority has. The control over the land determines the planning potential. Due to long-term contracts or the location of specific industry, it often proves difficult to undertake initiatives for the remaining area. In this sense, a determining factor is whether port sites have been let out to rent or lease or whether they have been sold. At present, it is not usual for port sites to be sold because this compromises the port authority's development options and because it may encourage unchecked land speculation. The careful management of the port area remains a difficult task for many port authorities. There is often a shortage of space which cannot be resolved in the short term. Moreover, the composition of the commercial enterprise plays a major role whereby increasing account must be taken of environmental and spatial planning criteria. In Hamburg and Kobe the framework of port location policy has been laid down in legislation: this clearly stipulates which enterprise can be located in which area.

In port planning, a distinction is traditionally made between *infrastructure* and *superstructure*. The generally applicable norm is that the infrastructure comes under the port authority or national government while the superstructure is financed directly or indirectly by the private sector. A number of reservations may be expressed about using this distinction as a port planning principle. In the first place, port authorities may undertake multiple activities and mix the costs and revenue of various activities so that less of the financial gain sharing from superstructure accrues to companies than would appear at first sight. Compare in this context, for instance, the service contracts which the port of Vancouver can negotiate with terminal operators, the simultaneous operation of the profitable airport and the port by the Port of Seattle, but also the fact that the HHLA container handling company in Hamburg is ultimately funded with government capital and can call on the city in times of need.

In the second place, the interpretation of the two concepts is not uniform in all ports. The financing of the Delta 2000-8 project in Rotterdam is an example of a unique construction. If, for financial reasons, a container stevedore is no longer able to bear the whole financial risk, the Rotterdam port authority can finance and pass on the costs for the *infrastructure-plus*. What were the superstructural components are then counted as part of the infrastructure for this specific supporting goal. In short, the distinction between infrastructure and superstructure is not a universal regulatory principle.

For the expansion of ports in Western Europe, it used to be felt that "the cart must be put before the horse". The economic growth in the past justified the construction of large-scale port areas. Unfortunately, the growth was not always genuine so that port investments burdened funds for many years after. In those days, too, a bright future was forecast for goods transport via the ports. New initiatives for capacity expansion have occurred in all Western European ports in the last seven years (chiefly container handling). The most important maxim for this still seems to be that the cart must be put before the horse. A world port cannot give no for an answer. Western Europe works on the principle that 'every supply creates its own demand'. The emphasis on growth is the main reason why the port of Antwerp, for example, lacks a concrete spatial economic vision for the Left Bank area. In the port of Rotterdam it appears to have been this sort of thinking in the past which led to the incompetent management of land grants because the land was available anyway. But it is due to this sort of growth thinking that the current expansion plan for the second Maasvlakte also looks as if it will not have to be borne by the market.

Finally, the collaboration between the Rotterdam port authority and the ECT container handling company on the Delta 2000-8 project can be questioned. Although this collaboration between market and port authority is coordinated, it is questionable whether it was based on the right market (that of the Rotterdam container handling). It might have been smarter to base it on the container shipping companies and container transport. After all, they are the ultimate conditioners of the logistic chains. Moreover, in the multi-year cooperation between the Rotterdam Municipal Port Management and ECT, the extent to which the RMPM has really been able to contribute to decisions about ECT's terminal concept is open to question. The fact is, the concept of dedicated terminals, automated vehicles and intermodal transport has had a major impact on the relationship with shipowners, the development of the Maasvlakte (use of space) and the necessary conditions with regard to hinterland transport (construction of the Betuwe Line, the shortsea project, inland shuttles). Particularly in the development of the limiting conditions the Rotterdam Municipal Port Management has had to play an important role; did it do this independently or had the Rotterdam port authority become an important political and financial resource for ECT? Finally, there is the question of whether an appropriate distance was kept in this close cooperation in view of the more general norm that the business community is supposed to stand on its own two feet. To what extent is the city of Rotterdam in a position to bring in Deense Maersk as a competitor to ECT without causing problems? How realistic and auditable was the city council's condition with regard to the financing of the infrastructure-plus for the Delta 2000-8 project, that they would only agree to it if the shipowners would bind themselves to the concept of dedicated terminals?

In the United States and Canada, on the other hand, it is customary for port expansion plans to be developed in close consultation with the market and other public authorities. Special examples in this context are the design of the new coal terminal in Los Angeles, in which no less than 41 companies are participating, and the joint construction of the Alameda Corridor by the competing ports of Los Angeles and Long Beach. Nothing will be built or developed until a user (approved by the authorities) has been found. There is thus more risk sharing than in Western Europe. The port authority will provide the financing for the preparation of the site. The size of the market segment will determine whether the port authority also takes a share in the financing of the superstructure. In Vancouver, for instance, the containers, the passenger terminal and the timber handling were formulated as policy spearheads. The more market-oriented approach found in American and Canadian ports is probably a result of the fact that they are only responsible for port planning and cannot rely on revenue from nautical management.

Port planning in Hong Kong occurs in an internationally unique way. The government issues long-term concessions for new terminals for which a centrally-determined *market division agreement* is used. The private sector finances both the infra- and the superstructure but wants as much certainty as possible regarding the commercial profitability. Against the backdrop of economic regulation and the huge financial risks, such an attitude is understandable but it remains to be seen whether the economic power of the container stevedores will work out favourably for the port in the long run. The fact is, their position can keep the port expensive for an unnecessarily long time. Moreover, there are still no general procedures and regulations with regard to the use of the environment and spatial planning. As long as the power of capital rules, these standards are unlikely to be introduced in the short term.

The port of Singapore invests and plans the port expansion and terminals entirely on its own. Due to its rosy prospects, Singapore does not have to worry about the return on capacity expansion. The success of the recently-launched regionalisation policy is dubious, however. It is understandable that new investment objects need to be found but it is not inconceivable that countries such as Malaysia, Indonesia, the Philippines, etc. will also try to make their ports as large as possible. The economic potential which they have for this is many times greater than the city-state will be able to produce in the long run.

The following table summarises the areas of attention for port planning.

Table 12.4. Areas of Attention for Port Planning

Port	Expected Control Relation Based on Current Norms and Values	Actual Control Relation	▪ Areas of Attention and Questions
Antwerp	government conditioning	government conditioning and market regulation	▪ how is the control over sold land regained? ▪ Which mechanism prevents investments in overcapacity and promotes the careful management of existing sites? ▪ Who oversees the formulation of a concrete plan for the long-term use of the port area?
Rotterdam	government conditioning	government conditioning	▪ how is the careless management of existing sites prevented? ▪ will the second Maasvlakte plan be borne by the market? ▪ in any collaboration with the business community, how can the appropriate distance be maintained (e.g. in protracted development projects such as Delta 2000-8)?
New York Vancouver Los Angeles	government conditioning	co-arrangements	▪ in any collaboration with the business community, how can the appropriate distance be maintained (e.g in the prefinancing of superstructure and entering into service contracts)?
Hong Kong	government conditioning	market regulation	▪ what safeguards are there to prevent the government being trapped in all policy areas by the finances of the business community which gives the latter a free hand?
Singapore	government monopoly	government monopoly	▪ which mechanism ensures that despite the numerous state participations justice is done to matters which concern the imperium? ▪ in the recent regionalisation policy, in what institutional way is the appropriate distance from other nations safeguarded?

12.4.3. PERCEPTIONS DERIVED FROM THE PORT SERVICES

The activities which fall into the category of port services have an essentially private character; they concern services to individual ships, for which the shipowners must pay a charge (towing services, bunker services, pilotage, stevedoring services, etc.). It is remarkable how the institutional format of these services can differ between one port to another. This is mainly due to ingrained tradition or to the port authority's (financial) objectives. In the service provision with a nautical character, such as e.g the pilotage of ships, additional rules of professional practice are often laid down for safety reasons or the activity is seen as a public responsibility. There now follows first an analysis of container handling and then a closer look at the pilotage of ships and the institutional format of the towing profession.

The Institutional Format of Container Handling
Container handling is a relatively young and innovative member of the port logistics family. Since the mid-sixties, every self-respecting port has become involved in the construction of container terminals. This was a highly capital-intensive affair for which, depending on the position of the port authority, a combining of private and/or public forces was necessary. In the port of Rotterdam, for example, this was achieved by the merger of various general cargo companies, whereas in Singapore, Durban and Seattle the port authority alone took care of the construction of container terminals and in Kobe a public-private cooperation was chosen. Thirty years on, a broad range of international experience with this form of logistics has been built up but the continuing development of techniques and logistical processes now occurs partly thanks to discoveries in the field of telematics and information technology. The ultimate aim of all these developments appears to be to acquire total certainty regarding the logistics process so that the sea cargo trade can be operated on the 'just in time' principle. This point has not yet been reached. In the field of international electronic data exchange, in particular, there is still a lot to learn. An interesting element in container logistics is that although market relations between shipowners seem to be in a constant state of flux, everything suggests that the trend towards scale increase has actually started and this will have implications for the relationship with port authorities worldwide.

  As already mentioned in the introduction to this subsection, it is remarkable how the institutional format of container handling activities varies. The question of how the learning capacity of these different formats can be appraised will now be considered. In Rotterdam, the provider concentration of container handling activities was established in a number of stages and has ultimately resulted in a private monopolist for the handling of deep-sea containers. Such an institutional format might seem surprising in view of the private character of this activity and the existing principle of market orientation in the Netherlands. The most important reason for this concentration was a more efficient use of public resources (e.g. less quays). Although such an aim may be relevant in the short term, in the long term it fails to take account of the learning capacity of control-based institutional relations (in this case: competition). It remains to be seen whether the competition with Antwerp and Hamburg, for example, will suffice to keep ECT on its toes in the long term. The intertwinement between ECT and the Rotterdam Municipal Port Authority described in the previous section would appear to justify the conclusion that the Rotterdam port authority is directly engaged in a private company activity. How does this relate to the general expectation that private activities in the modern economies of Western Europe should be left to the business community? Would the relationship with ECT not have been more sustainable if the principle of internal competition had been maintained, as is customary in Antwerp for example? Wouldn't a consistent disaggregation of the principle of internal competition also have prevented the city of Rotterdam from manoeuvring itself into a difficult position vis-à-vis the Deense Maersk?

The example of container handling in Antwerp makes it clear that it is possible for a port authority to actively implement the principle of internal competition. The distribution of concessions appears to be an appropriate instrument for achieving this. However, Antwerp has one temptation which could obstruct the principle of internal competition, i.e. if the privatised Port of Antwerp were to have shares in one of the container stevedoring companies. Then the same dubious situation would arise as in Hamburg, i.e. that the political administration could find itself in an embarrassing situation regarding the allocation of sites. On the one hand, there are good reasons to give preferential treatment to the public corporation because this would safeguard the public capital. On the other hand, it would be an excellent opportunity to give the private container stevedore the chance to keep the government corporation on its toes.

The existence of the HHLA government corporation in Hamburg is rather ambiguous. Due to the historical public funding and the employment objectives, the hands-off political influence can be understood to some extent although the experiences in other ports demonstrate that the creation of employment is highly relative. Argued from the private nature of the activity and the existing market principle, the monopoly position of the HHLA can be questioned. An earlier question, which also applies to the HHLA, is whether the competition with Rotterdam is sufficient to keep the public corporation on its toes in the long term. In the event of an economic downturn Hamburg still has the public committment to the corporation. Would it not have made more sense for the HHLA to have refrained from acquiring the Gerd Buss private container handling company and instead continued to operate in accordance with the principle of benchmarking?

In New York and Los Angeles, it is customary for the container shipping companies to be the terminal lessees. They take care of staffing and handling activities. In Vancouver and Seattle, on the other hand, terminal contracts are entered into with private stevedoring companies and occasionally with a major client such as e.g American President Lines in Seattle. In Seattle, it used to be customary for the Port of Seattle to provide the staff. This is now seen as an exclusively private activity. Although Vancouver and Seattle compete with each other, it is possible for Vancouver to enter into service contracts with a container stevedore. The port administration in Vancouver apparently opts for the provision of service and reliability rather than the lowest price. This slightly distorts the effect of competition. It remains to be seen whether the suppression of the price mechanism will activate the learning capacity of the private provider in the long term. There is a chance that after a time the mutual dependence between the port authority and the private provider will become so great that only serious dysfunctioning would induce the port authority to terminate the contract.

The port of Kobe has chosen to leave the leasing of international container handling facilities to a co-arrangement. In the Kobe Port Terminal Corporation the international shipowners have a twenty percent share and the national Ministry of Transport and the municipality of Kobe each have forty percent. In Kobe, international container handling is seen as a private activity and a number of the stevedores related to the shipping company carry out the actual handling operations. Due to the traditional links there is no question of competition but the financial spreading of risks and the private involvement in the port is clear. The KPTC is nevertheless a hybrid organisation. Given the fact that this mixing of public ownership in a private management organisation creates a tendency to turn terminals which were originally intended for private use into public terminals on occasion (if they are empty), the private control can be considered to be minimal. The private management organisation looks like a quasi-private farce in a predominantly publicly-controlled system.

In Hong Kong the container handling activities are offered under competition. Hong Kong is the most expensive container port in the world because it has virtually no competition from other ports and the supply of container facilities can barely keep up with the growth in container transport (partly due to political squabbles). Incidentally, the market division agreement with regard to the opening of new terminals does not correspond with free market thinking and this contextualises the competition between the various providers. The market principle is dominant, however. Nevertheless, it is difficult to theorise about the learning capacity of the container activities. On the one hand, despite the tremendous shortage of space, Hong Kong has managed to take care of the growth in the number of containers within the existing capacity. On the other hand, the massive demand, the strategic location, the weak competition position of nearby Chinese ports, and the ownership of the floating cranes and container facilities in other Chinese ports ensure that the container companies do not have to worry about their position; market shares are virtually guaranteed.

In Singapore and Durban, the container handling activities are seen as private activities on which port authorities make large profits because there is virtually no competition. This leads to great dissatisfaction among the international shippers. For the time being, the port authorities have chosen to split the public and private activities but this has not yet brought about market orientation. The learning capacity of the present set-up is thus also dubious. As long as there is clear economic growth and these public enterprises can display flourishing financial figures no-one will doubt their efficiency and effectiveness. The economic growth automatically forces the companies to expand. Its general function in the development of the country might justify the port authority's monopoly position. The actual test of its learning capacity can only take place when the competition increases or the economic growth stabilises. It is possible for such public corporations to continue hopping from one leg (general economic development) to the other (market orientation and efficiency) for a very long time.

A comparison of container handling in the United States makes it clear that there it is the *shipowner* who is the terminal lessee and not a private stevedore or a port authority itself. This means that the terminals in the United States do not have a multi-user character but are used exclusively by individual shipping companies and/or alliances. Where Western Europe has chosen to combine forces in a single major container handling company plus a few smaller ones, the United States believes in the principle of competition and basically that of a scale reduction in supply. Incidentally, in the shipowners market there is also a tendency to form oligopolies. Whereas the principle used to apply that the shore should be accessible to all, the container alliances have started a trend towards the *oligopolisation of the shore*. Although, in Rotterdam, the stevedore leases the terminal, experiments are already being carried out with the dedicated terminal concept to give the shipowner a more exclusive position. Such a development is possible in Europe because there is competition between the ports and the shipowners in their joint alliances are in a strong negotiating position. The port as a whole thus seems increasingly to be developing the characteristics of a toll good. After all, the traditional motivation for the public management of the port lay in the fact that there was a desire in principle to make the shore accessible to everyone. As a result of the scale increase in shipping, the shore is in fact no longer accessible to everyone but only for those enterprises which can afford it.

The fact that in Asia the trend towards oligopolisation of the shore has not continued is linked on the one hand to the tremendous growth in trade, so that there is considerable regional trade, in particular, which is not carried out by the global trunklines but by smaller shipping companies which own just a few ships. On the other hand, in Singapore the port has acquired too great a general economic importance and the state of Singapore is making a fortune from the flourishing container handling industry. In Hong Kong, the largest container companies are intertwined with other important economic sectors so that their position is not disputed. Furthermore in Japan, the government participates in the international terminals, there are public terminals and the handling tradition that a stevedore is bound to a shipping company still plays an important role.

The Institutional Format of the Pilots' Organisation

In most ports, the pilotage of ships is seen as a component of the safety policy. The appreciation of the pilots' profession naturally depends on the layout of the port and climatic conditions. Due to their specific, essential knowledge and expertise, the pilots traditionally form a separate and independent group within the port. Although port authorities work increasingly with radar-controlled traffic guidance systems and the developments in telematics are finding growing application in the shipping world, these techniques still do not function as comprehensive alternatives to the guidance of ships by pilots. In all ports, the technique is seen primarily as an additional aid. Rotterdam is the only port which is gradually going over to linking exemption from compulsory pilotage to the training in the use of the traffic guidance system. Rotterdam is also talking about introducing a tariff for the guidance of ships via the Traffic Guidance System.

In nearly all the ports in this study, the pilotage of ships is carried out by an independent, private association which is subject to more or less stringent conditions and professional regulations depending on the national viewpoint. As a rule, the justification for the independent institutional position of pilots appears to lie in their specific expertise, the guaranteeing of safety or simply in a tradition which has grown up over the years. In many cases, the independent position of the pilots is not disputed. A relevant question here, however, is what expectations are attached to the pilots' profession and how exactly does this position relate to the way in which government regulation is shaped? Although the pilots' costs only comprise a small part of the total running costs, in many cases the profession is judged on its efficiency.

The institutional format of the pilots' organisation in the port of Rotterdam displays ambivalent features which lead to a distinct lack of transparency and frustration for many of those involved. On the one hand, the pilotage of ships is deemed compulsory for reasons of safety whereby unmistakably public characteristics are ascribed to the nature of the operations. The public responsibility also finds expression in the formal power of the Ministry of Transport to set the pilotage tariffs. On the other hand, in 1988 the pilotage operations were privatised with the aim of raising the service level but which meant that even more private elements came to the forefront in the exercise and organisation of the pilots' profession. This ambivalence makes it awkward to assess the learning capacity because there is a free choice to hop from one leg to the other at will. Nevertheless, what this case does makes clear is that a very limited role must be allocated to the scope of government conditioning if at the same time more important political objectives are an underlying factor. After all, although the Ministry of Transport can formally determine the pilotage tariffs, it appears to limited by the Scheldt treaty.

The learning capacity of the dock pilots in the port of Antwerp must be given a very negative evaluation. There is a private monopolist, use of which in some cases is not compulsory but for which the shipowner nevertheless has to pay a certain sum (seventy percent of the normal amount). There is thus neither political control nor competition which gives the organisation a completely free hand. The reason for this institutional format seems to lie in the total lack of clarity regarding the status of the pilotage activities in the port. Why are some ships exempted from compulsory pilotage and yet have to pay for it? Which elements justify a private monopolist? In that case, would it not have been more logical for the pilotage in the port still to form part of the public service and be subject to direct political supervision?

In Hamburg, dissatisfaction with their work situation led the pilots to form an independent association. The acquiring of an independent status went hand in hand with the strict regulation of the profession. This government conditioning is expressed in the price and quantity regulations and in the training criteria. For the time being it seems to be working well but does assume the permanent active and direct involvement of the city-state to adequately respond to changing circumstances (supply-demand relation). It remains to be seen whether this can be sustained in the long term.

In Canada, the pilots are formally seen as a public responsibility. The profession comes under federal government corporations which have a regional public monopoly and in principle must be self-supporting. Such a construction does not improve the clarity of expectations. After all, in principle there is a market relationship with the shipowners but the government corporation is not judged on its efficiency because there is no competition and financial shortfalls are ultimately borne by the federal state.

The port authority of Los Angeles sees the provision of pilotage as a form of service to the shipowners. This assumes that it is not an exclusively public task whereas this service is offered exclusively by the port authority. This also provides the opportunity to absorb financial risks in the greater whole of the Harbor Department. The service level might be permanently activated by establishing a form of benchmarking with the private provision in Long Beach.

Pilotage in the port of Hong Kong is carried out by a private monopoly which is closely supervised by the director of the Marine Department (compulsory pilotage and tariffs). In view of the extremely busy nature of the port, close cooperation between the pilots and the traffic guidance system is seen as increasingly essential. Such a development might argue in favour of classifying the pilots under the Marine Department in the future. After all, the public character of the pilots is becoming increasingly clear in an ever more crowded port.

The Institutional Format of Towing Services

The towing or assisting of ships entering or leaving the port is seen in all ports as a private activity. This is not to say that it is always carried out under competition, however. Scale arguments are often used to advocate the exclusive public or private position. It is also sometimes alleged that for many newcomers their 'sunk costs' are too high to enter into competition with others. Empiricism demonstrates that these arguments may be disputed. An assessment of the learning capacity of towing services in the individual ports now follows.

In the port of Rotterdam for a number of years now the provision of towing services has taken place under competition. After having only two providers for years it proved possible for a third (Kotug) to gain a toehold. In response to the unrest which this caused, the Rotterdam port authority, in the interests of the permanent care for safety, laid down a number of additional (technical) conditions. As long as these conditions do not influence the competition between providers, the learning capacity of the towing market would appear to be optimal. A fairly similar situation has now arisen in Hamburg now that Kotug has also brought about competition there. In the ports of the United States, Canada and Hong Kong this practice had been customary for some time.

The towing services in the port of Antwerp are carried out exclusively by the Port Captain of the Port of Antwerp. The reasons for this are unclear. It remains to be seen whether this enterprising bureaucracy will be able to deliver market prices and whether in the long term it does not simply boil down to dilettantism (attractive ships, etc.). It might be more sensible in the existing institutional format to find a form of government conditioning with which artificial competition can be engineered. The same observations can also be made about Singapore and Durban. In Durban, the same situation exists as in Antwerp, and in Singapore a very limited form of benchmarking has been devised. Up to now, the private tugboats have only been allowed to turn out on the orders of the Singapore port authority at times when the port authority's capacity has been exceeded.

The following table shows a summary of the areas of attention for the port services.

Table 12.5. Areas of Attention for Port Services

Port	Expected Control Relation Based on Current Norms and Values	Actual Control Relation	Areas of Attention and Questions
New York Seattle Los Angeles Vancouver	market regulation	co-arrangements and market regulation	▪ how can the port authority as site leaser be kept on its toes? in the case of service contracts, how can the lessee be held responsible?
Rotterdam	market regulation	government conditioning, co-arrangements and market regulation	▪ is the competition with Hamburg and Antwerp enough to keep ECT on its toes?; ▪ how do the facilitating conditions of the RMPM vis-à-vis ECT relate to other private relations (e.g. Maersk)?; ▪ will central government acquire more influence over the pilotage tariffs now that the Scheldt is being dredged?
Antwerp	market regulation	government monopoly and market regulation	▪ how can resistance be offered to the port authority's temptation to take a share in the container handling?; ▪ what justifies the private monopoly position of the pilots in the port now that they obtain revenue without doing anything in return?; ▪ why is the towing of ships within the port reserved solely for the Harbour Captain's department?
Hamburg	market regulation	government monopoly and government conditioning	▪ is the competition with Rotterdam and Antwerp enough to keep the public container handling corporation on its toes?; ▪ can an effective government capacity steering of the pilots be sustained in the long term?
Kobe	government conditioning	co-arrangements	▪ doesn't the limited private control in the KPTC represent a rigid tie for companies which reduces its attractiveness?
Singapore and Durban	government monopoly	government monopoly	▪ to what extent does the organisational separation from nautical management also imply the possibility of future private participation? ▪ What does the separation mean for the competition in the towing market?
Hong Kong	market regulation	market regulation	▪ does the container oligopoly not make the port unnecessarily expensive?

12.5.   Summary

This chapter presented the general conclusions from the preceding analytical chapters by focusing on the crucial question of the learning capacity of ports. Just as in the empirical chapters, a comparison was made between the socio-economic significance and function of the port, the position of the port authority and the institutional format of various port activities. A comparison between different parts of the world provided the insight that fundamentally different meanings are ascribed to the function of ports. This turns out to have structural consequences for the way in which ports are managed and what may be expected of the port administration. In Western Europe, the ports are seen as industrial locations and as logistic links in numerous transport chains which must be managed as commercially as possible. In the United States, the port function appears to depend on the state in which the port is located. In Canada, the port administration is a federal responsibility and the emphasis lies on the promotion of trade. This emphasis on the trade function of the port is also the case for the ports in Asia and South Africa. A striking Asiatic case is that of Japanese port management where the ports need not make a profit but must integrate the functions of living, working and recreation as effectively as possible.

The function and significance of the port determine the expectations of the port administration's performance. In many cases, however, it is not specifically known which activities belong to the exclusive public or to the private responsibilities. In other words, in many cases it is difficult to discover the regulatory principles of the port administration. Since the port authority was the primary subject of research, a good picture of the various regulatory policies could be obtained. The position of the port authorities could be analysed from two perspectives. The first perspective concerned the autonomy of the port authority (territorial versus functional administration). The second perspective addressed the way in which political representatives are involved in port management (directly elected or appointed). This analysis provided some interesting findings. The port authorities of Rotterdam, Hamburg and Kobe are components of the territorial administration for which a directly elected political representative is responsible. On the other hand, the port authorities of New York, Singapore, Vancouver and Durban are examples of functional administration and appointed political representatives. In comparison with the first group, this group appears to enjoy a high degree of independence which should nevertheless be contextualised by the existence of a separate accounting system in Rotterdam and Kobe and the fact that Vancouver and Durban are clearly involved in cross-sectoral activities (municipal contributions and employment policy respectively).

The reason for a port authority's usually autonomous position should be sought in the necessity for independence. Port authorities always carry out a combination of public and private activities. The emphasis on one or both leads to distrust in implementing the other activity. The demand for accountability is highly relevant, particularly in such autonomous organisations because they tend to become introspective. In other words, there is a strong likelihood that there will be a disparity between the external expectations and internal perception of one's own position. Whether the competition between ports is enough to keep the port authority on its toes is doubtful. It is quite conceivable that a national government, using considerable financial clout, will create the conditions for the port so that essentially it becomes no longer a question of the competition between ports but of the competition between national states. The actual exploration of the position of the port authority and its furnishing of accountability only begins when individual port activities are subjected to closer analysis.

By taking the political value (public or private) of a port activity as fixed point of departure and comparing this with the actual institutional format, the learning capacity of an activity can be determined. In this way the learning capacity of the port as a whole can be contextualised. The nautical management activities have a public character. They are aimed at guaranteeing the safety of shipping. In all the ports, nautical management is looked after by a public authority. The requirement of a clear division between various public interests and checks and balances is not met in all ports, however. The ports of the United States, Canada, Japan and Hamburg, for example, where a national government organisation has exclusive responsibility and there is direct political control, do meet the requirement. In other ports, nautical management is linked with other public and private activities which are not specifically aimed at promoting safety. In Hong Kong, nautical management is seen as part of the total service provision where efficiency is the cardinal rule. The responsibility for nautical management is borne exclusively by a government monopoly. In Rotterdam, nautical management is linked to the port planning and the direct control of nautical management lies with the Rotterdam Municipal Port Management. The example of floating transhipment demonstrates that the distinction between wet and dry infrastructure is not a regulatory principle for the delineation of various public responsibilities. In the port of Antwerp at present, exclusive responsibility for nautical management and port planning is still incorporated in the privatised Port of Antwerp. There are no checks and balances whatsoever. Finally, in Singapore and Durban, environmental standards are still in their infancy or are awarded a different priority so that in those ports, too, no checks and balances exist.

On the one hand, the port planning activities involve general economic conditioning activities to make the port attractive to investors and cargo. On the other hand, they concern the more specific construction and financing of terminals which can have different use values (multi versus single user). The long-term contracts restrict the port authority's flexibility in the use of sites. Careful management of the land appears to be difficult. In this sense, there are marked differences between the approach in Western Europe and that of the United States. Where it is usual in Western Europe for 'the cart to be put before the horse' and to work on the principle of 'every supply creates its own demand', port authorities in the United States develop their plans together with the business community. Although the private investments in the port of Hong Kong are huge, the mechanism of port planning reveals that this port undergoes less competition that appears at first sight. After all, with the current growth figures, the market division agreement guarantees the substantial recovery of costs on the terminals.

The example of the co-arrangement for the realisation of the Delta 2000-8 project in Rotterdam made it clear that the distinction between infrastructure and superstructure is not a regulatory principle. The emphasis here lies instead on the financial risks which the government is running and the special position which the city of Rotterdam has apparently created for ECT.

Finally, an assessment was made of the learning capacity of the institutional format of container handling, pilotage and the towing of ships. In container handling and the towing profession in particular, which are seen in all ports as private activities, there are major institutional differences. In essence, these differences can be reduced to the various objectives which port authorities have and which can range from savings on public expenditure, to the traditional public financial ties and source of income, to the developing of a relatively new logistical concept, etc. As far as container handling was concerned, it became clear that there was once again a structural difference between Western Europe and the United States, namely that in the United States the terminal lessee is the shipowner and not an individual stevedore as is customary in Western Europe.

The pilotage of ships is in many cases an activity which comes under the safety policy of the port but is on the other hand an activity which shipowners expect to be carried out as efficiently as possible. The choice of a private independent position for the pilots, however, appears to make special demands on government regulation. In some cases, the regulation does not have the correct status so that the profession appears to acquire an ambiguous character and can no longer be induced to be made accountable for its actions.

Glossary

APL	American President Lines
CDR	Cooperative port Development Committee
DCMR	Rijnmond Environmental Protection Agency
DGSM	Directorate-General for Shipping and Maritime Affairs
DGV	Directorate General for Transport
DOT	Department of Transport
DSV	Shipping Department
DWT	Dead Weight Tonnage
EBB	Europoort/Botlek Foundation
ECT	Europe Combined Terminals
ESM	Exploitatiemaatschappij Schelde Maas
ESPO	European Seaport Organisation
EXCO	Executive Council
GATT	General Agreement on Tariffs and Trade
HOI	Port Reception Installation
IAPH	International Association of Ports and Harbors
IMO	International Maritime Organisation
HFLG	Hamburger Freihafen und Lagerhaus
HHLA	Hamburger Hafen- Und Lagerhas Aktiengesellschaft
ICTF	Intermodal Container Transfer Facility
ILA	International Longshoremen Association
KPTC	Kobe Port Terminal Corporation
Legco	Legislative Council
MARPOL	Marine Pollution convention
MOT	Ministry of Transport
MPA	Marine Port Authority
PAP	People's Action Party
PDB	Port Development Board
PHB	Port and Harbour Bureau
PSA	Port of Singapore Authority
RDP	Reconstruction and Development Plan
RIMH	Regional Environmental Hygiene Inspectorate
RMPM	Rotterdam Municipal Port Management
ROM	Regional Development Area
RSC	RailService Centra
RWS	Directorate General for Public Works and Water Management
SARWHU	South African Railway and Harbours Workers Union
SATS	South African Transport Services

SVZ	Shipping Union South
TCR	Tank Cleaning Rotterdam
TEN	TransEuropean Network
TEU	Twenty-foot Equivalent Unit
VBS	Traffic Guidance System
VPC	Vancouver Port Corporation

Summary

The Institutional Postition of Seaports[1]
An International Comparision

Motives and Purpose of the Study

In modern welfare societies, intertwinement between government and the private sector has increased in many ways. All too often it is no longer possible to trace clear-cut divisions of responsibilities between different organisations, although public and private sectors essentially operate according to different mechanisms and expectations. For this reason it is interesting to look for different state-market interaction categories and to value them in terms of supervision structures. This study concerns the academic question of governments' capacity to set conditions for other social actors (private companies and citizens). Although considerable regulation literature is already available, this study asks for permanent political attention to be devoted to the standardisation of government positions prior to embarking on policy actions.

Global seaports can provide useful examples for the consideration of this academic question as they consist of many separate and intertwined public and private activities. Many seaports are not as privatised as one might expect. Seaports in the United States, for examples, can be almost completely publicly owned, whereas seaports in Western Europe have to survive in a strongly competitive environment. Above all, port administration in different parts of the world varies widely. A third reason for investigating seaports is based upon their strong international trade elements and the ports' specific function that lead to permanent discussions on public and private tasks in global seaports. The discussion on the most appropriate position for a port authority has recently attracted worldwide interest due to privatisation and globalisation processes. A fourth reason is the lack of administrative academic literature on the various rationales underlying the different port administrative structures. The available literature on administrative port structures was written by economists and is of a highly descriptive nature. Recommendations are reasoned from existing and desirable property divisions and from the fluid element of scale. Such commercial criteria might be very useful for attaining goals but they do not provide insights into the normative frameworks of

---

[1] Summary of: Stevens, H. (1999). The Institional Position of Seaports, Ph.D.dissertation, Delft University of Technology, Delft, The Netherlands.

relationships within which port activities are performed. These economic theories do not throw light on the origin and nature of responsibilities for which public and private actors are held responsible. However, this way of reasoning is still customary in the international port community.

This study compares global seaports' institutional functioning in the context of their conditions for learning in a changing international environment. When the overall institutional seaport structure is consistent with the principle that defines the expected public and private role or responsibility, the seaport is able to *learn* by applying relevant external experience.

Reseach Method
To gain a clear picture of the operations of different port systems, international comparative research was necessary. The choice of seaports for this research was made based on the differences in their administrative structures. In Western Europe the research focused on the ports of Rotterdam, Antwerp and Hamburg. In North America the ports of New York, Vancouver, Seattle and Los Angeles were visited and in Asia the ports of Kobe, Hong Kong and Singapore were involved in the programme. Finally, the ports of Durban and Cape Town in South Africa were visited. The port of Rotterdam comprises a major part of this study. It formed the starting point to the research for getting acquainted with specific port activities and questions and provided a review point for the critical feedback after returning from the international trips.

In all the seaports, political representatives, chief executive officers and managers of the financial, marketing and legal/administrative departments were interviewed. In their pivotal role between public and private actors, port authorities were the major source of information. Whenever possible, conversations were held with representatives of the local Chamber of Commerce, specific interest groups from the port community or private companies.

Analytical Framework
In this study the concept of 'institution' has been sociologically defined: patterns of rules and/or norms on the basis of which each party can have sustainable expectations of the other's behaviour. An institution is characterised by sustainable possibilities for or limitations to action. In this study, the term institutional analysis refers to: seeking normative starting points for sustainable relationships between subjects.

With the help of theoretical insights from the disciplines of economics, sociology and law, four categories of *supervision relationships* between state and market have been formulated. A *government monopoly* reflects an institutional structure in which a government delivers services or goods and determines quality standards entirely on its own. The control relationship is typical of socialist societies. *Government conditioning* stands for the control relationship in which a government defines quality standards while paying, at least in part, for the service but without providing the service. Contracting out is an important example of this control relationship. In this book, the term *Co-arrangements* is used to refer to the direct cooperation between public and private actors. Finally, the control relationship of *Market regulation* refers to when private actors or consumers directly determine quality standards and completely pay for services and goods themselves. In this supervision relationship, a government determines the market conditions without intervening in the market process. This is generally the case for private goods.

The various port activities have been categorised as well. The category of *Nautical management* covers those activities involved in the safe and fast throughput of sea vessels: buoys, port state control, vessel traffic systems, dangerous substance control, etc. The activities pertaining to nautical management have a public nature: general functioning of rules and governmental responsibility. In terms of supervision relations, these activities require the division of different public responsibilities (supervision and enforcement) and 'checks and balances' (particularly for environmental affairs and spatial planning).

A second category, called *Port planning and infrastructure*, consists of activities demonstrating both a public (general use) and a private (specific use) character, for example the construction and redevelopment of port areas, spatial policies, the building of terminals, etc. Most port planning activities require huge investments which can only be financed by long-term loans (twenty years or more). Usually, private companies will not be able to make investments on this scale or timeframe so the investments necessarily have a public character.

The concept of *Port services* has been reserved for such private activities as the pilotage and towing of vessels, the work of linesmen and stevedores, the distribution of goods, etc. These activities are private in nature since the services are delivered to particular ships and paid for by the shipowners. Port services with a nautical character (pilotage) have generally been regulated for safety reasons or are seen as a government responsibility.

The distinction between different port activities, together with the categories of supervision relationships, facilitates the institutional analysis of each international seaport. Contrary to existing insights on port administrations, this analytical model includes the various port activities as its subject of research. This makes it possible to get a clear picture of the port authority's position in relation to private responsibilities. The well-known international distinction between *landlord port, toolport and service port* was not appropriate for formulating an answer to the research question. This distinction is based on the fluid division of means of production between public and private actors and does not make it clear how public and private actors will be able to address their mutual expectations. The general role division between the port authority and the private sector can be discovered by describing the history and socioeconomic functions of the seaport in question. By analysing the port authority's formal position, clear expectations of role divisions for different port activities can be identified. In challenging these expectations with empirical evidence, one can trace the (in)consistencies in the institutional seaport structure. The inconsistencies will provide the basis for formulating expectations about the seaport's learning.

Conclusions
The empirical chapters provided a description and analysis of each port's socioeconomic function and postion, the port authority's position and the institutional structure of various port activities. A comparison between ports in different parts of the world led to the conclusion that there are fundamental differences in the significance of seaports. This seems to have structural consequences for the port administration and the expectations of the port authority's performance. In Western Europe seaports are seen as combined industrial and logistical locations which have to be managed as profitably as possible. In the United States, the port function seems to be dependent on the state in which the seaport is situated. In Canada, seaport administration is a federal responsibility. Interest is focused on the trade function of the seaport. The same goes for seaports in Asia and South Africa. A special case is Japanese seaport administration. In Japan, seaports are not obliged to make a profit but must combine the functions of housing, work and recreation wherever possible.

The function of the port determines what is expected of the port administration. In many cases it is not explicitly understood which activities are either exclusively a public or a private responsibility. However, by analysing the port authority's position, a good picture of the actual responsibilities for different activities could be formed. Each port authority's position could be analysed along two dimensions. The first dimension concerned the port authority's autonomy (territorial versus functional administration). The other dimension concerned the way in which governments representatives were involved in the port administration (elected or appointed). This analysis resulted in some interesting findings.

The port authorities of Rotterdam, Hamburg and Kobe are part of a territorial administration for which an elected government representative is responsible. Conversely, the port authorities of New York, Singapore, Vancouver and Durban are examples of a functional administration with appointed government representatives. The latter group appears to enjoy greater autonomy. This must be further differentiated by the existence of a separate bookkeeping system in Rotterdam and Kobe and by the fact that Vancouver and Durban are engaged in cross-border activities (subsidies for municipalities and labour policies respectively).

The customarily autonomous postion of port authorities is based on the *necessity for independence*. Port authorities always perform a combination of public and private activities. An emphasis on one activity tends to result in distrust regarding the execution of the other. However, the question of accounting is very relevant for these internally autonomous organisations as they tend to become introverted. In other words, there is a high risk that external expectations and internal perceptions of their position no longer correlate. Whether competition between seaports will be sufficient to keep a port authority on its toes is dubious. It is possible for a national government to set the conditions for the seaport so that competition is state-based rather than seaport-based. Real research into a port authority's position can only begin when the separate port activities have been thoroughly analysed.

By taking the political value (public or private) of a port activity as the starting point and comparing this with the actual institutional structure, one can develop expectations about the learning capacity of that specific activity. In this way, the learning of the seaport as a whole can be differentiated.

The nautical management activities have a public character. They are focused on guaranteeing the ships' safety. In all the seaports investigated, nautical management is performed by a government agency. Not every port can satisfy the requirement of a clear division of the various public interests and 'checks and balances'. This is the case in the seaports of the United States, Canada, Japan and Hamburg where a national government department has exclusive responsibility and there is direct governmental supervision. In the other seaports, the nautical mangement is linked with other public and private activities which are not particularly focused on guaranteeing safety. In Hong Kong, nautical management is perceived as part of the total port service. Efficiency seems to be the ultimate performance yardstick. Responsibility for nautical management is exclusively determined by means of a government monopoly. In Rotterdam, nautical management is linked to port planning and direct control over nautical affairs rests exclusively with the Port of Rotterdam. In Antwerp, the privatised Port of Antwerp has exclusive responsibility both for nautical management and port planning activities. There are no checks and balances at all. And in Singapore and Durban, the development of environmental standards is still in its infancy.

Port planning activities are focused, on the one hand, on setting general economic conditions for making the port more attractive for investors and cargo and on the other, on building terminals with different user values (multi-versus single user). The location is the most important but also the most difficult element of port planning. The long-term nature of contracts reduces the port authority's options for making flexible use of the location. It appears very difficult to make optimal, effective use of the location. In this respect, there are important differences between the approach taken in Western Europe and that of the United States. In Western Europe, the maxim is ' don't put the cart before the horse' and it is common for there to be large reserves, whereas in the United States port authorities develop their projects in close cooperation with private companies. Finally, the learning capacity of the institutional structure of container activities, pilotage and the towing of vessels was assessed. The container sector and the tug profession in particular differ widely from one seaport to another, although in every port they are seen as private activities. These differences can be explained by the individual port authority's goals (savings on public expenses, traditional public relationship and source of income, development of a new concept, etc.). Here again, there is a fundamental difference between Western Europe and the United States. In the United States, a shipping company rents a terminal whereas in Western Europe individual, private stevedoring organisations own the terminal.

In many cases, pilotage is seen as part of the port safety policy. However, the shipowners expect pilotage to occur as efficiently as possible. Particular government conditions need to be set if a choice is made to give pilots an independent organisational position. In some cases this regulation does not have the right status. The character of the pilotage profession thus becomes ambiguous and can no longer be held accountable for providing an efficient service.

Bibliography

Agranoff, R. L. (1986). *Intergovernmental management. Human services problem solving in six metropolitan areas*, New York.

Alten, F. von (1995). *The role of government in the Singapore economy*, Frankfurt am Main.

American Association of Port Authorities (1995). *Seaports of the Americas. 1995 AAPA directory*, Coral Gables.

Argyris, C. en D.A. Schön (1978).*Organizational learning: a theory of action perspective*, Reading (Mass.)

Banks, J.S. en B.R. Weingast (1992). "The political control of bureaucracies under assymmetric information, in: *American Journal of Political Science*", 36:2, pp. 509-524.

Bardach, E. en R.A. Kagan (1982). *Going by the book. The problem of regulatory unreasonableness*, Philadelphia.

Baudelaire, J.G. (1986). *Port Administration and Management*, Tokyo.

Behrendt, D.K. (1994a). "Die Hamburger Hafenordnung. Bilanz nach 20 Jahren", in: *Hansa*, 1, pp. 71-75.

Behrendt, D.K. (1994b). The Hamburg Government's role in the port, internal document.

Bekemans, L. en S. Beckwith (red.) (1996). *Ports for Europe. Europe's maritime future in a changing environment*, Brussel.

Bendor, J., S. Taylor en R. van Gaalen (1987). "Politicians, Bureaucrats and assymmetric information", *American Journal of political science*, 31:4, pp. 796-828.

Ben-Ner, A. en B. Gui (red.) (1993). *The nonprofit sector in the mixed economy*, Michigan.

Berle, A.A. en G.C. Means (1932). *The modern corporation and private property*, New York.

Bish, R.L. en R. Warren (1972). "Scale and monopoly problems in urban government services", *Urban affairs quarterly*, september, pp. 97-122.

Blonk, W. A.G. (1994). "Developments in EU maritime transport policy", *Marine Policy*, 6, pp. 476-482.

Booz-Allen & Hamilton (1995). Organizational review of the Los Angeles Harbor Department. Final report, McLean.

Branch, A.E. (1986). *Elements of port operation and management*, London.

BST Associates (1991). Washington ports and transportation study. Final report, Olympia.

Cargo Systems (1994). "Waiting for the rush", 21:6, pp. 47-48.

Chamberlin, J.R. en J.E. Jackson (1987). "Privatization as institutional choice", *Journal of Policy analysis and management*, 6:4, pp. 586-604.

Charter International (1994). *Port of Hong Kong*, London.

Chen, M. (1995). *Asian management systems. Chinese, Japanese and Korean styles of business*, London.

Chihaya, M. (1981). "Kobe Port Island", *Maritime Asia*, pp. 19-20.

Chubb, J.E. en T.M. Moe (1988). "Politics, markets and the organization of schools", *American Political Review*, 82:4, pp. 1065-1087.

Chung Pui Hoong, J.T. (1984). Legislative checks on the Singapore bureaucracy: roles of the Public accounts committee and the Estimates Committee, Singapore.

Clavell, J. (1975). *Taipan*, Londen.

Coase, R.H. (1937). *The nature of the firm*, Economica 4, pp. 386-405.

Coase, R.H. (1990). *The firm, the market and the law*, Chicago.

Containerisation International (1986)." Ports promote variations on an expansionary Japanese theme", November, pp. 62-67.

343

Containerisation International (1991). "Port development: Hong Kong style", december 1991, pp. 71-77.

Containerisation International (1995). "Singapore Shift", October, pp. 79-81.

Coser, L.A. (1978). *Gulzige instituties*, Deventer.

Crichton, J. (1991). "Portnet", *Containerisation International*, 25: 2, pp. 45-50.

Daryanani, R. (ed.) (1995). *Hong Kong 1995. A review of 1994*, Hong Kong.

Denning, M. and D.J. Olson (1981). Public enterprise and the emerging character of state service provision: application to public ports, paper for the Annual Meeting of the American Political Science Association, Seattle.

Denning, M. (1983). A Politico-economic survey of the major west coast ports. Part I: The California ports, Seattle.

Dinten, W.L. van (1989). *Bedrijfskundige analyse en synthese. Op zoek naar de kern van de zaak.* (in Dutch) Oratie EUR, Rotterdam.

Dinten, W.L. van (1991). *Over helpen en gebruiken, collegestof Bedrijfskundige analyse en synthese*, (in Dutch) Rotterdam.

Dinten, W.L. van (1996). "Organiseren in een democratie", in: W.L. van Dinten (red.) *Democratie, dimensies en divergenties; van Descartes via Darwin naar Guehenno*, (in Dutch)'s-Gravenhage.

Doig, J.W. (1993). "Expertise, Politics, and Technological Change. The search for mission at the Port of New York Authority", *Journal of the American Planning Association*, 59:1, pp. 31-44.

Downs, A. (1967). *Inside bureaucracy*, Boston.

Dredging and Port Construction (1984). "Sea pollution in Singapore Waters", September, pp. 17-18.

Driel, H. van (1990). *Samenwerking in haven en vervoer in het containertijdperk*, (in Dutch) Delft.

Driel, H. van (red.)(1993). *Ontwikkeling van bedrijfskundig denken en doen: een Rotterdams perspectief*, (in Dutch) Rotterdam.

Economic Bureau Kobe City Government (1993). *Economic overview of Kobe*, Kobe.

Economic Development Board (1993). *Economic Development of Singapore*, Singapore.

Economic Development Board (1995). *Singapore*, Singapore.

Eggertson, T. (1990). *Economic behavior and institutions*, Cambridge.

Erdmenger, J. (1996). "Seaports in the Trans-European Transport Network", in L. Bekemans en S. Beckwith (red.) *Ports for Europe. Europe's maritime future in a changing environment*, Brussel.

Eucken, W. (1950). *Economische orde in een vrije maatschappij*, (in Dutch) Voorburg.

Fleming, D.K. (1987). "The port community: an American view", *Maritime Policy and Management*, 14:4, pp. 321-337.

Fligstein, N. en P. Brantley (1992). "Bank control, owner control, or organizational dynamics: Who controls the large modern corporation?", *American Journal of Sociology*, 98: 2, pp. 280-307.

Frankel, E.G. (1987). *Port planning and development*, New York.

Geelhoed, L.A. (1983). De interveniërende staat. Rapport van de projectgroep Beleidsinstrumenten in directe zin, (in Dutch) 's-Gravenhage.

Geelhoed, L.A. (1985). *Crisis en economisch recht. Wetgeving en economische crisis: crisis in economische wetgeving?* (in Dutch), Zwolle.

Geelhoed, L.A. (1993a). "Deregulering en mededinging: het Europese integratieproces en het coördinatievraagstuk", in J.J.M. Kremers (red.). *Inspelen op Europa, uitdagingen voor het financieel-economisch beleid van Nederland*, (in Dutch) Schoonhoven, pp.173-184.

Geelhoed, L.A. (1993b). "Deregulering, herregulering en zelfregulering", in Ph. Eijlander, P.C. Gilhuis en J.A.F. Peters (1993) (red.). *Overheid en zelfregulering. Alibi voor vrijblijvendheid of prikkel tot aktie?*, (in Dutch) Zwolle.

Goey, F.M.M. de (1990). *Ruimte voor groei: Rotterdam en de vestiging van industrie in de haven 1945-1975*, (in Dutch) Rotterdam.

Goey, F.M.M. de en H. van Driel (1993). "De relatie tussen overheid en bedrijfsleven in de Rotterdamse haven", in H. van Driel (red.) *Ontwikkeling van bedrijfskundig denken en doen: een Rotterdams perspectief*, (in Dutch) Rotterdam.

Goss, R.O. (1979). *A comparative study of seaport management and administration*, London.

Goss, R.O. (1990a). "Economic policies and seaports: 1. The economic function of seaports", *Maritime Policy and Management*, 17:3, pp. 207-219.

Goss, R.O. (1990b). "Economic policies and seaports: 2. The diversity of port policies", *Maritime Policy and Management*, 17:3, pp. 221-234.

Goss, R.O. (1990c). "Economic policies and seaports: 3. Are port authorities necessary?", *Maritime Policy and Management*, 17:4, pp. 257-271.

Goss, R.O. (1990d). "Economic policies and seaports: 4. Strategies for port authorities", *Maritime Policy and Management*, 17:4, pp. 273-287.

Griffiths, D. (1993). "South Africa: port development. Eyes on the prize", *Port Development International*, 19:7/8, pp. 30-33.

Griffiths, D. (1995). "Singapore: port privatisation", *Container Management*, Sept, pp. 4-5.

Gupta, A.K. en L.J. Lad (1983). "Industry self-regulation: an economic organizational and political analysis", *Academy of Management Review*, 8:3, pp. 416-425.

Hart,H. W. ter (1985). *Hong Kong*, 's Gravenhage.

Hazewinkel, H.C. (1974). *Geschiedenis van Rotterdam*, (in Dutch) Zaltbommel.

Hazewinkel, F. (1978a). *Zeehavenbeleid van Rotterdam 1967-1975. De jongste geschiedenis van de ontwikkelingen rond Europoort, Maaslvlakte en de Gouden Delta. Jaaroverzichten*, (in Dutch) Rotterdam.

Hazewinkel, F. (1978b). *Zeehavenbeleid van Rotterdam 1968-1975. De jongste geschiedenis van de ontwikke-lingen rond Europoort, Maasvlakte en de Gouden Delta. Facetten van het beleid*, (in Dutch) Rotterdam.

Heaver, T.D. (1994). Port efficiency, competition and cooperation, paper presented to the Joint Conference of the Korea Maritime Institutie and the International Association of Maritime Economists, Vancouver.

Heaver, T.D. (1995). "The implications of increased competition for port policy and management", *Maritime Policy and Management*, 22: 2, pp. 125-133.

Hegge, M.C. en Y.M. de Muynck (1995). "Private naast publieke taken: marktwerkings- en mededingingsaspecten", *Tijdschrift Privatisering*, jrg. 2, nr 7, pp. 4-5. (in Dutch)

Hellingman, K. en K.J.M. Mortelmans (1989). *Economisch publiekrecht, rechtswaarborgen en rechts-instrumenten*, (in Dutch) Deventer.

Henig, J.R. (1990). "Privatization in the United States: theory and practice", *Political Science Quarterly*, 104:4, pp. 649-670.

Hoed, P. den, W.G.M. Salet en H. van der Sluijs (1983). *Planning als onderneming; Voorstudies en achtergronden, (in Dutch) V34 WRR*, 's-Gravenhage.

Hoed, P. den (1995). *Bestuur en beleid van binnenuit*, (in Dutch) Amsterdam/Meppel.

Hofstede, G.H. (1995). "Principes van internationaal-vergelijkend onderzoek", (in Dutch) in A.F.A. Korsten, A.F.M.Bertrand, P. de Jong en J.M.L.M. Soeters (red.), *Internationaal-vergelijkend onderzoek*, 's-Gravenhage.

Hong Kong Marine Department (1995a). *Handbook, Hong Kong*.

Hong Kong Marine Department (1995b). *Performance pledge*, Hong Kong.

Hong Kong Marine Department (1996). *Statistical tables 1995*, Hong Kong.

Hooydonk, E. van (1996). *Beginselen van havenbestuursrecht. Onderzoek naar de grondslagen van de havenbestuurlijke autonomie*, (in Dutch) Antwerpen.

Humes, S. (1991). *Local governance and national power: a worldwide comparison of tradition and change in local government*, New York.

Institute of Shipping Economics and Logistics (1977; 1984; 1991; 1996; 1999). *Shipping Statistics Yearbook 1977; 1984; 1991; 1996*, Bremen.

International Association of Ports and Harbors (1995). Reports of the IAPH technical committees for trade affairs group to the 19th IAPH biennial conference Seattle/Tacoma, Tokyo.

Ircha, M.C. (1993). "Institutional structure of Canadian ports", *Maritime Policy and Management*, 20:1, pp. 51-66.

Jacobs, J. (1994). *Systems of survival*, New York.

Japanese Ministry of Transport (1995a). *Ports and Habors in Japan*, Tokyo.

Japanese Ministry of Transport (1995b). *Reconstruction work and development plan*, Tokyo.

Japanese Maritime Safety Agency (1995). *To keep the sea safe, clean and enjoable for future generations. Maritime Safety Agency*, Tokyo.

Jaques, B. (1993). "South Africa", *Seatrade Review*, August, pp. 29-45.

Jong, W.M. de en H. Stevens (1995). Intermodal transport by co-production. The building of infrastructural services for combined container transport, TRAIL-communications nr 95/6, Delft.

Kagan, R.A. (1990). *Patterns of port development. Government, intermodal transportation, and innovation in the United States, China, and Hong Kong*, Berkeley.

Kaufman, F.X., G. Majone en V. Ostrom (1986). *Guidance, control and evaluation in the public sector*, Berlin/New York.

Kemperink, G.N.H. (red.) (1995). *Publieke taak, private markt. De gevolgen van privatisering voor de publieke taakstelling,*(in Dutch*)* Deventer.

Kettl, D.F. (1993). *Sharing power. Public governance and private markets*, Washington D.C.

Kingdon, J.W. (1984). *Agenda's, alternatives and public choices*, Boston-Toronto.

Kinnock, N (1996). "Address", in L. Bekemans en S. Beckwith (red.) *Ports for Europe. Europe's maritime future in a changing environment*, Brussel.

Klink, H.A. van (1995). *Towards the borderless mainport Rotterdam: an analysis of functional, spatial and administrative dynamics*, Rotterdam.

Kobe Port Terminal Corporation (1994). *Port of Kobe. Leading wharves in the world*, Kobe.

Kreukels, A.M.J. (1991). Rotterdam: port and city; spatial mutations and urban plans, paper voor het seminar 'Port Cities in Europe; social mutations; spatial mutations; Barcelona, Genoa, Hamburg, Liverpool, Marseilles, Rotterdam', Utrecht

Kreukels, T. en E. Wever (1996). "Dealing with competition: the port of Rotterdam", *Tijdschrift voor Economische en Sociale Geografie*, 87:4, p. 293-309.

Krimpen, C. van (1991). *Ports between government control and private enterprise, paper voor Scientific and Technical Conference on Organisation of Works in Ports in the Conditions of Market Economy and Private Property*, Rotterdam.

KwaZulu-Natal Marketing Initiative (1996). *Investing in KwaZulu-Natal*, Durban.

Lewis, D.K. (1992). *Hong Kong. A completely new port*, Hong Kong.

Lim Kay Hwan (1976). *Decision making in a statutory board in Singapore: containerisation in the Port of Singapore Authority*, Singapore.

Lindblom, Ch.E. (1977). *Politics and markets. The world's political economic systems*, New York.

Lipsky, M. (1980). *Street-level bureaucracy: dilemmas of the individual in public services*, New York.

Lloyd's List International (1992). "South African Maritime Affairs", October, pp. 1-40.

Low, J. (1987). *Privatization in Singapore: the political and administrative implications*, Singapore.

Maritime Administration Office of Port and Intermodal Development in cooperation with The American Association of Port Authorities (1994). Public Port Financing in the United States, Wasgington DC.

Miners, M. (1995). *The government and politics of Hong Kong*, Hong Kong.

Mintzberg, H. (1983). *Structure in fives: designing effective organizations*, Englewood Cliffs.

Mors, J.T. (1986). *The potential for privatising harbours*, Johannesburg.

Mortelmans, K.J.M. (1985). *Ordenend en sturend beleid en economisch publiekrecht*, (in Dutch) Deventer.

Nagorski, B. (1972). *Port problems in Developing Countries*. International Association of Ports and Harbours, Tokyo.

Nalliah Pillai, Ph. (1983). *State enterprise in Singapore: legal importation and development*, Singapore.

Nationale Havenraad (1996). *Jaarverslag 1995*, 's-Gravenhage.

Neelen, G.H.J.M. (1994). *Principal-agent relations in non-profit organizations*, Enschede.

Nelson, R.R. (1987). "Roles of government in a mixed economy", *Journal of policy analysis and management*, 6:4, pp. 541-557.

Nonet, Ph. en Ph. Selznick (1978). *Law and society in transition*, New York.

Noort, J.W.P.P. van den (1990). *Pion of pionier. Rotterdam - Gemeentelijke bedrijvigheid in de negentiende eeuw*, (in Dutch) Rotterdam.

Noort, J.J. van (1993). *Van Hoek naar haven. Veertig jaar scheepsbegeleiding in de grootste haven ter wereld*, (in Dutch) Alkmaar.

Nijkamp, P. en S.A. Rienstra (1995). "Private sector involvement in financing and operating transport infrastructure", *Annals of Regional Science*, 29, pp. 221-235.

Nyawo, S. (1995). Port planning and development in the port of Durban, speech for the International Conference on Port Planning and Development, Durban.

Olson, D.J. (1980). Regionalism in port development: The Pacific Northwest experience, paper for the Regional Port Institutions Seminar, Seattle.

Olson, D.J. (1984). Reorganizing public enterprise: principles and practices, paper for the Annual Meeting of the American Political Science Association, Seattle.

Pearce, J.A. en S.A. Zahra (1991). "The relative power of CEOs and boards of directors: associations with corporate performance", *Strategic Management Journal*, 12, pp. 135-153.

Pearson, T. (1995). *African Keyport. Story of the port of Durban*, Rossburgh.

Philips, K. (1991)."South Africa bulk ports and trades prospects in the 1990s", *International Bulk Journal, International report - South Africa*, 11:11, pp. 6-27.

Planning Department Hong Kong en Port Development Board (1995). *Port Development Strategy. Second Review*, Hong Kong.

Port and Harbor Bureau Kobe City Government (1994). *Port of Kobe*, Kobe.

Port and Harbor Bureau Kobe City Government (1995). *Port of Kobe. Statistical Report 1994*, Kobe.

Port Development Board (1994). *Annual Report*, Hong Kong.

Port Development Board (1995). *Annual Report*, Hong Kong.

Port of Seattle (1990). *Purposes & objectives of the Port of Seattle*, Seattle.

Port of Seattle (1993). *Harbor handbook. A guide to Port of Seattle Marime facilities and services*, Seattle.

Port of Seattle (1994). 1995-99 *Capital improvement plan and draft plan of finance*, Seattle.

Port of Seattle (1995). *Pioneers and partnerships. A History of the Port of Seattle*, Seattle.

Port of Seattle (1996). *Annual Report 1995*, Seattle.

Port of Singapore Authority (1995). *Annual report 1994*, Singapore.

Port of Singapore Authority (1996). *Annual report 1995*, Singapore.

Portnet (1995a). *The port of Durban. Handbook & directory 1995/96*, Durban.

Portnet (1995b). *Ports of South Africa*, Florida.

Portnet (1995c). *Port reorganisation in South Africa*, Johannesburg.

Portnet (1995d). *The gateway to southern Africa*, Johannesburg.

Posner, R.A. (1974)."Theories of economic regulation", *The Bell Journal of Economics and Management Science*, 5: 2, pp. 335-358.

Posthuma, F. (1972). "Het havenbedrijf der gemeente Rotterdam", *G.E. van Walsum (red.) Rotterdam Europoort*, (in Dutch) Rotterdam.

Prime Minister's Office (1994). *Competitive salaries for competent & honest government. Benchmarks for ministers & senior public officers*, Singapore.

Quah, J.S.T. (1985). "Statutory boards", in J.S.T. Quah, C.H. Chee and S.C. Meow, *Government and politics of Singapore*, Singapore.

Robinson, J. (1969). *Economics of imperfect competition*, London.

Ross, R.L. (1988). *Government and the private sector. Who should do what?*, Santa Monica.

Rotterdam Municipal Port Management (1976). *Enige aspekten van het havenbeheer*, (in Dutch) Rotterdam.

Rotterdam Municipal Port Management (1989). *Strategische visie*, (in Dutch) Rotterdam.

Rotterdam Municipal Port Management /Europe Combined Terminals (1990). *Delta 2000-8. Naar een grootschalig containeroverslag- en goederendistributiecentrum op de Maasvlakte*, (in Dutch) Rotterdam.

Rotterdam Municipal Port Management (1991). *Ontwerp Havenplan 2010 Toekomstbeeld van Mainport Rotterdam*, (in Dutch) Rotterdam.

Rotterdam Municipal Port Management (1992). *Marktgegevens in de Hamburg - Le Havre range*, (in Dutch) Rotterdam.

Rotterdam Municipal Port Management (1993a). *Overslag in de havens van de Hamburg - Le Havre range in 1992,*(in Dutch*)* Rotterdam.

Rotterdam Municipal Port Management (1993b). *Havenplan 2010*, (in Dutch) Rotterdam.

Rotterdam Municipal Port Management en Gemeentewerken (1994). *10 Jaar Project Onderzoek Rijn. Het baggerspeciebeleid toen, nu en in de toekomst*, (in Dutch) Rotterdam.

Rotterdam Municipal Port Management (1995). *Milieu met beleid. Integratie* milieu- en veiligheidsbeleid in de bedrijfsvoering van het Gemeentelijk Havenbedrijf, (in Dutch) Rotterdam.

Rotterdam Municipal Port Management (1996a). *Bedrijfsplan 1997-2000*, (in Dutch) Rotterdam

Rotterdam Municipal Port Management (1996b). *Eindrapportage Inbreiding ruimtegebruik*, (in Dutch) Rotterdam.

Rotterdam Municipal Port Management (1996c). *Jaarverslag 1995*, (in Dutch) Rotterdam.

Rotterdam Municipal Port Management (1996d). *Container Yearbook 1996*, Rotterdam.

Rotterdam Municipal Port Management (1997a). *Haven in cijfers*, (in Dutch) Rotterdam.

Rotterdam Municipal Port Management (1997b). *Jaarbericht 1997 (in Dutch)* , Rotterdam.

Saitua, R. (1995). "De veranderende rol van de gemeentelijke overheid in de Rotterdamse havenclusters", (in Dutch) in D. Jacobs en A.P. de Man (red.) *Clusters en concurrentiekracht: naar een nieuwe praktijk in het Nederlandse bedrijfsleven?*, Alphen aan den Rijn.

Salet, W.G.M. (1987). *Ordening en sturing in het Volkshuisvestingsbeleid. Voorstudies en achtergronden WRR, V59*, (in Dutch) 's-Gravenhage.

Salet, W.G.M. (1989). "Behoed de volkshuisvesting voor tripartisering: nieuwe rollen voor huurders en verhuurders in de jaren negentig", *Beleid & Maatschappij*, nr 3, pp. 133-144.(in Dutch)

Salet, W.G.M. (1994). *Om recht en staat. Een sociologische verkenning van sociale, politieke en rechtsbetrekkingen*, (in Dutch) 's-Gravenhage.

Sappington, D.E.M. en J.E. Stiglitz (1987). "Privatization, information and incentives", *Journal of policy analysis and management*, 6: 4, pp. 567-582.

Sato, H. (1981). "The development of ports in Japan", *Ports and Harbors*, October 1981, pp. 9-11.

Savas, E.S. (1982). *Privatizing the public sector. How to shrink government?*, Chatham.

Savas, E.S. (1987). *Privatization. The key to better government*, Chatham.

Schmalensee, R. (1979). *The control of natural monopolies*, Massachusetts.

Schreuder, C.A. (1994). *Publiekrechtelijke taken, private rechtspersonen. Verzelfstandiging en privatisering in de vorm van vennootschappen en stichtingen*, (in Dutch) Deventer.

Scott, R. (1987). "Hong Kong," *Fairplay*, 302:5414, pp. 15-16.

Scott, W.R. (1995). *Institutions and organizations*, Thousand Oaks.

Seatrade Review (1995). *South Africa Report. Shiprepair. Ringing the changes*, June, pp. 151.

Selznick, Ph. (1992). *The moral commonwealth. Social theory and the promise of community*, Berkeley.

Singapore Ministry of Information and the Arts (1995). *Singapore 1995*, Singapore.

Singapore Ministry of Trade & Industry (1991). *The strategic economic plan. Towards a developed nation*, Singapore.

Singapore Port Workers Union (1986*). The port worker and his union. The first 40 years of Singapore Port Workers Union*, Singapore.

Singapore Prime Minister's Office (1994). *Competitive salaries for competent & honest government. Benchmarks for ministers & senior public officers*, Singapore.

Slauerhoff, J. (1996). *Alleen de havens zijn ons trouw*, (in Dutch) Amsterdam.

Standing Committee on Transport (1995). *A National Marine Strategy*, Ottawa.

Stevens, H. (1996). Opportunities and shortcomings of port privatisation. The port of Rotterdam in an international perspective, congresverslag Vlaamse Ingenieursvereniging Antwerpen.

Stevens, H. (1999). "Containeroverslag in de Rotterdamse Haven", in M.W.M. van Twist en W.W. Veeneman (red.) *Markt en Infrastructuur*, (in Dutch) Utrecht.

Stichting International Contract Research (1996). *Zuid-Afrika; nieuwe verhoudingen verkend*, (in Dutch) Leiden.

Stigler, G.J. (1971). "The theory of economic regulation", *The Bell Journal of Economics and Management Science*, 2:1, pp. 3-21

Suykens, F. (1987). "Over havens en steden",(in Dutch) *Economiche en sociaal tijdschrift*, 41:5, pp. 747-766.

Suykens, F. (1994). "Het Europees havenbeleid", (in Dutch*) Tijdschrift voor vervoerwetenschap*, nr 4, pp. 293-303.

Suykens, F. (1995). The European scene, voordracht gehouden tijdens de tweejaarlijkse bijeenkomst van de International Association of Ports and Harbors in Seattle.

Suykens, F. (1996). "The future of European ports", in: L. Bekemans & S. Beckwith (red.) *Ports for Europe. Europe's maritime future in a changing environment*, Brussel.

Swann, D. (1981). *The retreat of the state. Deregulation and privatisation in the UK and US*, New York.

The Dock and Harbour Authority (1994). *Hong Kong port development*, 75: 855, pp. 179-182.

The Port Authority of New York/New Jersey (1994*). More ships. More trains. More trucks. More planes. More from the Port of New York & New Jersey*, New York.

The Port Authority of New York/New Jersey (1995a). *Port Department. First Quarter 1995 Performace Report*, New York.

The Port Auhtority of New York/New Jersey (1995b). *Budget 1995*, New York.

The Port of New York & New Jersy (1996). *Port guide NY/NJ*, New York.

Transnet (1995). *Annual report 1995*, Johannesburg.

Truong, N. (1976). *The role of public enterprise in national development in Southeast Asia: problems and prospects*, Singapore.

United States Department of Transportation (1996). *A report to congress on the status of the public ports of the United States 1994-1995*, Washington.

Vakgroep Haveneconomie Erasmus Universiteit Rotterdam (1993). *The ins and outs of mainports, rapport studieproject 1992-1993 Hongkong&Singapore*, Rotterdam.

Vancouver Port Corporation (1994). *Port Plan. Land use management*, Vancouver.
Vancouver Port Corporation (1995a). *Vancouver International Handbook*, Vancouver.
Vancouver Port Corporation (1995b). *Submission of the Vancouver Port Corporation to the House of Commons Standing Committee on Transport. Marine Policy Review*, Vancouver.
Vancouver Port Corporation (1996). *Annual Review 1995*, Vancouver.
Vasil, R. (1992). *Governing Singapore*, Singapore.
Veerman, T. F. B. (1995). *Handleiding Handhaving Milieuwetgeving*, (in Dutch) Rotterdam.
Verhoeff, J.M. (1981). "Zeehavenconcurrentie: overheidsproduktie van havendiensten", (in Dutch) *Tijdschrift voor Vervoerswetenschap*, 17: 3/4, pp. 181-202.
VIA Port of New York - New Jersey (1995a). *Access is everything*, 47:, 1.
VIA Port of New York - New Jersy (1995b). *New York/New Jersy trade stats 1994*, 47: 4.
Vickers, G. (1965). *The art of judgement. A study of policy making*, London.
Walsum, G.E. van (red.)(1972). *Rotterdam Europoort 1945-1970*, (in Dutch) Rotterdam.
Washington Reserach Council (1990). *Washington's public ports*, Olympia.
Weerdt, J. de (1990). *De wereld van de zeevrachtvaart*, (in Dutch) Schoonhoven.
Williamson, O.E. (1975). *Markets and Hierarchies; Analysis and antitrust implications*, New York.
Winkelmans, W. (1984). *Paradoxen in de zeehaveneconomie, inaugurale rede EUR*, (in Dutch) Rotterdam.
Winkelmans, W. en E. Poelvoorde (1996). "Het havenbeleid in Vlaanderen: op een nieuwe economische leest?",(in Dutch) *Tijdschrift Vervoerswetenschap*, nr 1, pp. 55-66.
Winship, C. en S. Rosen (1988). "Introduction: Sociological and economic approaches to the analysis of social structure", *American Journal of Sociology*, 94 (supplement), pp. S1-S16.
Wolf, Ch. (1978). "A theory of nonmarket failure: framework for implementation analysis", *Journal of Law and Economics*, april, pp. 107-139.
Wolf, Ch. (1993). *Markets or governments. Choosing between imperfect alternatives*, London.
Worldport LA (1995). *Annual financial statement*, Los Angeles.
Wormmeester, (1992a). "Concrete voorstellen voor de gemeenschappelijke innovatie van mainport Rotterdam", (in Dutch) in Teisman, G.R. en R.J. in 't Veld, *Over effectieve structuren tussen overheid en bedrijfsleven*, 's-Gravenhage.
Wormmeester, (1992b). *Met het oog op morgen. Ontwikkelingstraject infrastructuur goederenvervoer*, (in Dutch) Rotterdam.

Official Documents

Act respecting the Canada Ports Corporation 1992
Canada Ports Corporation Operating By-Law A-1
Charter of the City of Los Angeles, Article XI Harbor Department.
Havenverordening Rotterdam (in Dutch).
Statuten Gemeentelijk Havenbedrijf Antwerpen, goedgekeurd 24 juni 1996 en gewijzigd 9 september 1996. (in Dutch)
The House of Commons of Canada, Bill C-44 Ministry of Transport, November 20 1996
The Port of Singapore Authority Act.
The Basic Law of the Hong Kong Special Administrative Region of the People's Republic of China.

Index